I0095332

Winning the War of the Flea

Irregular

Stability Operations (SO)

Security Force Assistance (SFA)

Foreign Internal Defense (FID)

Unconventional Warfare (UW)

Counterterrorism (CT)

Counterinsurgency (COIN)

Warfare

Winning the War of the Flea

The American Way of Irregular Warfare

Robert Ball, Lieutenant Colonel,
U.S. Army (Retired)

Sheffield Ford III, Major,
U.S. Army (Retired)

Paul LeFavor, Master Sergeant,
U.S. Army (Retired)

BLACKSMITH 2013 PUBLISHING

"In war many roads lead to success, and they do not all involve the opponent's outright defeat." – Clausewitz

"Do not put a premium on killing. To capture the enemy's army is better than destroying it. To win a hundred victories in a hundred battles is not the acme of skill. To subdue the enemy without fighting is the acme of skill." – Sun Tzu

Winning the War of the Flea: The American Way of Irregular Warfare

By Robert Ball, Sheffield Ford, and Paul LeFavor

Copyright © 2025 Blacksmith Publishing

ISBN 978-1-956904-34-5

Printed in the United States of America

Published by Blacksmith LLC
Fayetteville, North Carolina

www.BlacksmithPublishing.com

Direct inquiries and/or orders to the above web address.

All rights reserved. Except for use in review, no portion of this book may be reproduced in any form without the express written permission from the publisher.

This book does not contain classified or sensitive information restricted from public release. The views presented in this publication are those of the author and do not necessarily represent the Department of Defense or its components.

While every precaution has been taken to ensure the reliability and accuracy of all data and contents, neither the author nor the publisher assumes any responsibility for the use or misuse of information contained in this book.

Additionally, as a disclaimer, this book is based on open source information and the experiences of the authors. The views and analysis in this book do not represent the views of any government agency or organization.

Contents

Foreword ix

Introduction 1

Part One: Setting the Stage

1 – From Clausewitz to Kilcullen 5

2 – Two Sides of the Same COIN 22

3 – Five PLAIN Laws of Irregular Warfare 36

Part Two: Spanish-American War to World War II

4 – The Accidental Guerrilla War 49

5 – They Remained: Resistance in the Philippines 80

6 – War in the Shadows: Resistance in Europe 122

Part Three: The Cold War

7 – The Hukbalahap Rebellion 141

8 – War Without Fronts: The Vietnam War 160

9 – The Cold War Next Door 197

Part Four: Global War on Terrorism

10 – The Afghanistan War 220

11 – Operation Enduring Freedom – Philippines 268

12 – The Iraq War 279

Part Five: Great Power Competition

13 – U.S. Efforts in African States 319

14 – The Syrian Civil War 339

15 – The War in Ukraine 355

16 – Winning the War of the Flea 370

Conclusion 391

Annex A: Case Study Matrix 393

Annex B: Select List of Axioms 394

Annex C: The Twenty-Seven Articles of T.E. Lawrence 395

About the Authors 403

Bibliography 406

Acknowledgments 409

Recommended Reading List 410

Index 412

Dedicated to Ed Brodie. Green Beret, mentor to many, and one heck of a model American.

Foreword

"Whoever wishes to foresee the future must consult the past." – Machiavelli

Perception is influenced by interpretation. While many have attempted to decipher the riddle of irregular warfare (IW), perceptions of it are not unlike the six blind men in the famous poem by John Godfrey Saxe:

It was six men of Indostan, to learning much inclined,
Who went to see the Elephant, (Though all of them were blind),
That each by observation, might satisfy his mind.

In the poem, each blind man touches a different part of the elephant. Each believes his own experience best captures its essence. One of the blind men touch its tusk and says it's like a snake. Another touches it's leg and concludes that elephants are like trees, and so on to the sixth blind man. The blind scholars then fall into an endless debate about who is right:

And so these men of Indostan, disputed loud and long,
Each in his own opinion, exceeding stiff and strong,
Though each was partly right, and all were in the wrong!

This goes to show there is no knowledge of facts and things that are not influenced by our interpretive activity. It is no wonder then that many have sought to understand the complexities of irregular warfare through the lens of their own experiences – a very influential perspective indeed.

Since July 4, 1776, America has been at war 232 out of 249 years. Statistically, America has been at war 93% of the time, and at peace for less than 20 years total since its inception. All told, the United States has been involved in 118 military conflicts, and 11 formally declared wars. Nearly every one of these conflicts and wars have involved some sort of irregularity or asymmetry. By current doctrine, irregular warfare, is defined as *"the overt, clandestine, and covert employment of military and non-military capabilities across multiple domains by state and non-state actors through methods other than military domination of an adversary, either as the primary approach or in concert with conventional warfare."*[1] The U.S. Army has codified is doctrinal approach to IW via its various publications: ATP 3-18.1, Special Forces Unconventional Warfare; FM 3-24 Counterinsurgency; FM 3-07.1, Security Force Assistance; ATP 3-05.2. Foreign Internal Defense; Joint Publication 3-26 Counterterrorism; and FM 3-07 Stability Operations.

While there is one definition of irregular warfare, along with doctrinal publications covering all six IW activities, there are various schools of thought regarding *how* the complexity of irregular warfare is to be applied.

And that's why I like this book! It offers timeless lessons regarding the nature of irregular warfare, showing the good, the bad, and the ugly of America's long history with this asymmetrical aspect of war. Beginning with the Spanish-American War (1898), this book canvases over 125 years of historical examples of the American way of irregular warfare. Without some sense of historical continuity, American warriors will have to relearn the lessons of history each time we face a new conflict.

I was a young man, and young in the Army, when I decided to volunteer for Special Forces training. For me I was not only finding a new career, but I was also finding my way in life and was learning very fast. During a part of my training, in late 1992, I was very fortunate to be in a large classroom at Camp Mackall when a retired Master Sergeant from 10th Special Forces Group spoke to my class. He told us how he had, through certain terms gone to Afghanistan to teach and mentor mujahideen fighting against the Soviet Union. His stories of several trips to do this work inspired me to learn about Unconventional Warfare (UW), a type of irregular warfare. Studying and practicing irregular warfare for many years as a noncommissioned officer and as an officer in Special Forces gave me a foundation to work with as the War on Terror came into full swing.

I entered the war as a Special Forces Operational Detachment Commander, otherwise known as an ODA Team Leader. Our ODA was

[1] FM 1-02.1, *Operational Terms* (Washington D.C.: Department of the Army, 2024), 42.

conducting Direct Action missions while conducting Foreign Internal Defense (FID) with the Afghan Soldiers and Government Officials. I personally felt that we, Special Forces, was missing an opportunity to do more. It was during a visit in 2006 from our CJSOTF-A (Combined Joint Special Operations Task Force – Afghanistan) Commander with the SOC-CENT (Special Operations Command – Central) Commander to our ODA's Fire Base in Kandahar that I was asked by the General, "What would you do differently if you could?" Using a large map on a wall I explained how I would get our ODAs out of the fire bases and into areas where we could create stability for one village, then two, then multiple villages, while integrating Afghan forces and police to maintain the stability. Then I would move the ODAs to an adjacent area to do the same to connect the stabilized areas. The plan was simple. Utilizing multiple ODAs spread across larger areas, we could slowly connect the villages to the larger cities through irregular warfare, integrating FID and conventional forces, while building confidence in both the civilian populace and the Afghan government forces. Both commanders liked the idea. Then, on a follow-on deployment, the CJSOTF-A Commander asked if I had written down the idea from the year before. I ended up briefing the CJSOTF-A J3 and J3 Sergeant Major on the idea, they too liked the idea. I called it "Operation Outposts."

In 2008, I was part of a four-man element that was sent to Pakistan. For 32 months, I remained in Pakistan, part of a team that did an amazing amount of work. That is its own story. However, for me it was the highlight of my career and the opportunity to conduct irregular warfare in many layers. I learned a lot about the people of that region, and how to work with them to fight common enemies. I looked forward to the opportunity to return to Afghanistan to use the experience and continue the mission.

After returning from Pakistan in 2011, and preparing to return to Afghanistan, our unit was briefed on a new initiative, entitled, Village Stability Operations (VSO). The program called for ODAs to establish Village Stability Platforms (VSPs), from the ODAs would isolate and secure the villages from the Taliban, integrating Afghan forces and police to maintain the stability. It was really cool to learn that behind the initiative was my former CJSOTF-A Commander and J3 along with a gentleman from the RAND Corporation. While I was in Pakistan, that same person from RAND came for a visit. We discussed "Operation Outposts." The plan I had presented in 2006. He found the idea interesting, and I pointed out that it really wasn't anything new. The U.S. had done something similar in settling the western United States with our calvary, by establishing outposts to support settlements. As history shows, this concept has likewise been used to stabilize unsettled or conquered areas throughout many wars. Feeling as though I may have had an influence on the development of VSO through my idea of "Outposts" sold me on the idea, as many of my colleagues and ODAs leadership struggled with "why?" So, as the poem of the six blind men goes, what does right look like in this environment.

We, the Soldiers of a Special Forces Company Headquarters with eight ODAs, combined our knowledge from training, reading, and experience to successfully conduct VSO in central and northern Kandahar in 2012. Our campaign plan was built in conjunction with conventional U.S. forces, and both Afghan Special Operations and Afghan Conventional forces. Using irregular warfare, we dominated and denied the enemy influence over the local populace. Although not successful in all VSP areas of operation, collectively there was success over more areas than not, connecting people with markets, medicine, education, and government – creating stability.

Although some may argue with what my idea of irregular warfare is, and how I may have applied it during my career, I enjoy the continued conversation and sharing of experiences to continue the mission and provide tools to our current warfighters making them more effective.

This book warrants my strongest recommendation for you to learn of others' experiences in irregular warfare. It is a must-read for warfighters operating in the most political, complex and ambiguous environments. May it be another tool to make you and your team better as you execute the foreign policy objectives of the United States. Charlie Mike! Continue the Mission!

De Oppresso Liber

Major Sheffield Ford III, U.S. Army (Ret.)

Introduction

"The guerrilla fights the war of the flea, and his military enemy suffers the dog's disadvantages: too much to defend; too small, ubiquitous, and agile an enemy to come to grips with. If the war continues long enough—this is the theory—the dog succumbs to exhaustion and anemia without ever having found anything on which to close its jaws or to rake with its claws." – Robert Taber

From the Philippine Insurrection to the War in Afghanistan, America's warriors have experienced the accuracy of Robert Taber's analogy: "The guerrilla fights the war of the flea, and his military enemy suffers the dog's disadvantages." The flea is the guerrilla. The dog is the government or occupying power of the host nation. For over seventy years America has fought the war of the flea from Vietnam to Afghanistan, and everywhere else in between. Experience has taught us that irregular warfare (IW) is more difficult than operations against an enemy that fights according to the conventional paradigm. Though few have stood up to the United States in a conventional fight, when it comes to this most challenging form of war, America's track record has proven to be less consistent.

Famously, Solomon declared, "Of making many books there is no end, and much study is toilsome to the soul" (Ecclesiastes 12:2). This book investigates the American way of irregular warfare. Why another book on this topic? No doubt much ink has been spilt. A host of theorists and practitioners have produced many honored volumes. But what is well-known is not necessarily understood. And as history shows, what is learned must often be relearned. Aside from the accuracy and usefulness of the host of volumes on IW, what is lacking is a single book that brings lessons learned together. As will be argued, history reveals there is an American way of irregular warfare. In fact, regarding this most challenging form of war, it will be argued that America has a distinct formula for success; one that has been largely forgotten.

To manage your expectations, let it be said up front that we do not claim to be experts on irregular warfare, or experts on anything in particular. Even so, we were privileged to serve in some of America's

1

finest fighting units. During our Army service, we took part in actions that spanned the globe from Afghanistan to Bosnia and canvassed all six IW activities, which include: Foreign internal defense (FID), counterterrorism (CT), counterinsurgency (COIN), security force assistance (SFA), stability operations (SO), and, in a more limited sense, unconventional warfare (UW). Our experience is primarily at the tactical level. We took part in combat operations in Iraq, Afghanistan, and the Philippines, peacetime stability operations (SO) in places like Bosnia, and foreign internal defense (FID) missions on five continents. Across the spectrum, in every theater, we witnessed dogged American stamina, professionalism, and ingenuity in the tackling of a myriad of difficult missions.

What we discovered from our experiences is this: echoing the sentiment of General Cleveland, in each conflict "American tactical brilliance was followed by strategic muddling and eventual failure."[1] Certainly, it cannot be overemphasized, the tactical brilliance often redeemed failed policy, or at least forestalled catastrophe. The feeling we had was that in every conflict and peacetime operation we simply reinvented the wheel. More to the point, within the particular units we served in at the time, the unanimous consensus was we all felt we were untrained and unprepared for most of the situations we faced.

In our humble opinion, the U.S. Military fails to adequately hand down hard-earned lessons-learned in its institutional memory. It is hoped that this volume will serve as a prescription. With that being said, we hold no expectation that our analysis constitutes the silver bullet remedy to irregular warfare success. Nonetheless, it is hoped that this volume will arm America's irregular warfare practitioners with conceptual tools that will enable them to win the war of the flea and bring glory and honor to our standard.

Our goal will be to gain a thorough understanding of the American way of irregular warfare by tracing out the most prominent IW operations in U.S. history. Beginning with definitions, our method will be to explore the various concepts associated with insurgency and counterinsurgency, briefly surveying some of the more important theorists. From there, the

---

[1] Charles T. Cleveland, *The American Way of Irregular Warfare: An Analytical Memoir* (Santa Monica, CA: Rand, 2020), iii.

discussion will move to approaches for both insurgency and counterinsurgency. This will be the focus of chapters one through three. In chapters four through fifteen, we will survey the various approaches to irregular warfare America has employed through the lens of history. These chapters will cover every major conflict since 1898. Finally, chapter sixteen concludes the study with some practical applications from all that will be surveyed, with the aim of offering a formula for IW success as we have now returned to the Great Power Competition.

Central to our agenda, we will argue that there are five plain laws that govern irregular warfare and predict with a high degree of certainty whether or not success will be achieved. As will be argued, these laws are: Political objective(s), legitimacy, adaptability, influence, and native face. IW is a complex competition for contested terrain – the people. By adhering to these five PLAIN laws, and adaptively employing the ten tactical lines of effort that will be introduced, IW practitioners stand the best chance of victory.

While at first glance this all may seem self-evident, however, as borne out by history, many have ineptly bungled their way through conflicts. In the recent past, America's forces have prosecuted a version of counterinsurgency that is both excruciatingly frustrating and expensive in the loss of blood and treasure.

Regarding counterinsurgency, while we hold no naïve expectation that our own analysis will resolve the long-standing impasse between the major interpretive traditions, of whether which approach is the most sound, nonetheless, we hope this contribution will move the dialogue in new directions; in particular, a holistic one.

Heeding David Galula's dictum, the pitfall of dogmatism is inherent in any effort at abstraction, an overall effort will therefore be made to demonstrate that people themselves are the contested *land*. As such, IW is a fight over people.

Part One

Setting the Stage

1

From Clausewitz to Kilcullen

"Wisdom begins with the definition of terms." – Socrates

Green Berets training Civilian Irregular Defense Group (CIDG) personnel. South Vietnam, 1962. Courtesy AP.

As stated by Carl von Clausewitz, the end state in traditional war involves three things: destroying the enemy's armed forces, occupying his land, and taking from him his will to fight. By contrast, it's the third end state of the Prussian's dictum that is most often crucial to irregular warfare success. This is because the battlefield of IW includes the hearts and minds of the affected population. Victory therefore may or may not include the outright defeat of the enemy's armed forces. Adding to the complexity, as history shows, in a counterinsurgency (COIN) environment, traditional and irregular warfare (IW) may occur simultaneously. Likewise, a COIN force must apply discriminative violence, if they hope to win. As this is the case, to conduct military actions without proper analysis of their political effects will at best be ineffective and at worst aid the insurgent. A COIN force could win every battle but lose the war. Consider America's wars in Vietnam and Afghanistan as ready examples. Success in IW therefore requires a paradigm shift in strategy, tactics, and mindset. Success depends not necessarily on occupying political space, but rather on the abstract psychological metrics of legitimacy and influence. This makes IW more art than science, and more about anthropology than kinetics.

This book canvases American irregular warfare, which is defined as *"the overt, clandestine, and covert employment of military and non-military capabilities across multiple domains by state and non-state actors through methods other than military domination of an adversary, either as the primary approach or in concert with conventional warfare."*[1]

The Twelve Irregular Warfare Activities

There are twelve IW activities. These include:

IW Missions

- Unconventional Warfare (UW)
- Foreign Internal Defense (FID)
- Counterinsurgency (COIN)
- Counterterrorism (CT)
- Security Force Assistance (SFA)
- Stability Operations (SO)

IW Operations

- Military Information Support Operations (MISO)
- Civil Affairs Operations (CAO)
- Civil-military operations (CMO)
- Counter Threat Finance (CTF)
- Counter Threat Networks (CTN)
- Security Cooperation (SC)

---

[1] FM 1-02.1, *Operational Terms* (Washington D.C.: Department of the Army, 2024), 42.

This book will concentrate on the six IW missional activities, defined below:

Unconventional warfare (UW) is defined as "activities conducted to enable a resistance movement or insurgency to coerce, disrupt, or overthrow a government or occupying power by operating through or with an underground, auxiliary, and guerrilla force in a denied area."[2] Anyone familiar with military doctrine understands its evolutionary nature. The 1955 definition of UW was, "operations conducted in time of war behind enemy lines by predominately indigenous personnel responsible in varying degrees to friendly control or direction in furtherance of military and political objectives. It consists of interrelated fields if Guerrilla Warfare, evasion and escape and subversion against hostile states."[3] Additionally, an insurgency may be defined as "the organized use of subversion and violence to seize, nullify, or challenge political control of a region.[4]

Foreign internal defense (FID) is defined as "activities that support a Host Nation's internal defense and development (IDAD) strategy and program designed to protect against subversion, lawlessness, insurgency, terrorism, and other threats to their internal security, and stability, and legitimacy."[5] FID is considered a subset of counterinsurgency as FID is essentially the military strategy of supporting a foreign government in combating an insurgency within its own territory, the core focus of counterinsurgency operations.

Counterinsurgency (COIN) is a type of warfare that involves a range of tasks and capabilities beyond those of conventional conflict. As such, it is defined as "a blend of civilian and military efforts designed to end insurgent violence and facilitate a return to peaceful political processes. COIN is a war for legitimacy in the eyes of the effected populace."[6] As

---

[2] Ibid.
[3] Department of the Army, Field Manual 31-21 Guerilla Warfare (Washington, D.C.: U.S. Government Printing Office, 23 MAR 1955), 2
[4] Joint Publication 3-24, Counterinsurgency, ix.
[5] JP 3, V-3.
[6] Joint Publication 3-24, Counterinsurgency.

will be argued, success requires coordination across all lines of effort, winning the support of the local population, depriving insurgents of sustenance and intelligence, while adaptably adjusting approaches to fit the evolving situation and local dynamics.

Counterterrorism (CT) are "actions taken directly against terrorist networks and indirectly to influence and render global and regional environments inhospitable to terrorist networks.[7] CT and COIN are related activities. While COIN focuses on defeating a larger, organized insurgency that may have significant popular support within a population, CT primarily targets individual terrorist groups or cells. These targeted groups or cells are often an insurgency's leadership.

Security force assistance (SFA) is the unified action to generate, employ, and sustain local, host-nation, or regional security forces in support of a legitimate authority (FM 3-07). It may be asked, what is the difference between FID and SFA? While SFA generally focuses on helping a partner nation build capabilities to address both internal and external threats, FID focuses efforts on assisting a country in defending against internal threats like insurgencies or rebellions within their borders. Additionally, FID is often executed through smaller-scale operations conducted by Special Operations Forces.

Stability operations (SO) encompass "various military missions, tasks, and activities conducted outside the United States in coordination with other instruments of national power to maintain or reestablish a safe and secure environment, provide essential governmental services, emergency infrastructure reconstruction, and humanitarian relief."[8] Stability operations are military missions that aim to strengthen legitimate governance and restore or maintain the rule of law, support economic development, and foster national unity, assisting the host government to assume responsibility for civil administration. Stability operations can take place before, during, and after combat operations.

---

[7] Ibid.
[8] JP 3-07, Stability Operations.

FID

COIN

CT

SFA

SO

Support of a resistance movement or insurgency **against a** nation state or occupying power.

UW

Support of a nation state **against** a resistance, insurgency, or terrorists.

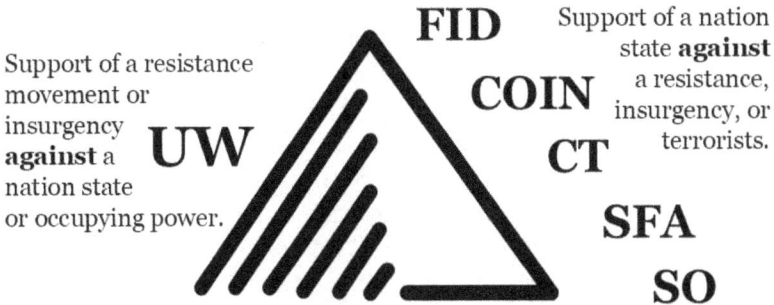

Six Core Activities of Irregular Warfare.

Carl von Clausewitz once famously observed, "war is an extension of politics by other means." He further opined that "war is an act of violence to impose our will on our enemy." As such, traditional war uses organized violence to achieve political ends, which normally includes the gaining of political space. Moreover, traditional or "conventional" war, in large measure, is uniformed organized violence. That is, one normally can visibly identify who is friend or foe by either uniform type, which side of the front line one is on, along with rear areas, and location of enemy headquarters, etc. However, irregular warfare requires an entirely different way of thinking, that is, in order to prosecute it successfully.

Considering its broad scope, IW touches upon every form of human relation and, in the last analysis, it is a contest to win the adherence of the population. With this in mind, it is easy to see why legitimacy and influence should be considered the two prime movers of IW. In such a contest, when countering insurgency, military force is but one instrument. One would presume this to be self-evident. However, as history testifies, mistakes in COIN have been legion.

Likewise, though many COIN campaigns have led to the successful demise of insurgent leaders and their organizational frameworks, due to the course in which these campaigns have run, in the end, there is no better state of the peace. The cause submerged itself into the background, waiting for more opportune times. Presenting two sides of the same coin, insurgency and counterinsurgency are complex forms of irregular warfare (IW). This chapter is a literature review of these two forms of warfare.

In his book *On War* (1832), Clausewitz referred to insurgency as a "people's war." Yet, he was not a fan of such wars, believing a people's war to be a legalized form of anarchy, and as much a threat to the social order at home as it was to the enemy. Interestingly, as noted by Liddell Hart, Clausewitz failed to make any reference to the most striking example of insurgency during the wars of his time, that of the Spanish guerrilla war against Napoleon. After all, it was this conflict which brought the term "guerrilla," meaning "little war," into military usage.[9] Perhaps this was because, in Clausewitz's estimation, wars by an armed populace could only serve as a strategic defense.[10] Given his limited view of irregular warfare, he likewise failed to appreciate the irregulars' ability to assist the operational commander as a source of information.

Clausewitz's contemporary Antoine Henri de Jomini, also briefly delved into insurgency in his *Art of War* (1838). While both military philosophers discussed certain aspects of insurgency, regarding counterinsurgency, their approach may be said to be solely enemy-centric. This highlights a phenomenon regarding theories of guerrilla warfare. As Ian Beckett has noted, such theories "invariably reflect the experience of their author and are specific to a particular point in time and to particular circumstances."[11] This indeed holds true.

The most practical wisdom regarding theories of irregular warfare comes from history's theorist-practitioners. In the more recent past, heading such a list is John S. Mosby, the Confederate Ranger, guerrilla leader, and scourge of the Union Army of the Potomac during the American Civil War (1861-1865). In his book *War Reminiscences* (1889), Mosby recounts how his partisan Rangers kept Lee's Army of Northern Virginia well-informed of enemy movements, harassed the Federals as much as possible, and wreaked havoc on their supply lines and depots.

There were several key elements to Mosby's success. First, he enjoyed the moral and logistical support of the people of Northern Virginia. Second, his enemy was unable to guard everywhere. Third, was his speed and audacity with which he used his guerrilla tactics. All these points

---

[9] B.H. Liddell Hart, *Strategy* (London: Faber & Faber, 1954), 362.
[10] Carl von Clausewitz, *On War, Everyman's Library* (Princeton, NJ: Princeton University Press, 1993), VI.26, 578.
[11] Ian Beckett, *Modern Insurgencies and Counter-Insurgencies: Guerrillas and Their Opponents Since 1750* (New York, NY: Routledge, 2001), 70.

were to be noted by later guerrilla warfare practitioners. In his *War Reminiscences* Mosby recounts:

The military value of the species of warfare I have waged is not measured by the number of prisoners and material of war captured from the enemy, but by the heavy detail it has already compelled him to make, and which I hope to make him increase, in order to guard his communications, and to that extent diminishing his aggressive strength.[12]

Following Mosby, albeit with no known understanding of Mosby's guerrilla war or works, was Thomas E. Lawrence, aka "Lawrence of Arabia." In his autobiographical account *Seven Pillars of Wisdom* (1926), Lawrence gives an extended account of his role in the revolt of the Arab tribes against the Ottoman Turks during World War I (1914-1918). Of all that may be said of Lawrence, his genius lies in his ability to quickly size up the various challenges and then conceptualize a doable strategy for the Arab revolt. T. E. Lawrence's famous advice to those who would study his exploits:

Do not try to do too much with your own hands. Better the Arabs do it tolerably than that you do it perfectly. It is their war, and you are to help them, not to win it for them.[13]

Numerically inferior, Lawrence determined that the Arabs must adopt a "strategy of detachment." Following the tactics of earlier practitioners, Lawrence observed:

Most wars are of contact, both forces striving to keep in touch to avoid tactical surprise. Our war should be a war of detachment: we were to contain the enemy by the silent threat of a vast unknown desert, not disclosing ourselves till the moment of attack. This attack need only be nominal, directed not against his men, but against his materials: so, it should not seek for his main strength or his weaknesses, but for his most accessible material.[14]

Lawrence correctly deduced that defending everywhere, the Turks would defend nowhere. This itself reflects what was for Clausewitz the

---

[12] John S. Mosby, *War Reminiscences* (New York: Dodd, Meade & Co, 1889), 45.
[13] Thomas E. Lawrence, *Twenty-Seven Articles*.
[14] Thomas E. Lawrence, *Seven Pillars of Wisdom* (New York: Doubleday, 1926), 131.

most characteristic feature of an insurgency: ubiquity. To be nebulous and elusive; to "exist everywhere and nowhere."[15]

Lawrence therefore determined that in order to contain the revolt, the Turks would need some 600,000 troops in fortified posts consisting of twenty men, every four-square miles. For this daunting task, in a hostile environment, the Turks had a force of only 100,000. Over against this "strategy of detachment," British Army HQ in Cairo wanted Lawrence to attack the 10,000-strong Turkish garrison in Medina. Gaining an epiphany during a bout with dysentery, Lawrence instead applied the indirect approach to the problem. He surmised that the best strategy would, "let the Turks in Medina die on the vine." Lawrence understood that the Hejaz rail line served as the Turkish garrison in Medina's lifeline. Rather than a direct approach, Lawrence bypassed the garrison and explosively interdicted the rail line which served as its vital link.

Lawrence argued for guerrilla warfare of which he advanced six principles:

1. A successful guerrilla movement must have an unassailable base. 2. The guerrilla must have a technologically sophisticated enemy. 3. The enemy army of occupation must be sufficiently weak in numbers so as to be unable to occupy the disputed territory in depth with a system of interlocking fortified posts. 4. The guerrilla must have at least the passive support of the populace, if not its full involvement. 5. The irregular force must have the fundamental qualities of speed, endurance, presence and logistical independence. 6. The guerrilla must be sufficiently advanced in weaponry to strike at the enemy's logistics and signals vulnerabilities.[16]

As a seventh principle, Lawrence argued that a guerrilla force can operate alongside a conventional one. This he successfully demonstrated in support of British General Allenby's Palestine Campaign. Arab guerrillas, operating in the Turkish rear areas, and in concert with Allenby's forces, tied up tens of thousands of enemy troops, leading to the Turkish defeat.

One of the most influential of all irregular warfare theorists is Mao Zedong. From his meagre beginnings as the son of hardworking peasants in Hunan Province, Mao gained political power through the barrel of a

---

[15] Ibid., VI.26, 581.
[16] Ibid., 196.

gun. Mao was a founding member of the Chinese Communist Party. After the fallout with Nationalist Chiang Kai-shek in 1927, Mao began to put into practice his revolutionary theories.

Mao's experiences in guerrilla warfare led him to pen several works. In 1937, he wrote *On Guerrilla Warfare*. In it, he blends philosophy, political science, and war to trace out the relationship between the insurgent and the civil population. Mao argued for guerrilla warfare as a viable, more mobile form of warfare to be employed by revolutionaries in a political and military style. For Mao, guerrilla troops should have no conception of defense or battle lines but rather like a fish they are to "swim in the sea of the people."[17] Moreover, because guerrillas cannot hope to defeat a conventional enemy in the open field, Mao advocated a strategy of pitting one man against ten and a tactic ten against one.

According to Mao, the insurgent should prepare for the long haul. By way of this protracted struggle, Mao stressed that insurgency is the traditional strategy of the weak resisting the strong. Thus, in his strategy, one that reflects Lawrence's philosophy of detachment, "If the enemy advances, we retreat. If he halts, we harass. If he tires, we attack." This axiom of Mao in essence sums up Sun Tzu's third chapter in the *Art of War;* "fight the enemy's strategy, not his forces."[18] Mao further emphasized identifying with the population and blending guerrilla and conventional military tactics to achieve success. His three basic principles are as follows:

1. Yield any town or terrain you cannot hold safely.
2. Limit yourself to guerrilla warfare as long as the enemy has numerical superiority and better weapons.
3. Organize regular units and pass over to the general counter-offensive only when you are sure of the final victory.[19]

For Mao, the underlying cause or ideology in an insurgency is of the utmost concern. His ideology comes by way of a three-phased approach:

---

[17] Shu Guang Zhang, *Mao's Military Romanticism: China and the Korean War, 1950-1953* (Lawrence, KS: University Press of Kansas, 1995), 13.
[18] Sun Tzu, *The Art of War* (New York, NY: Oxford, 1963), 77.
[19] Franklin Mark Osanka, *Modern Guerrilla Warfare: Fighting Communist Guerrilla Movements, 1941-1961* (New York: The Free Press of Glencoe, 1962), 259.

strategic defense, strategic stalemate, and strategic offense.[20] This first phase, strategic defense, is a prerevolutionary one. The goal of this phase is to expand the organization and establish key infrastructure. During this phase, limited force is to be applied to prepare the way for guerrilla action. For Mao, a popular support base is imperative. As he argued, "Identifying with the people is the single most important task of the guerrilla." Noting the aim of this phase, Ian Becket observes:

The careful political preparation of this first phase was designed to convince the peasantry that their lives could be improved only by supporting the communists, and it was essential if the guerrillas were to survive, for the guerrilla 'must be in the population as little fishes in the ocean.'[21]

At the appropriate time, the strategic defense would give way to the strategic stalemate. This phase would be characterized by guerrilla warfare in which minor attacks would demonstrate that the communists were in competition with the government. Last would be the strategic offensive or war of movement. In this phase, regular units would meet and defeat government forces in battle to achieve a strategic end state. If unsuccessful, the insurgency would be forced to revert back to a previous phase to regroup and build strength.

Beginning in 1927, the Chinese Civil War continued unabated until Mao and Kai-shek agreed to an uneasy truce to defend China against Japanese invaders. Following the defeat of the Japanese in 1945, the Chinese Civil War resumed, and Mao and his movement were able to use their rural foundation to outmaneuver and eventually overwhelm the Nationalists. With his overthrow of Chiang Kai-shek, Mao's theories have become something of a gold standard and have been successfully applied around the globe. They have long been used because they provide an essential link between the ideological and the practical, having both a political and a military wing. This strategy requires a high level of organization, indoctrination, action along multiple lines of effort, and leadership to direct the shifting of phases according to circumstances.[22]

---

[20] Mao Zedong, *Selected Military Writings of Mao Tse-Tung* (Beijing: Foreign Language Press, 1966), 210.
[21] Ian Beckett, *Modern Insurgencies and Counter-Insurgencies*, 74.
[22] FM 3-24.2, *Tactics in Counterinsurgency*, 2-17.

Riding this wave of success, Soviet leader, Nikita Khrushchev promoted what he deemed "national wars of liberation." As we will see later in chapter eight, Mao's approach was successfully employed by North Vietnamese General Nguyen Giap, first against French and later American forces. Nuancing Mao, Che Guevarra in *Guerrilla Warfare* (1961) argued for a "Focoist approach." This approach entails a small group of guerrillas operating in a rural environment where the grievances of the local population can be easily exploited to build the conditions necessary to overthrow a government. This was the approach used by the Sandinistas in Nicaragua.

On the flip side, many counterinsurgency works have been written by those with firsthand knowledge of its complexity. Among the first modern COIN theorists is the French Colonel David Galula. Drawing on his experience as a French military officer who served in the Algerian War (1956-58), Galula wrote *Counterinsurgency Warfare: Theory and Practice* (1964). Galula states succinctly that "insurgency and counterinsurgency are two different aspects of the same conflict."[23] Moreover, following Clausewitz, Galula correctly stressed the military action is to be secondary to the political one, stating, "counterinsurgency is 80 % political action and only 20 % military." Emphasizing a people-centric view, Galula argued that the people are the primary objective of a counterinsurgency campaign.[24] His dictum encapsulates one of the axioms of this book: *The people themselves are the contested land.*

Another French colonial infantryman, Colonel Roger Trinquier, shared his experiences in Indochina and Algeria. In his book, *Modern Warfare: A French View of Counterinsurgency* (1964), Trinquier underscored the imperative for a COIN force to have a coherent political end state. Trinquier witnessed firsthand the appalling effects of not having one, as French forces were left to devise their own. Trinquier also stressed the absolute importance of learning and adapting to insurgent tactics.

---

[23] David Galula, *Counterinsurgency Warfare: Theory and Practice* (Westport, CT: Praeger, 2006), xiv.
[24] Ibid, 86.

Perhaps one of the most influential theorists of counterinsurgency was British Sir Robert Thompson. Based on his success in countering Communist insurgents during the Malayan Emergency (1948-1960), Thompson was regarded as an expert on countering insurgency. He emphasized winning the hearts and minds of the people, and gearing up for the long haul, he states, "It is a persistently methodical approach and steady pressure which will gradually wear the insurgent down."

Also following a people-centric view, in his 1967 *Counter-Insurgency Operations: Techniques of Guerrilla Warfare*, British Army Colonel Julian Paget identified five essentials for COIN operations: These are civil-military understanding, joint command and control, good intelligence, mobility, and training. Paget argued that while insurgent leaders are to be sought and eliminated, and the insurgent cause to be undermined and diminished, for the counterinsurgent to win, the affected population must be won over. Pointedly, he states:

There are two basic requirements to be met before the support of the local population can be won by the counter-insurgent forces, either in the short or in the long term. Firstly, the government must demonstrate its determination and its ability to defeat the insurgents, for no one likes backing a loser, particularly in an insurgency. Secondly, the government must convince the populace that it can and will protect its supporters against the insurgents, for no one likes being shot as the reward for loyalty.[25]

In a similar vein, Edward Lansdale, an Air Force officer and CIA agent, advised Philippine Defense Minister Ramon Magsaysay during the Huk Rebellion (1945-54). Gaining a firsthand impression of the people, Lansdale devised a holistic COIN approach that led to the defeat of the Huks. In his book, *In the Midst of Wars* (1971), Lansdale argued for multiple approaches to uproot an insurgency. Central to his argument was the understanding that a cause *is* an insurgency. By having its cause undermined, an insurgency can lose its political base.

Witnessing the Cuban Revolution and the U.S. efforts in Vietnam, Robert Taber offered a COIN primer with the book *War of the Flea* (1965). Tracing out earlier ideas, Taber argues that separating the insurgent from the population is an essential key to the

---

[25] Julian Paget, *Counter-Insurgency Operations: Techniques of Guerrilla Warfare* (New York: Walter and Company, 1967), 176.

counterinsurgent's victory. As history shows, separating the insurgent from the population involves both psychological as well as physical means. As demonstrated by Taber, the *how* to such a task is as important as the *what*. Additionally, Taber argues that much like traditional war, guerrilla warfare is an extension of politics.

More recent theorists of COIN have argued for various additional tenets to be stressed. For example, John A. Nagl in *Learning to Eat Soup with a Knife* (2002), argues that adaptability is the most important ingredient for success. Accordingly, the counterinsurgent must possess the ability to learn and adapt. Showcasing the British success in Malaya, Nagl proposes that victory resulted from their ability to adapt as a "learning institution."[26] By comparison, Nagl claims that the American military's penchant for conventional war caused it to fail to become a "learning institution," thereby inhibiting its ability to modify its counterinsurgency approach.

Arguably, one of the most influential of today's COIN theorists is retired Australian Army Lieutenant Colonel David Kilcullen. In *Counterinsurgency* (2010), Kilcullen recapitulates most of what earlier COIN theorists have said, arguing that fighting insurgency requires an utterly different way of thinking and training that large, peacetime armies aren't equipped for. Systematizing the best COIN practices, Kilcullen offers the classic approach, observing that success requires gaining and maintaining the hearts and minds of the local people. This is his emphasis in *The Accidental Guerilla* (2009), where Kilcullen argues, "Effective counterinsurgency provides human security to the population, where they live, 24 hours a day. This, not destroying the enemy, is the central task."[27]

In *Out of the Mountains* (2015), Kilcullen takes a prescient look at how global trends are shaping the environment for future conflicts. In it, he argues that military power must be tailored to social, political, and economic realities. Then, in *Blood Year* (2016), Kilcullen analyzes the threat that ISIS poses to the Middle East and argues that even if the West

---

[26] John A. Nagl, *Learning to Eat Soup with a Knife: Counterinsurgency Lessons from Malaya to Vietnam* (Westport, CT: Praeger, 2002), 11.
[27] David Kilcullen, *The Accidental Guerrilla: Fighting Small Wars in the Midst of a Big One* (New York: Oxford University Press, 2009), 486.

kills all of ISIS, there remains an underlying problem. Under the context of countering a global insurgency, David Kilcullen rightly observes: "paraphrasing Clausewitz, to wage this war effectively we must understand its true nature: not mistaking it for, or trying to turn it into, something it is not."[28]

Most extensively, the research and analysis of the think-tank RAND has long been a help to policymakers and leaders, giving understanding of the complexities of insurgency and counterinsurgency. In our own analysis, we are deeply indebted to such works as *How Insurgencies End*, *Keys to Successful Counterinsurgency Campaigns Explored*, and *Victory Has a Thousand Fathers*. When it comes to IW, RAND's analysis reveals many gems and key factors, among which is the axiom: "Good counterinsurgency practices tend to run in packs."

In the recent past, former U.S. Army Special Forces Major Jim Gant authored the paper *One Tribe at a Time: A Strategy for Success in Afghanistan*. Gant argued that the United States should leverage the Pashtun tribal system in Afghanistan by creating Tribal Engagement Teams that would embed themselves at the village level and work with locals to protect and advance both American and Afghan interests. Embraced by senior Special Forces leadership, Gant's winning strategy evolved into what became known as Village Stability Operations (VSO). The bottom-up approach provided stability for rural villages within key districts and empowered the local-tribal government structure. As will be argued in chapter eleven, this COIN strategy could have won the war in Afghanistan.

Success or failure in insurgency and COIN often depends upon several crucial factors. A highly useful metric has been devised by Gordon McCormick to demonstrate this.

---

[28] David Kilcullen, *Countering Global Insurgency*.

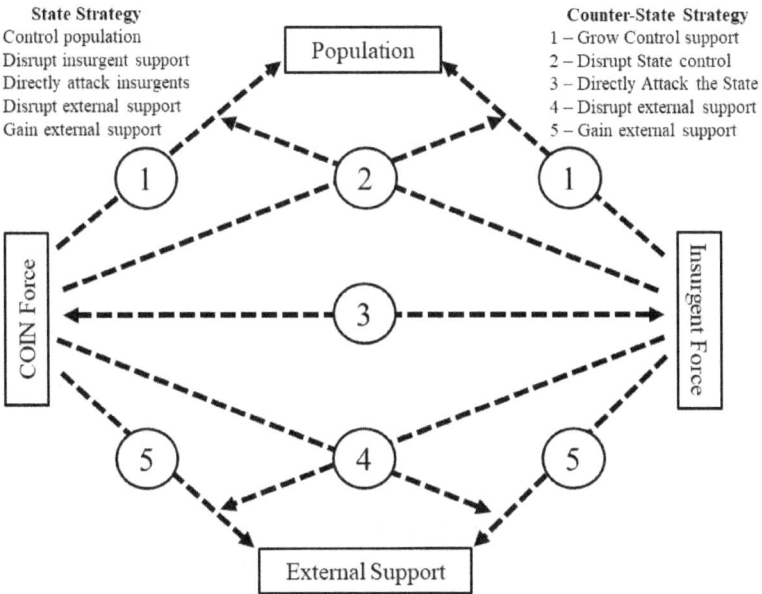

McCormick's Magic Diamond.

McCormick identifies four factors: These are, the population, the insurgent, the counterinsurgent, and the international community. By the conceptual framework of the "magic diamond," we may better understand the complexities of irregular warfare. In the diagram, the counterinsurgent/insurgent demonstrates the first action by gaining support over the affected population (1). In the second action, the counterinsurgent/insurgent disrupts their opponent's control over the population (2). In the third action, the counterinsurgent/insurgent conducts direct action against the opponent (3). The counterinsurgent/insurgent works to disrupt their opponent's relations with the international community (4). Last, the counterinsurgent-insurgent establishes relations with the international community.

In a well-researched and detailed analysis, Max Boot addresses this "complex set of circumstances and motives" in his epic history of guerrilla warfare entitled *Invisible Armies*. Out of his narrative, Boot analyzes important insights that are relevant to today's counterinsurgents.

As an in-house discussion, there is debate as to which approach in counterinsurgency is best: people-centric, enemy-centric, intelligence-centric, and leader-centric. Most theorists posit either a population or leader-centric approach. In *A Question of Command* (2009), Mark Moyar argues that counterinsurgency is "leader-centric" warfare in which those with superiority in certain leadership attributes usually win.[29] Moyar argues that certain leadership attributes (initiative, flexibility, creativity, judgment, empathy, charisma, sociability, dedication, integrity, and organization) produce the ideal counterinsurgency commander and quality of leadership that determines the outcome of any counterinsurgency campaign.

More recently, in his *The American Way of Irregular War: An Analytical Memoir* (2020), Lieutenant General (Retired) Charles Cleveland traces his career in the Special Forces and offers his recommendations for the restructuring of America's approach to irregular warfare. According to Cleveland, when it comes to IW, America has been largely successful at the tactical level. However, over the past four decades, lacking effective irregular warfare-fighting concepts and population-centric doctrine, the United States has failed to achieve its strategic objectives in nearly every population-centric military campaign. Voicing a hundred years' worth of collective COIN wisdom, Cleveland states, "Population-centric conflicts are irregular warfare contests that are won and lost by controlling and influencing populations rather than occupying territory."[30] The present need, argues Cleveland, is for America's irregular warfare-fighting capability to match her nuclear and conventional capabilities. We couldn't agree more. The hope is for this volume to help fill that void.

Summary and Implications

The literature we have surveyed reveals several salient tenets of irregular warfare. Principally, every theorist argues for a population-

---

[29] Mark Moyar, *A Question of Command: Counterinsurgency from the Civil War to Iraq* (New Haven, CT: Yale University Press, 2010), 3.

[30] Charles Cleveland, *The American Way of Irregular War: An Analytical Memoir* (Santa Monica, CA: RAND, 2020), xiv.

centric approach that is buttressed by a doable political end state. Secondly, the complexity of this subset of warfare has led every IW theorist to underscore the importance of learning and adaptability. Lastly, while not clearly articulated by every pundit, legitimacy in IW operations carries the day with the affected populace. As will be seen, these tenets comprise the very foundation for success.

For Further Reading:

1. *On Guerrilla Warfare* by Mao Zedong.
2. *The Accidental Guerrilla* by David Kilcullen.
3. *A Question of Command* by Mark Moyar.

2

Two Sides of the Same COIN

"For always, no matter how powerful one's armies, to enter a conquered territory one needs the goodwill of the inhabitants." – Machiavelli

Strategic Hamlet Program, South Vietnam 1961. Courtesy AP.

As two sides of the same coin, insurgency, and counterinsurgency (COIN) are complex forms of irregular warfare. The nature of their complexity is underscored by the fact that defining them is often "plagued with semantic and definitional difficulties and misconceptions."[1] This difficulty is alleviated somewhat by using functional definitions. Current U.S. doctrine defines insurgency as "an organized movement aimed at the overthrow of a constituted government through the use of subversion and armed conflict."[2] In its simplest form, insurgency is waged by those who are inferior in numbers,

---

[1] Napoleon D. Valeriano and Charles T. R. Bohannan, *Counter-Guerrilla Operations: The Philippine Experience* (New York, NY: Praeger, 1962), 3.
[2] U.S. Army, *FM 3-24, Counterinsurgency* (Washington, D.C.: Government Printing Office, 2006), 1-1.

equipment, and financial resources, making it impossible to meet their opponents in conventional battle. However, as an asymmetric threat, insurgencies employ guerrilla warfare, conventional warfare, terrorism, or all of them combined. Yet, despite limited resources, an insurgency can have an enormous impact. For example, while everyone knows the American War of Independence ended shortly after Lord Cornwallis surrendered British forces to George Washington at Yorktown, few realize he was in Yorktown because his army was mauled by guerrillas in the Carolinas.

As another example, its common knowledge that Napoleon sent his 600,000-man *Grande Armée* into Russia in June of 1812 only to see it decimated in its retreat that winter. Fewer, however, know that in 1813 Napoleon sent an even larger force of some 670,000 to conquer Spain, only to see it frittered away in its *little war* against guerrillas in less than two years. Even fewer realize that on the eve of the Battle of Waterloo, Napoleon had to send some 30,000 troops to contain the royalist insurgents in Western France (Vendée), leaving him with only 72,000 to face Wellington.[3] An insurgency can have an enormous impact!

On the flip side is counterinsurgency (COIN), which is defined as "military, paramilitary, political, economic, psychological, and civic actions taken by a government to defeat insurgency."[4] COIN is essentially a competition with the insurgent for the right to win the hearts, minds, and acquiescence of the population. One of the aims of this book is to understand how these two forms of irregular warfare, insurgency and counterinsurgency, interact. To that end, this chapter is an effort to provide a framework of this phenomenon of human history, with the goal of lessening its complexity.

Insurgency

As Christopher Paul has poignantly stated, insurgency has been the most prominent form of armed conflict since 1949.[5] Insurgency is the

---

[3] George Lefebvre, *Napoleon from Tilsit to Waterloo* (New York: Columbia University, 1969), p. 363.
[4] Joint Publication 1-02.
[5] Christopher Paul, *Victory Has a Thousand Fathers* (Santa Monica, CA: RAND, 2010), xiii.

traditional strategy of the weak resisting the strong. According to current Army doctrine, there are five general root causes for insurgencies: identity, religion, occupation or exploitation, economic failure, and corruption and repression.[6] Yet, paradoxically, in many instances, the more concrete the objectives, the easier the insurgency can be defeated.[7] Insurgency is essentially a political phenomenon brought on by particular drivers of instability and grievances, real or perceived. As such, it has taken many forms over time, and each one is unique. As borne out by history, there are essentially three drivers of instability: land grievances, lack of state services, and monetary disputes.

Giving voice to one of these drivers of instability or other grievances, insurgents employ both violent and nonviolent tactics to achieve their objectives. Early on, an insurgency normally uses terrorism and guerrilla warfare to gain momentum and acceptance. Insurgents use military tactics, such as ambushes, sabotage, and raids to harass larger and less mobile traditional forces. Following the tactic of the weak versus the strong, insurgents strike at vulnerable targets and then melt back into the populace. Successful insurgents may often use nonviolent tactics in conjunction with violent tactics, such as subversion and propaganda; the two most prevalent forms of nonviolent warfare.

According to current U.S. doctrine, an insurgency is "an organized movement aimed at the overthrow of a constituted government through use of subversion and armed conflict." An insurgency is typically an internal struggle within a state, not between states. It is normally a protracted political and military struggle designed to weaken the existing government's power, control, and legitimacy, while increasing its own power, control, and legitimacy. Because insurgents initially lack the resources necessary to confront the incumbent government in conventional warfare, they tend to adopt an irregular approach that uses violence and other means to wrest the support of the population away

---

[6] FM 3-24.2, *Tactics in Counterinsurgency*, 1-18.
[7] Napoleon D. Valeriano and Charles TR Bohannan, *Counterguerrilla Operations* (New York: Praeger, 1962), 18.

from an established authority.[8] All insurgencies therefore use a combination of force and popular appeal.[9]

As current U.S. doctrine holds, there are specific physical and environmental conditions that allow for a successful resistance or insurgency. The three main conditions are: (1) a weakened or unconsolidated government or occupying power, (2) a segmented population, and (3) favorable terrain from which an element can organize and wage subversion and armed resistance.

Six Major Insurgent Approaches

An insurgency must be understood before it can be defeated. As articulated by *FM 3-24 Counterinsurgency*, current U.S. doctrine understands there to be six common insurgent approaches/strategies. These are:

1. Urban
2. Military-focused
3. Protracted popular war
4. Identity-focused
5. Conspiratorial
6. Composite & Coalition

In the urban approach, insurgents attack government targets with the intention of causing government forces to overreact against the population. This was the strategy employed by the National Liberation Front in Algeria. They launched a series of bombings and attacks, causing significant civilian casualties, in order to shock the French into negotiations. A nuance to this is a military-focused strategy which believes that military action can create the conditions necessary for success. Che Guevara and Fidel Castro both used this approach.

Arguably, the most followed is the protracted popular war approach of Mao Zedong. As previously stated, this theory posits a winning strategy

---

[8] David C. Gompert and John Gordon IV, *War by Other Means: Building Complete and Balanced Capabilities for Counterinsurgency* (Santa Monica, CA: RAND, 2008), iii-iv.
[9] Ibid., xxix.

broken down into three distinct phases: latent or incipient, guerrilla warfare, and war of movement (strategic defense, strategic stalemate, and strategic offense). Each phase builds upon the previous one and continues activities from the previous phases. A classic example of this approach is the one that was executed by the communist insurgents in the Chinese Civil War and the Vietnam War.

The <u>identity-focused</u> approach mobilizes support based on the common identity of religious affiliation, clan, tribe, or ethnic group. This was the approach employed unsuccessfully by the Tamil Tigers in Sri Lanka.

In <u>conspiratorial</u> strategy, insurgents work at subverting the government from within. This approach often involves a few leaders and a militant cadre. This was the approach used by the Bolshevik Party to seize power in Russia in 1917. Lastly, the <u>composite and coalition</u> approach combines various strategies to serve the purposes of the different groups. This approach has been used by Al Qaeda and ISIS with successful results. Regardless of the approach, insurgencies are often a coalition of disparate forces united by their common enmity for the government.

Looking at all these approaches, central to any insurgency is the gaining of political power through the galvanizing of the people's dissatisfaction. To be successful, an insurgency must develop unifying leadership, doctrine, organization, and strategy. Though insurgencies take many forms, most share certain common attributes. An insurgent organization normally consists of five elements: Leaders, combatants, political cadre, auxiliaries, and the mass base, that is, the bulk of the membership. The proportion of each element relative to the larger movement depends on the strategic approach the insurgency adopts.[10]

Seven Insurgent Dynamics

While the ultimate goal or end state desired by an insurgency may vary, all insurgents use organized violence to affect change within a state.[11] Insurgencies often arise when the populace perceives that the

---

[10] Taber, *The War of the Flea*, viii.
[11] Williams, *Insurgency Terrorism: Attitudes, Behavior and Response*, 14.

government is unable or unwilling to redress their issues or the demands of important social groups. These groups band together and begin to use violence to change the government's position. As this occurs, generally speaking, there are seven dynamics that are common to most insurgencies. These dynamics may be remembered by the acronym: **PEE OOIL**

P – Phasing and timing. Insurgencies often progress through three phases in their efforts: latent or incipient, guerrilla warfare, and war of movement. As an example, communist insurgents in Vietnam waged "a classic, peasant-based, centrally directed, three-stage, Maoist model insurgency, culminating in a conventional military victory."[12]

E – Environment and Geography. These two are the cultural and demographic factors which influence insurgent doctrine, decisions, as well as tactics that are most often suited to the terrain. Successful insurgencies have taken root in environments as diverse as deserts and jungles. Whatever the terrain, "dispersion is an essential condition of survival and success on the guerrilla side."[13]

E – External support. Examples such as Vietnam, Nicaragua, Afghanistan, and Iraq demonstrate how external support can accelerate events and influence the outcome of insurgencies. Likewise, an absence of such support has led to the isolation and eventual demise of island or peninsular insurgencies, such as that of the Tamil Tigers, and the Malayan National Liberation Army.

O – Objectives. Insurgencies normally seek to achieve one of three objectives: overthrow the existing government, expel whom they perceive to be "occupiers," or create a region that they can exploit where there is little or no governmental control. As stated by *FM 3-24.2, Tactics in Counterinsurgency*, the central issue in an insurgency is the reallocation of power.

---

[12] Jeffery Record, *Iraq and Vietnam: Differences, Similarities and Insights*, 2.
[13] B.H. Liddell Hart, *Strategy* (London: Faber & Faber, 1954), 365.

O – Organizational and operational patterns – Insurgent organizational and operational patterns vary widely between one province or urban area and another. Different insurgent groups using different methods may form loose coalitions when it serves their interests. Such as the initial Sunni and Shi'a wings of the Iraqi insurgency. While each insurgency organization is unique, they often follow common patterns or strategies. Knowing these helps COIN practitioners predict the tactics and techniques insurgents may employ against the government.

I – Ideology. To win, an insurgency must be able to justify its actions and have a program that explains what is wrong with society. An insurgency's ideology must therefore promise great improvements after the government is overthrown or an occupying power is ousted. According to most political analyses, the ideological content of a revolution is generally considered a necessary condition for setting a revolution in motion, as it provides a unifying framework, motivates people to action, and articulates the goals and grievances that drive the revolutionary movement. In short, an insurgency must have a cause. Understanding the root cause(s) of the insurgency is therefore essential to a COIN victory. According to Bernard Fall, over against the common misconception, "Unlike machinegun bolts, ideologies are not easily interchangeable." Each insurgency will have to be fought on its own ground.

L – Leadership. An insurgency requires leadership to provide direction, guidance, coordination, and organizational cohesiveness. The goal of insurgent leadership is to replace the government's legitimacy with that of its own.

In the last analysis, the insurgent must acquire at least the passive support of the native population in order to survive and develop. As will be made clear by the following case studies, without the consent and active aid of the people, the guerrilla would be merely a bandit and could not long survive. If, on the other hand, the counterinsurgent could claim this same support, the guerilla would not exist, because there would be no war, and no revolution. The cause would have evaporated; the popular impulse toward radical change – cause or no cause – would be dead.[14]

The Seven Phases of a U.S. Sponsored Insurgency

Unconventional warfare (UW) is defined as "activities conducted to enable a resistance movement or insurgency to coerce, disrupt, or overthrow a government or occupying power by operating through or with an underground, auxiliary, and guerrilla force in a denied area."[15] UW will vary in its application as each environment is unique but will generally pass through seven distinct phases. While a fuller explanation of these phases is beyond the scope of this volume, current U.S. doctrine organizes UW into the following seven phases:

Phase I Preparation – Intelligence preparation, planning, and shaping activities.

Phase II Initial Contact – U.S. Government agencies coordinate with a resistance or government in exile.

Phase III Infiltration – Special Forces (SF) unit infiltrates and establishes communications with a resistance movement.

Phase IV Organization – SF unit organizes, trains, and equips resistance cadre into three basic components: guerillas, underground, and auxiliary.

Phase V Build Up – Development of infrastructure: SF unit assists cadre with expansion into effective organization.

Phase VI Employment – Resistance forces conduct combat operations.

Phase VII Transition – UW forces revert to national control.

---

[14]Taber, *The War of the Flea*, 12.
[15] Ibid.

Counterinsurgency (COIN)

Counterinsurgency is a competition with an insurgency for the right to win the hearts, minds, and acquiescence of the population. COIN strategies typically fall into either a population-centric approach, where the primary focus is on winning over the affected population and isolating the insurgents, or an enemy-centric approach, which prioritizes directly defeating the insurgent forces through military means. A population-centric strategy emphasizes building relationships with the affected population, providing essential services, and addressing their grievances to prevent them from supporting the insurgents. The enemy-centric strategy prioritizes actively engaging and neutralizing the insurgent forces through military operations, often with a focus on intelligence gathering and targeted strikes. Arguably, success in COIN requires a balance between these two strategies.

Regardless of the approach, the people must perceive that the legally constituted government of their country is working in their best interest. For a counterinsurgency to work, the government must be perceived as legitimate. Such legitimacy must be perceived through a careful coalescence of political, social, psychological, economic, and military measures.

Six Major Counterinsurgency Approaches

Successful COIN requires an adaptive force led by agile leaders. While every insurgency is different because of distinct environments, root causes, and cultures, all successful counterinsurgency is based on common principles. Additionally, as aforementioned, a successful COIN approach must apply discriminate use of force in order to separate the insurgency from the population. Separating the insurgency from the population both psychologically and physically is an important, if not essential, mechanism to aid in their defeat. In light of the most successful counterinsurgencies, there are arguably six main counterinsurgency approaches:[16]

1. Development/Reforms
2. Amnesty and Rewards
3. Pacification
4. Border Control
5. Resettlement
6. Shape-Clear-Hold-Build-Transfer

1. Development/Reforms. Arguably, a comprehensive strategy that addresses societal needs is more effective than conventional military efforts alone. In the opinion of the RAND researchers, this COIN approach, when applied correctly, receives the highest recommendation as it focuses on winning the hearts and minds of the effected populace. As previously noted, there are essentially three drivers of instability: land grievances, lack of state services, and monetary disputes. When a government applies this approach, the aim is to reduce the perceived grievances. Successful examples of this approach include that of the U.S.-led counterinsurgency campaign during the Salvadoran Civil War which brought stability following the twelve-year conflict.

2. Amnesty and Rewards. The amnesty and rewards approach is among the most successful. It's an attractive option, observes Christopher Paul, that reduces "the need for a fight to the finish."[17] Amnesty and rewards is a cost-effective option. In many successful COIN campaigns, amnesty for insurgents has been a key component. This approach, however, is not a stand-alone option but rather is adjunct with other approaches to accelerate the end of an insurgency. A textbook case is that which was used against the Huks in the Philippines (see chapter 7).

---

[16] Other COIN approaches consist of cost-benefit, psyops, strategic communications, search and destroy, etc. As will be argued, these approaches may be best understood as supplementary nuances to these six main approaches.

[17] Christopher Paul, Colin P. Clarke, and Beth Grill, *Victory Has a Thousand Fathers: Sources of Success in Counterinsurgency* (Santa Monica, CA: Rand Corporation, 2010), 54.

3. *Pacification.* As noted by many, pacification is a broad-based approach that employs various population-centric techniques.[18] When a government uses this approach, it seeks to assert its influence over the affected area through security efforts, the implementing of programs aimed at the distribution of food and medicine, while coordinating such efforts with the military and civil agencies. Perhaps the most famous example of this approach was the U.S. pacification efforts during the Vietnam War. Among these efforts were the Strategic Hamlet Program, which was successfully employed by the British during the Malayan Emergency; it nonetheless failed to take root in South Vietnam for various reasons (see chapter 8).

According to Christopher Paul, there are three main focuses to this approach: a perception of security among the affected population, along with various short-term investments and improvements, as the COIN force expands secure areas under their control.[19]

4. *Border Control.* Being more stick than carrot, when applying this approach, a government seeks to siphon off the external support insurgents may receive to bolster their cause. As history shows, insurgencies often enjoy cross border sanctuaries, whereby they may regroup and equip themselves indefinitely. As observed by Jonathan Owen, "effective border security strikes directly at the center of gravity of a cross-border insurgency by turning the cross-border sanctuary into a critical vulnerability that can be cut and exploited."[20] As will be argued, border control is most effective when it's part of an overall COIN strategy.

5. *Resettlement.* Population resettlement has contributed to the defeat of insurgent movements by "draining the swamp," i.e., physically separating them from the population. Arguably, resettlement works best when used in conjunction with other approaches, while left on its own, it's a poor COIN practice. History shows this approach has been the key element in eradicating insurgencies from South Africa to Malaya.

---

[18] Richard A. Hunt, *Pacification: The American Struggle for Vietnam's Hearts and Minds* (Boulder, CO: Westview Press, 1995), 11-15.
[19] Christopher Paul, *Victory Has a Thousand Fathers,* 39.
[20] Jonathan Owen, *#Fail: Why the U.S. Lost the War in Afghanistan* (Fayetteville, NC: Blacksmith, 2017), xvii.

6. *Shape-Clear-Hold-Build-Transfer* – As advocated by current U.S. doctrine (FM 3-24 *Counterinsurgency*), this approach emphasizes seven lines of effort: establish civil security, establish civil control, support Host Nation forces, support to governance, restore essential services, support economic and infrastructure development, and conduct information engagement.

The COIN force begins by *shaping* the environment and information to their advantage, concentrating on population centers and the reduction of insurgent influence over them. In *clearing*, the COIN force destroys, captures, or forces the withdrawal of insurgents to secure a physical and psychological environment that establishes or re-establishes government control.

The COIN force then *holds* the contested areas and *build* defensive networks, radiating security and influence from this cleared area. Build tasks include improving infrastructure and roads, reestablishing services, schools, and facilities. Finally, in *transition*, the COIN force hands over operations to host nation security forces

Ultimately, there are no cookie-cutter approaches to counterinsurgency. COIN practitioners must therefore be adaptive to fit their particular strategy and tactics to specific circumstances and then be flexible enough to adapt them as often as necessary to win. There is therefore no standard set of metrics, benchmarks, or operational techniques that apply. Nonetheless, while every insurgency is different, all successful COIN operations are based on common principles. The following table illustrates the six COIN campaigns and their outcomes that are discussed in this volume:

Conflict	Strategy	Main Approach	Result:
Philippine Insurrection	Hybrid[21]	Resettlement	COIN Win
Hukbalahap Rebellion	Hybrid	Resettlement	COIN Win
Vietnam War	Enemy-centric	Pacification	COIN Loss
Afghanistan War	Population-centric	Clear-Hold-Build	COIN Loss
OEF Philippines	Population-centric	Pacification	COIN Win
Iraq War	Population-centric	Clear-Hold-Build	COIN Win

---

[21] By hybrid is meant population, enemy, and intel-centric.

In light of history and reason, success in COIN requires an operational design that: (1) Identifies the root cause(s) of an insurgency and its leadership; (2) applies a discriminate use of force to separate the insurgency from the population, both psychologically and physically; and (3) stabilizes the affected area by identifying with the people, working with them to create a better state of the peace, winning their allegiance, and, if possible, redress the grievances that led to the insurgency. Conceptually, this is a three-legged stool, and leadership is the seat.

While at first glance this all may seem self-evident, however, as borne out by history, it is not. Likewise, though many COIN campaigns have led to the successful demise of insurgent leaders and their organizational frameworks, due to the course in which these campaigns have run, in the end, there is no better state of the peace, the cause having submerged itself into the background, waiting for more opportune times.

As will be argued, the Shape-Clear-Hold-Build-Transition framework is the best operational approach to successful counterinsurgency.

Summary and Implications

Winston Churchill once described the Soviet Union as "a riddle, wrapped in a mystery, inside an enigma." It's fair to say that many see COIN in precisely the same way. To be sure, there are complexities, yet the basic tenets are clear enough. Moreover, the annals of history include many COIN campaigns that were highly successful. They are, nonetheless, unacceptable by modern standards. Take for example Caesar's brutal methods in his Gallic Wars, which, in large measure consisted of counterinsurgency. Caesar's tactics were as resourceful and decisive as they were savage. Caesar's conquest of Gaul was his greatest achievement. Yet, as J.F.C. Fuller observes:

Had he heeded a maxim to be found in a book obtainable in his day, might not he have accomplished all he did more speedily, at lesser cost, and at greater advantage? The maxim reads: 'But he who acts in a harsh and savage manner, immediately after becoming master of a city…makes other cities hostile, so that the war becomes laborious for him and victory difficult to attain...For nothing makes men so brave as the fear of what they will suffer if they surrender. [22]

As history shows, Russia's Afghanistan and Chechnyan experiences would fall into this category. In Afghanistan, with little understanding of Afghan culture, or a desire to gain the people's consent, the Russians' intent was to wipe out opposition with overwhelming firepower. They likewise applied the same approach in Chechnya. In both conflicts, Russian mistakes were legion. Rather than attempt to separate the insurgents from the population, either physically or psychologically, they "tried to extirpate the population with artillery fires." [23]

For Further Reading:

1. *FM 3-24, Counterinsurgency.*
2. *FM 3-24.2, Tactics in Counterinsurgency.*
3. *Victory Has a Thousand Fathers* by Christopher Paul.

---

[22] J. F. C. Fuller, *Julius Caesar*, 165.
[23] Robert M. Cassidy, *Russia in Afghanistan and Chechnya: Military Strategic Culture and the Paradoxes of Asymmetric Conflict* (Carlisle, PA: Strategic Studies Institute), 2003.

3

Five PLAIN Laws of Irregular Warfare

"In war many roads lead to success, and they do not all involve the opponent's outright defeat." – Clausewitz

Iraqi soldiers seize a weapons cache in Basra, Iraq, April 2008. Courtesy of AP.

Irregular warfare is a type of combat among state and nonstate actors that uses indirect and asymmetric tactics to weaken an opponent's power, legitimacy, influence, and will to fight. As an example of a state actor, a government may support an insurgency in a neighboring country via covert operations to destabilize a regime or use propaganda to sway public opinion. Likewise, nonstate actors, such as insurgents, fight against a government, using terrorist and guerrilla attacks to achieve political goals. As a contest over the hearts and minds of the affected population, IW uses a variety of tools, including economic coercion, information operations, such as propaganda, cyber operations, covert action, and espionage.

36

Complex and multidimensional, IW can occur alongside conventional warfare, providing a significant line of effort to an overall campaign, and can include large-scale combat operations (LSCO). With all its complexity, can it be said that there are laws of IW? Arguably, there are five PLAIN laws of irregular warfare, which may be remembered by the acronym: **PLAIN**

P – Political Objective
L – Legitimacy
A – Adaptability
I – Influence
N – Native Face

P – Political Objective: <u>Political objectives are the goals that a war is fought to achieve</u>. This obvious tenet is often overlooked. Without defined political goals, a COIN force can easily become trapped in a cycle of simply suppressing violence without addressing the root cause(s) of an insurgency. Likewise, without a cause, an insurgency cannot exist. This is the principal imperative. French theorist David Galula rightly argued that, "a cause is an insurgency." Accordingly, Mao opined, "Political power grows out of the barrel of a gun."

For belligerents on both sides of the coin, the justification, or cause for the war, should be seen as the need to achieve "a better state of the peace." In order for the war to serve political ends, policy should be a guide for how operations develop from planning to execution. Military actions may meet their objectives but if they are not conducted toward a political objective the results they achieve may be entirely counterproductive. For the counterinsurgent, success requires an operational design that identifies and undermines the root cause of an insurgency. They must, as Sun Tzu suggests, "attack the enemy's strategy." For this, he said, "is of supreme importance in war."[1]

Likewise, on both sides of the coin, a vision must be cast to what a better state of the peace looks like. According to FM 3-24, the U.S. Army's field manual on counterinsurgency, "Resolving most insurgencies

---

[1] Sun Tzu, *The Art of War*, Translated by Samuel Griffith (New York: Oxford, 1971), 77.

requires a political solution; it is thus imperative that counterinsurgent actions do not hinder achieving that political solution."[2] As borne out by history and reason, a government that ignores this first law defeats itself more often than they would be defeated by the insurgency they are combating. Failing to adhere to this law is a failure to address root causes that led to the insurgency.[3]

Axiom 1: A cause is an insurgency.
Axiom 2: The political objective should be to achieve a better state of the peace.
Axiom 3: An actionable policy should be undergirded by a doable goal.

L – Legitimacy: Legitimacy is the qualitative condition of the governing authority, as is perceived by the people who give their consent.[4] In large measure, IW is a contest for legitimacy (Galula). Legitimacy includes legality, level of popular support, and international acceptance. IW practitioners must gain and maintain a condition of perceived competence (Galula). For counterinsurgents, this is achieved through diminishing the grievances that brought on instability.

As Napoleon once remarked, in war, the moral is to the physical three to one. He meant success often depends upon the human aspect as well as the numerical and technical. The litmus test for a governing authority is how it uses power. In the context of counterinsurgency, a governing authority is legitimate if it can secure the population and separate it from the insurgency in accordance with the law. Just how this separation comes about is a delicate balance between the age-old dichotomy of either a population-centric or enemy-centric approach. Population-centric conflicts cannot be fought with military concepts and doctrine designed for the physics of conventional war and instead require approaches that blend anthropology, economics, history, sociology, and

---

[2] U.S. Army, *FM 3-24, Counterinsurgency* (Washington, D.C.: Government Printing Office, 2006), 1-22.
[3] Ben Connable and Martin C. Libicki, *How Insurgencies End* (Santa Monica, CA: RAND, 2010), 152.
[4] John Locke, *Two Treatises on Government.*

an understanding of when and where the reality of the physics of war applies.[5]

To quote the British Army field manual on countering insurgency, the principal focus for the security line of operation in counterinsurgency must be:

The security of the population, rather than security for the forces themselves, or attrition of insurgents. Experience continues to demonstrate that until the local population starts to believe it is secure it will not start to support its government, nor will it begin to provide the intelligence and information crucial for effective counterinsurgency.[6]

As is self-evident, for a counterinsurgency effort to succeed it must neutralize insurgents. Albeit, success in COIN requires a discriminate use of force to separate the insurgency from the population, both psychologically and physically while securing the affected population.

As for consent, the people must perceive that the governing authority is competent and is acting in accordance with the law and acceptable values and norms. The people must sense the governing authority is acting in the people's best interest. To quote again the British Army field manual on countering insurgency, "It is axiomatic that gaining consent will reduce support available to the insurgent, and this will weaken the insurgency. The insurgent will not give up support easily and a fierce backlash should be anticipated when operations get underway to secure the population."[7]

It's important to remember that consent is conditional. In many ways, it may take the form of willing acceptance or passive compliance. Galula writes,

Having attained the support of the population it is imperative to remember that this support is conditional. What you do matters, and support can be lost if your actions are unfavorable to the population.[8]

---

[5] Charles T. Cleveland, *The American Way of Irregular War: An Analytical Memoir* (Santa Monica, CA: Rand Corporation, 2020), xviii.

[6] British Army Field Manual, Volume 1 Part 10, October 2009, 3-8.

[7] Ibid., 3-12.

[8] David Galula, *Counterinsurgency Warfare: Theory and Practice*, 86.

Not only should COIN practitioners consider all actions in light of the rules of engagement, but they should also further consider the perception of the manner in which they are carried out. The neutral majority must be won over and the hostile minority neutralized by an appropriate and discriminate use of force.

Axiom 1: Legitimacy involves gaining and maintaining a condition of perceived competence.
Axiom 2: Legitimacy is achieved through diminishing the motives that brought on the instability.
Axiom 3: The moral is to the physical three to one. In UW, success often depends on the human aspect as much as it does the numerical and technological.

A – Adaptability: Irregular warfare is a learning competition that requires adaptive approaches to complex and changing threats.[9] Two crucial components of adaptability are cultural intelligence (CQ) and competence. Success in IW involves an inversion of the traditional war principle of dominating political space. As such, the people are the contested *land*, and the center of gravity (COG). Mixing metaphors, Mao referred to the people as the *sea* in which the insurgent swims.

Culture is a learned behavior and consists of the set of opinions, beliefs, values, and customs that define a society. Integrated into every facet of a society, culture includes things like social behavior, language, and religion. Arguably, the principal component and starting point in any war is knowing the terrain over which it will be fought. The same is true regarding irregular warfare, albeit a further dimension is added – the people are the contested land. As IW is conducted on these two playing fields, understanding human terrain is key. As a complex undertaking, COIN cannot be conducted effectively without having a detailed understanding of human terrain. Cultural intelligence recognizes and adapts to cultural differences and arguably includes three basic competencies: CQ knowledge, CQ drive, and CQ capability.

CQ knowledge is understanding how culture influences the way people think and behave, as well as one's level of familiarity with how cultures

---

[9] John Nagl, *Learning to Eat Soup with a Knife.*

are similar and dissimilar. CQ drive is the motivation to learn and adapt to new and diverse cultural settings. CQ capability is being able to draw from one's understanding of the target culture to solve complex problems, along with being able to act appropriately when interacting: "Failing to understand the human terrain can lead to misunderstandings or, worse, bring soldiers into conflict with the local population."[10]

Underscoring the importance of competence, in IW, the side that learns faster and adapts more rapidly—the better learning organization— usually wins. *Irregular warfare is therefore a learning competition; making the ability to shift gears and change one's approach a key ingredient to success.* Underscoring John Nagl's emphasis, prosecuting IW therefore involves adapting tactics, operations, and strategy to fit specific circumstances, and then being prepared to change them as often as necessary to win. To quote the U.S. Army's field manual on counterinsurgency, "An effective counterinsurgent force is a learning organization. Insurgents constantly shift between military and political phases and approaches.

In addition, networked insurgents constantly exchange information about their enemy's vulnerabilities – including information exchanged with other insurgents in distant theatres. "A skillful counterinsurgent is able to adapt at least as fast as the insurgents."[11] Underscoring this necessity of adaptability, Sun Tzu compared war to the flow of water, and observed that the one able to gain victory by modifying his tactics in accordance with the enemy situation "may be said to be divine."[12]

Axiom 1: The people themselves are the contested *land*.
Axiom 2: People are the center of gravity (COG).
Axiom 3: The people are the "sea" in which the insurgent swims.

I – Influence: Influence is the capacity to affect perception. Insurgents/Counterinsurgents influence perception by actively managing how the affected population views their presence. All

---

[10] British Army Field Manual, Volume 1 Part 10, October 2009, 3-5.
[11] U.S. Army, *FM 3-24, Counterinsurgency* (Washington, D.C.: Government Printing Office, 2006), 1-26.
[12] Sun Tzu, *The Art of War*, Translated by Samuel Griffith (New York: Oxford, 1971), 101.

information afforded to the affected population must be carefully crafted according to a narrative that suits the political goal. Influence involves vying with an opponent to retain or regain control of the support base – the people. Every action requires a message that fits the narrative (messaging battle). The goal in influence competition involves winning over fence-sitters and neutralizing nay-sayers.

To regain or retain credibility, a narrative is to be wrapped in a blanket of truth to gain the compliance of the affected population with norms and a worldview that is favorable to the host government. A strategic narrative is generated to be leveraged at the tactical level. For example, during the American Revolution, Thomas Paine wrote a pamphlet entitled *Common Sense*. In it, he advocated that independence from Great Britain was moral, arguing "The cause of America is in a great measure the cause of all mankind." Such narratives are more art than science, and build trust, emphasize the righteousness of a cause, and foment nationalistic fervor.

Regarding the capacity to influence, there are arguably four essential aspects: leadership, intelligence, identification with the people, and integration. Influence is the essence of what leadership is and does. It's the ability to influence others to obtain their obedience, respect, confidence, and loyal cooperation. As the central warfighting function, good leadership generates the support and compliance of the people. As history is replete with examples of disastrous military operations and failed counterinsurgency campaigns, both of which have costly price tags in terms of blood and treasure, the importance of good leadership cannot be overstated.

Intelligence drives operations. In counterinsurgency, intelligence is more important than firepower. Without intelligence, "COIN forces are at risk of conducting unfocused operations which waste resources, including time, resulting in alienation of the local population while creating tensions with partners, regional and international audiences."[13]

While intelligence is self-explanatory, as borne out by history, identifying with the people is not. Consider the Greek guerilla army of National People's Army of Liberation (ELAS) under the leadership of the Communist Party of Greece (KKE), while the mistakes of the ELAS were

---

[13] British Army Field Manual, Volume 1 Part 10, October 2009, 3-15.

legion, chief among them was their alienation of the population. As considered by Mao, identifying with the people is therefore the single most important task of the guerrilla; the same holds true for the counterinsurgent.

As intelligence drives operations, integration of military and political efforts ensures actions are comprehensive and unified. By definition, COIN is a blend of civilian and military efforts. Effective COIN therefore integrates the elements of national power and wields them together with police, security forces, those of nongovernmental organizations (NGOs), intergovernmental organizations (IGOs), and everything else in between, to include the integration of intelligence, and psychological operations into activities.

To reiterate, IW is a contest of control over the support base – the people. Such control includes security, infrastructure, essential services, public administration, economic stability, civil order, including border control. As Epictetus would tell us, some things are up to us, others are not. A great many things play crucial roles in the outcome of a COIN effort, but many are beyond the control of those who prosecute it. Such elements include media, private companies, local leaders, and religious groups. Counterinsurgents do best to focus on what they can control, remain aware of the influence of such groups, and prepare to work with, through, or around them.

Axiom 1: A narrative must be wrapped in a blanket of truth to gain compliance.
Axiom 2: Every action requires a message that fits the narrative (messaging battle).
Axiom 3: Fence-sitters must be won over and nay-sayers neutralized.

N – Native Face: <u>Counterinsurgents/insurgents cannot win as an outside power</u>. For counterinsurgents, an indigenous force must be compatible with the affected population and provide connectivity with the host nation government. This entails that the COIN force finds local solutions to local problems; all COIN is local. In this effort, if the counterinsurgent military force ratio is lopsided to the indigenous, chances are counterinsurgents will appear as a foreign-sponsored enterprise, leading to what David Kilcullen calls the "accidental guerrilla syndrome." Likewise, IW practitioners must heed T.E. Lawrence's axiom, "It's better the indigenous force does it tolerably than for you to do it perfectly.

Employing local forces can be more cost-effective than deploying large foreign military contingents. However, there are various challenges to consider when using indigenous security forces. There may be loyalty concerns. Ensuring their loyalty to the government and not being influenced by the insurgency can be difficult. Additionally, they may require substantial training and proper equipment to be effective in combat operations. Counterinsurgency campaigns involve a large influx of funds. As such, there could be a risk of corruption, especially if proper oversight mechanisms are not in place.

Axiom 1: Find local/national solutions to local/national problems.
Axiom 2: Counterinsurgency is local (Galula).
Axiom 3: It's better the indigenous force does it tolerably than for you to do it perfectly (Lawrence).

Ten Irregular Warfare Lines of Effort

By adhering to the five PLAIN laws, and adaptively employing the ten tactical lines of effort (LOEs), IW practitioners stand the best chance of victory. The ten LOEs may be remembered by the acronym: **DISSECTION**

D – Diminish or Deepen the Motives. IW is a learning competition. Success therefore requires a COIN force to work with the host government to diminish the motives that contributed to the cause of the insurgency. Likewise, competent insurgents work to deepen motives that demonize the host nation. Such motives are grievances of various sorts, real or perceived.

I – Intelligence. Effective IW requires carefully considered actionable intelligence, gathered and analyzed at the lowest possible levels, and disseminated and distributed throughout the force.

S – Security Forces. Not surprisingly, a COIN force must stabilize the affected area. Under martial law, COIN forces can impose curfews, cordon off areas, and issue passes. How that is accomplished is crucial to success as security forces in COIN are the connecting link between the affected population and the government. Insurgents actively work to degrade this capability.

S – Strategic Communications (SC). Every action requires a message that fits a carefully crafted narrative designed to manage perception. IW practitioners employ SC as a unifying messaging effort to align a population with norms and a worldview that is favorable to their cause. As a foremost example from the Philippine War, General Bell's circulars ensured local commanders could achieve unity of purpose while communicating a national rule of law.

E – Essential Services and Economic Infrastructure. Effective counterinsurgency requires close cooperation between military forces, government ministries, and humanitarian organizations to provide essential services like healthcare, education, and infrastructure development. As a levering motive, insurgents seek to diminish COIN efforts by providing their own parallel infrastructure.

C – Civil Control. In IW, civil control is crucial to fostering legitimacy and winning over the local population. The aim of IW practitioners should be to gain the support of the population rather than control of territory.

T – Tangible Support Reduction or Rise (TSR). Insurgents rely on foreign (external) support for recruits, finances, weapons, material, and sanctuary, which counterinsurgents must target or negate. As a RAND study suggests, "tangible support trumps popular support."[14]

I – Integration. Effective COIN requires integration of military, police, security forces, and political efforts to ensure actions are comprehensive.

---

[14] C. Paul, et al, *Victory Has a Thousand Fathers* (Santa Monica, CA: RAND, 2013), 14.

Competent insurgents work through social, political, and economic systems to undermine the host nation.

O – <u>Overall Quality of Force</u>. The litmus test for IW practitioners is how it uses power in accordance with the law. By emphasizing restraint, preventing excessive force, and minimizing collateral damage, such overall quality of counterinsurgent security forces helps to prevent alienation of the civilian population, which can fuel further insurgency.

N – <u>Native Government</u>. Counterinsurgency forces should avoid being perceived as occupation forces to prevent what David Kilcullen calls the "accidental guerrilla syndrome" from taking place. For their part, insurgents work to paint the COIN force with the brush of foreign invaders. As such, social, political, and economic programs should be coordinated and administered by the host nation's leaders.

Correlation of the 5 PLAIN Laws and the 10 IW Lines of Effort:

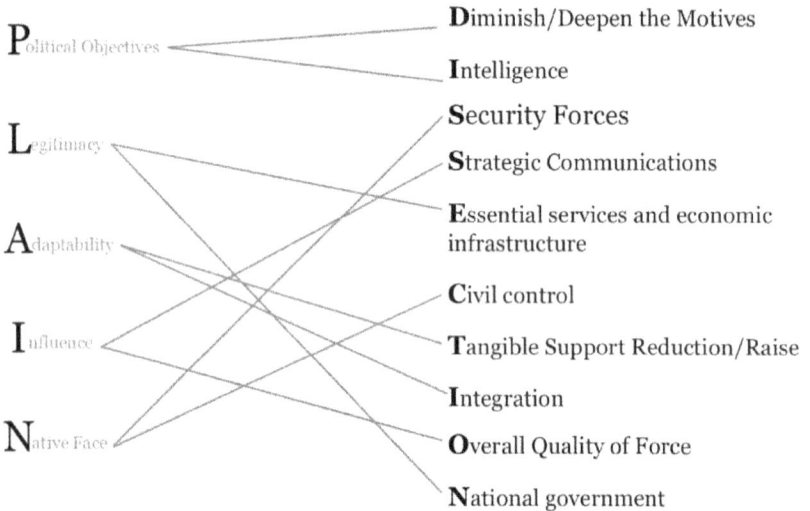

Political Objectives

Legitimacy

Adaptability

Influence

Native Face

Diminish/Deepen the Motives

Intelligence

Security Forces

Strategic Communications

Essential services and economic infrastructure

Civil control

Tangible Support Reduction/Raise

Integration

Overall Quality of Force

National government

Summary and Implications

As a violent struggle for legitimacy and influence over the relevant populations, success in IW requires victory in the abstract: an accumulation of more legitimacy and influence than the enemy. The aim of this book will be to flesh out the five PLAIN laws of IW by way of the case studies that follow.

For Further Reading:

1. *Learning to Eat Soup with a Knife* by John Nagl.
2. *The American Way of Irregular War* by Charles Cleveland.
3. *How Insurgencies End* by Ben Connable and Martin C. Libicki.

Part Two

Spanish American War
to World War II (1898-1945)

4

The Accidental Guerrilla War

"It's one thing to conquer a nation. Yet it's another thing entirely to set up a working administration."

Marines halt on a road march during the Philippine Insurrection. Courtesy AP.

In a forgotten chapter of history, the United States successfully invaded Spanish-held Cuba, Puerto Rico, and the Philippines, and, in just eight months, deprived Spain of its overseas empire. Initially, the U.S. Army governed these Islands until the American Government could determine their future disposition. While the Army's nation-building efforts were relatively peaceful in Cuba and Puerto Rico, this was decidedly not the case in the Philippines.[1] There, in the island archipelago, the United

---

[1] Andrew J. Birtle, *U.S. Army Counterinsurgency and Contingency Operations Doctrine, 1860-1921* (Washington, D.C.: Center of Military History, 1998), 99.

States soon found itself emmeshed in a war that lasted from February 4, 1899, to July 2, 1902. In terms of blood and treasure, the enterprise cost the U.S. $400 million and claimed 4,200 American lives. For the people the U.S. was liberating, the metrics were also grim. Some 20,000 Filipino combatants were killed, and as many as 200,000 Filipino civilians died from the ravages of war, famine, and disease.

The annals of history include many counterinsurgency campaigns that were highly successful, albeit they are deemed unacceptable by modern standards of Western culture. Notwithstanding it's successes, as a whole, the U.S. Army's prosecution of COIN during the Philippine Insurrection would most likely fall into this category. Nonetheless, it would serve as a model for America's counterinsurgency theory and practice for the next forty years. The goal of this chapter is to identify the ingredients of this model, and while avoiding its excesses, ascertain which elements may be used today.

Manifest Destiny

As America would discover, the insurgency in the Philippines evolved largely from unfortunate economic, political, and social conditions that had prevailed for hundreds of years. The Philippine Islands had been a colony of Spain since 1565. Over the centuries, there had been numerous uprisings that failed to overthrow the occupying Spanish. In the more recent past, a secret revolutionary movement named *Katipunan* was formed by Andreas Bonifacio in 1892. The movement's central aim was the expulsion of the Spanish to gain independence through armed force.

Chief among their grievances included that of the Spanish colonial land system. Known as *encomienda*, this system granted huge tracts of land to Spanish elites, or *principales*, while the Filipinos who lived on these land grants became the tenant farmers for their Spanish masters. In the passage of time, these Spanish masters became mestizos, literally "mixed" Spanish and indigenous descent. They owned vast areas of the best agricultural land in the Philippines. For the most part, the *encomienda* system was a ruthless economic exploitation of the agricultural peasantry which would later give a prominent voice to Communist movement in the archipelago (Chapter 7).

Katipunan gave a voice to the centuries-old grievances of the Filipino people who had been dominated by their Spanish overlords. The revolution began in August 1896 when the Spanish colonial government discovered the secret society. Hundreds of Filipino suspects were arrested, imprisoned, and many were executed, including the movement's intellectual founder, Jose Rizal. What followed was a mixture of pitched battles and ambushes mainly in the eight provinces of Luzon. Chief among the revolutionary leaders was Emilio Aguinaldo, a Chinese Filipino mestizo, who came from a well-to-do family that enjoyed relative wealth and power in the city of Cavite El Viejo (now Kawit), where he was mayor. Aguinaldo emerged as the movement's leader after he had Bonifacio assassinated.

Despite being outnumbered, Aguinaldo capably led a resistance that forced some 2,000 Spanish troops to leave the province of Cavite. Following a year of fighting ending in a stalemate, Aguinaldo signed an agreement with the Spanish. Under the accord, Aguinaldo agreed to leave the Philippines and to remain permanently in exile on condition of the promise of liberal reforms, and a substantial financial reward of $400,000. While Aguinaldo and his cadre waited, the promised reforms never materialized.

As the year 1897 came to an end, war between the United States and Spain loomed on the horizon. Due to its close proximity of about 90 miles to Florida, the defining issue was Cuba's independence. In a manner similar to the Philippines, Cubans armed themselves against their Spanish overlords. To crush this rebellion, Spanish general and colonial administrator Valeriano "Butcher" Weyler authorized public executions, mass exiles, and the forced concentration of about a million people into protected zones. Having separated the people from the *insurrectos*, Weyler then ordered the destruction of farms and crops. Under these repressive measures, it is estimated that 250,000 perished in such camps.

As reports of these death camps reached American ears, public opinion swayed in support of the rebellion in Cuba. Pressure began to weigh on President McKinley to do something. Then on February 15, 1898, when the U.S. Navy armored cruiser *USS Maine* mysteriously exploded and sank in Havana Harbor, political pressures pushed McKinley into a war that he had wished to avoid. Not everyone in the American government shared McKinley's stance on the matter. In fact, many saw a war with Spain as a unique opportunity to expand America's ideals and realize her manifest destiny – a spirit of the time which envisioned America's mission as carrying her form of culture and government globally.

Among those who aspired imperialist ambitions was the Assistant Secretary of the Navy, Theodore Roosevelt. For his part, Roosevelt perhaps said more to give shape to the arguments for war with Spain and did more to replace European primacy in world affairs than anyone of his time. To this end, on February 25, just ten days after the sinking of the *Maine*, Roosevelt sent a coded telegram to Commodore George Dewey: "Order squadron to Hong Kong. Keep full of coal. In the event of declaration of war with Spain, your duty will be to see that the Spanish squadron does not leave the Asiatic coast, and then offensive operations in the Philippine Islands." As observed by Gregg Jones, "In a stroke of breathtaking audacity (and questionable authority), Theodore Roosevelt had just set in motion America's conquest of the Philippines."[2]

American intelligence efforts were also ongoing as to the ability of the Filipino revolutionaries to rise against Spain. To leverage this untapped resource, U.S. officials courted Filipino exiles. For instance, while in Singapore, Aguinaldo met U.S. Consul E. Spencer Pratt. Inscrutably, Pratt strongly inferred, if not promised, Aguinaldo Philippine independence in exchange for his assistance in the war against Spain.

As pressure for war with Spain mounted, McKinley continued to negotiate for Cuba's independence. Spain however refused McKinley's proposals, and on April 20, 1898, Congress declared war. Then, as Dewey's Asiatic Fleet steamed toward Manila, Aguinaldo made preparation to return to the Philippines and fight alongside the Americans. In the predawn hours of May 1, 1898, the American squadron entered Manila Bay, and in one of the most one-sided decisive naval victories in history, Dewey destroyed the Spanish Pacific Squadron under Rear Admiral Patricio Montojo. Before noon, American cruisers sank twelve Spanish antiquated warships, killing 381 Spanish with the loss of only one American.

---

[2] Gregg Jones, *Honor in the Dust: Theodore Roosevelt, War in the Philippines, and the Rise and Fall of America's Imperial Dream* (New York: New American Library, 2012), 42.

Dewey's surprise victory opened a second front in the war with Spain, however, he lacked the ground troops necessary to defeat the Spanish garrison in Manila. Toward this end, two days later, Secretary of War Russell Alger ordered troops to the Philippines. However, it would be months before they would arrive. Faithful to the promise, after the battle, Dewey permitted Aguinaldo to come back to the islands aboard the *McCulloch*, which reached Manila Bay on May 19, 1898. Aguinaldo's intent was to resume the revolution and to fight alongside the Americans. Dewey was impressed with Aguinaldo and sympathized with his struggle. Things took a turn on May 26, 1898, when Dewey received instructions from the Secretary of the Navy. As ordered by John D. Long, Dewey was to avoid allying U.S. forces with the revolutionaries.[3]

This foreboding turn of events notwithstanding, Aguinaldo was not one to leave things to chance. On June 12, 1898, he managed to get himself named the provisional president. By September, a revolutionary assembly met and ratified Filipino independence. With Filipino revolutionary spirit running high, the parceled out Spanish troops became easy prey to Aguinaldo's forces. Now swelling to some 40,000 men, Filipino revolutionaries armed themselves with captured Spanish Mauser rifles. These Model 1893 7.65 mm bolt-action rifles were superior to the rifle many American troops were carrying at the time. Without clear policy objectives, American-Filipino relations proceeded on vague assumptions.

The Army's choice of commander for this second front with Spain was Major General Wesley Merritt. Having served under Generals Sheridan and Sherman, Merritt was a capable field commander, who, in addition to the Civil War, had a lot of experiences in the Indian campaigns of the American West. At the Presidio in San Francisco, receiving a force of some 13,000 men from the new Volunteer Regiments, Merritt quickly trained and equipped his force into an orderly formation which constituted the Army's new VIII Corps. Assisting him was his able second in command Brigadier General Elwell S. Otis, another veteran of the Civil War and Indian campaigns. Before striking out, Merritt tried in vain to get detailed instructions from McKinley. Merritt wasn't sure if he was to

---

[3] Brian McAllister Linn, *The U.S. Army and Counterinsurgency in the Philippine War, 1899-1902* (Chapel Hill, NC: University of North Carolina Press, 1989), 7.

"subdue and hold all Spanish territory in the islands, or merely to seize and hold the capital."

Merritt departed San Francisco for a month-long voyage with the following instructions: "establish a strong garrison to command the harbor of Manila." In grave oversight, McKinley failed to give Merritt instructions as to what relationship, if any, he was to have with the Filipino insurgents.[4] McKinley's failure to clearly define political objectives would create complex difficulties.

Back in Luzon, before U.S. troops arrived, Aguinaldo had issued the Philippine Declaration of Independence, proclaimed a revolutionary government, and, then with a force of some 15,000 regulars, laid siege to the Spanish garrison in Manila. Under orders to not recognize Aguinaldo's government, Merritt faced the dilemma of how to defeat the Spanish without provoking an open conflict with the Filipino *insurrectos*. His solution was a secret arrangement with General Fermín Jáudenes, the commander of the Spanish garrison. In a mock battle, Jáudenes would surrender the city to American forces. Dewey and Merritt intentionally kept Aguinaldo in the dark as his army encircled Manila in some fourteen miles worth of trenches. With the taking of Manila, Filipinos imagined an independent country of their own. Under the supposed auspices of protecting the city from looting, Merritt convinced Aguinaldo to keep his forces out of the attack. Merrit, having moved Aguinaldo's forces aside, the American assault commenced on August 13, 1898. After putting up a token resistance, with "enough gunfire and casualties to satisfy both Spanish and American honor," the Spanish garrison in Fort Santiago and the city promptly surrendered.[5]

As the dust was settling, Philippine independence failed to materialize. This came as a shock to many Filipinos who claimed the U.S. had promised them a nation of their own. Merritt was then relieved by Otis on August 30 to advise the U.S. delegation that would eventually meet in Paris to sign the treaty that would end the war against Spain. Learning of the upcoming treaty, Aguinaldo designated Felipe Agoncillo to represent Filipinos. However, McKinley rejected Agoncillo, and the

---

4 John Gates, *Schoolbooks and Krags* (Westport, CT: Greenwood Press, 1973), 6.

5 Linn, *The U.S. Army and Counterinsurgency in the Philippine War, 8.*

Treaty of Paris was held on December 10, 1898, without Filipino representation. As stipulated in the treaty, Spain relinquished its claim to Cuba and sold the Philippines and Guam to the U.S. for $20 million. Less than two weeks later, on December 21, 1898, McKinley proclaimed American sovereignty over the Philippines. This decision, which was later approved by the Senate, set off an intense and emotional debate across the U.S. Many, like Roosevelt, saw the decision as America's duty. Others believed America, which itself had once been an overseas colony, had no right to take the islands. In response, Aguinaldo declared the First Philippine Republic on January 23, 1899.

The Burden of Victory

With the new year of 1899, an ambiguous and volatile relationship between Filipinos and Americans persisted. As American and Filipino troops faced each other along Manila's trench lines, the first weeks of 1899 were tense. The spark that set off hostilities was ignited on the evening of February 4, 1899, when Private William Grayson from Nebraska fired on an insurgent patrol resulting in three Filipino soldiers killed. The next day, Brigadier General Arthur MacArthur's 2nd Division routed Aguinaldo's forces, driving them from Manila.

Disgruntled Filipinos flocked to Aguinaldo's banner in droves. While Filipino nationalists viewed the situation as a continuation of their struggle for independence, the U.S. regarded it as an insurrection. Hence, there were multiple names for the same conflict: Philippine Insurrection or Philippine War. At the outset of hostilities, Aguinaldo tried to stop the war by sending Otis an emissary. In response, Otis countered by saying, "fighting, having begun, must go on to the grim end."[6]

Having no clearly defined political objectives, Otis sought the unconditional surrender of the *insurrectos*. As for U.S. policy, having gotten rid of the Spaniards, it seems little thought was given beyond that. McKinley was quoted as saying, "The truth is I didn't want the

---

[6] Stuart C. Miller, *Benevolent Assimilation: The American Conquest of the Philippines, 1899–1903* (Yale, CT: Yale University Press, 1982).

Philippines, and when they came to us, as a gift from the gods, I did not know what to do with them."

The annexation of the Philippines presented the United States with an unusual set of problems and challenges for which it had limited experience. Following the U.S. victory in Manila, American forces easily captured the islands of Iloilo, Cebu, Bacolod, and Jolo. However, as American commanders were soon to realize, all these areas would require garrisons to prevent the guerrillas from reclaiming them.

During the first months of 1900, as observed by Mark Moyar, Otis ordered U.S. commanders to concentrate on civic, rather than military, action. He and a good many other Americans wrongly believed that Filipinos would support the U.S. once they saw the benefits of American rule.[7] Otis began with Manila, launching a civic action campaign to clean up the city. Once hailed as the Pearl of the Orient, it was now a "starving, filthy, disease-ridden city" that had now swollen to include some 70,000 refugees.[8] Otis established schools, implemented sanitation measures, and stood up a 3,000-man garrison to protect the city. As an able administrator, Otis hoped these measures would be enough to convince the Filipinos of the benevolent intent of the United States.

On a fact-finding assignment, McKinley appointed the First Philippine Commission led by Dr. Jacob Schurman, president of Cornell University. Created January 20, 1899, the five-man panel included Otis, Dewey, Charles Denby, the former minister to China, and Dean Worchester, an authority on the Philippines. The commission was McKinley's first attempt at figuring out what to do with the Philippines. Being ousted from Manila, Aguinaldo established a formal government and established his capital at Malolos, some twenty-five miles north of Manila. While U.S. forces were advancing against him, there were however some promising developments in peace talks. In April, the commission met with Colonel Manuel Arguelles, Aguinaldo's representative, to whom they presented McKinley's offer.

---

[7] Mark Moyar, *A Question of Command: Counterinsurgency from the Civil War to Iraq* (New Haven, CT: Yale University Press, 2010), 72.
[8] *Honor in the* Dust, 98.

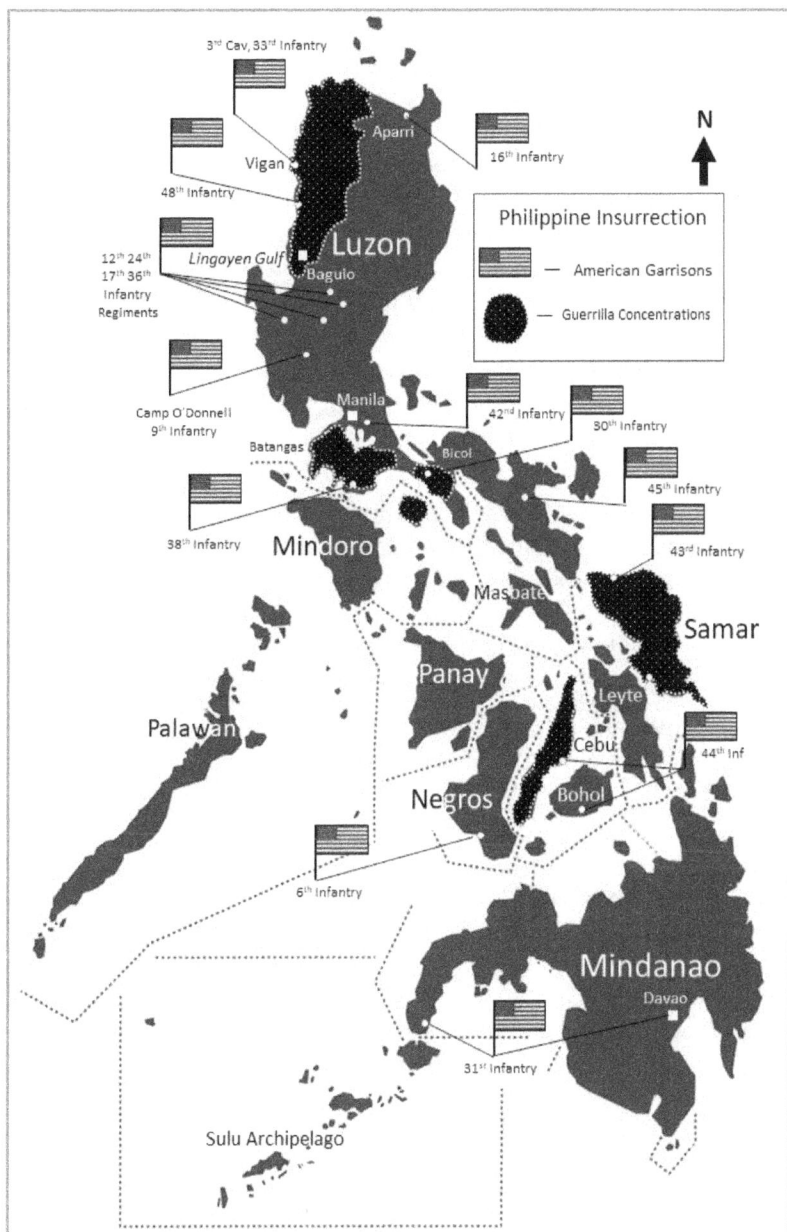

Philippine Islands depicting U.S. garrisons, 1900.

The government McKinley proposed consisted of a governor-general appointed by the president, a cabinet appointed by the governor-general, and a general advisory council elected by the people. This was well received. Aguinaldo convened a council which voted unanimously to cease fighting and accept this proposal. However, an internecine squabble between those loyal to Aguinaldo and General Antonio Luna, field commander of the revolutionary army, fractured the fledgling peace talks. Confronted with this development, and perhaps convinced the United States did not intend to create an independent Philippine state, Aguinaldo withdrew his support from the peace table. Now, fully committed to all-out war with the United States, the four main aims of the First Philippine Republic were as follows: National emancipation, representative government, individual liberty, and separation of church and state.

The commission acknowledged Filipino aspirations for independence but deemed them not ready for it. On November 2, 1899, the commission submitted a preliminary report containing the following statement:

Should our power by any fatality be withdrawn, the commission believe that the government of the Philippines would speedily lapse into anarchy, which would excuse, if it did not necessitate, the intervention of other powers and the eventual division of the islands among them. Only through American occupation, therefore, is the idea of a free, self-governing, and united Philippine commonwealth at all conceivable. And the indispensable need from the Filipino point of view of maintaining American sovereignty over the archipelago is recognized by all intelligent Filipinos and even by those insurgents who desire an American protectorate. The latter, it is true, would take the revenues and leave us the responsibilities. Nevertheless, they recognize the indubitable fact that the Filipinos cannot stand alone. Thus, the welfare of the Filipinos coincides with the dictates of national honor in forbidding our abandonment of the archipelago. We cannot from any point of view escape the responsibilities of government which our sovereignty entails, and the commission is strongly persuaded that the performance of our national duty will prove the greatest blessing to the peoples of the Philippine Islands.[9]

---

[9] Dean Worcester, *The Philippines: Past and Present* (New York: Macmillan, 1914),

Things being as they were, the commission therefore recommended "prosecution of the war until the insurgents submit."[10] In response, McKinley vowed to send all necessary force to suppress the insurrection.

As lampooned by the movie *War Machine*, the burden of victory could be surmised by saying, it's one thing to conquer a nation, yet it's another thing entirely to set up a working administration. "No one," the report said, "can foresee when the diverse peoples of the Philippine Islands may be molded together into a nation capable of exercising all the functions of independent self-government."[11] Divided by geography, religion, language and race, the seven million people that inhabited the Philippines was (and remains) a diverse people. Based on the findings of the First Philippine Commission, McKinley believed the Filipinos were incapable of self-government. Yet what kept American presence there was he simply did not want another foreign power to take over the islands.[12] With the aim of creating an American Protectorate, McKinley directed Otis to "win the confidence, respect, and admiration" of the inhabitants of the Philippines. Just how Otis was to do that while prosecuting a war until the insurgents submitted was yet to be determined.

A Learning Competition

There is a time-honored axiom of war which states that war is "an extension of politics." In a lesser known but no less important statement, without policy, says Clausewitz, war would be an unbridled, "absolute manifestation of violence," that would, "of its own independent will, usurp the place of policy." In the absence of clearly defined political objectives, this latter axiom underscored the mission-creep nature of American actions. With nearly 20,000 troops at his disposal, Otis launched a spring offensive north and east of Manila that routed the Philippine rebel army. He then set out to crush what remained of the rebellion. As he pushed outposts farther and farther from Manila,

---

[10] Frank H. Golay, *Face of Empire: United States-Philippine Relations, 1898-1946* (Manila: Ateneo de Manila University Press, 1997), 50.

[11] Report of the Philippine Commission to the President, vol. 1: 113.

[12] H. W. Brands, *Bound to Empire: The United States and the Philippines* (New York: Oxford University Press, 1992): 24–25.

resistance increased. Lines of supply became their favorite targets. Telegraph lines were cut. Small bands of troops were ambushed. In the first thirteen months of the war, the Army reported 1,026 engagements, with 245 Americans killed, 490 wounded, and 118 captured, versus 3,854 rebels killed, 1,193 wounded, and 6,572 captured.[13] Otis's goal was to fix the rebels in place long enough to produce a crushing victory.

By mid-November 1899, after suffering heavy losses in attempts at fighting in accordance with conventional tactics, Aguinaldo ordered Luna to disperse their forces into the mountainous interiors of the islands and conduct guerrilla warfare. Yet, Otis failed to recognize this paradigm shift. Moreover, he had other problems. Of the 20,000 troops at his disposal, some 3,000 had to remain in Manila to serve as provost guard, while 15,000 were volunteers who were eligible for discharge beginning February 14, 1900. This huge turnover threatened his ability to sustain offensive momentum. And so, while Otis received additional troops, they did little more than replace the thousands of homeward bound volunteers.[14]

Keenly aware of American public opinion against the annexation of the Philippines, Aguinaldo hoped to forestall any decisive U.S. gains until the American presidential elections in November 1900. It seemed likely that a Democratic victory would mean recognition of Filipino independence. McKinley also recognized the Philippines as a political liability unless he could bring the situation there firmly under control.

As a guerrilla army, Aguinaldo's forces took on a decentralized organization into several military zones throughout the Philippine Islands. Operating in bands numbering somewhere between 20 and 200, and being supported by thousands of auxiliary forces, semi-autonomous regional commanders led ambushes and raids, harassing and exhausting American forces. With his force swelling to some 100,000 guerrillas, Aguinaldo now enjoyed a three-to-one advantage over the Americans. Additionally, many of the guerrillas were better armed than their American counterparts. The .28 caliber Spanish Mausers had a greater long-range accuracy than the Springfield rifles issued to most of the

---

[13] Asprey, Robert B. *War in the Shadows: The Guerrilla in History.* Vol. 1. 2 vols. Garden City, NY: Doubleday, 1975), 192
[14] Gates, *Schoolbooks and Krags*, 102.

60

American volunteer units. Tactics used by guerrillas included ambushing American supply columns and attacking sentries with bolos.

American patrols also encountered punji-laden traps and tripwires. The guerrillas also tore down telegraph lines and then ambushed the parties sent to repair them.[15] Such hit-and-run tactics produced an ever-increasing number of casualties. For instance, typical weekly metrics included thirty-four killed. Of which five were killed in action, six drowned, seven succumbed to typhoid fever, four to malaria, four to dysentery, and one committed suicide.[16]

Despite his gains at atritting U.S. forces, Aguinaldo was only able to prolong the war. His strategy was one of protraction. His aim was to exhaust the American Army in the field and for the American public back home to demand for a U.S. withdrawal. For its part, the United States was epitomizing a phenomenon that so often characterizes counterinsurgency. Otis was winning all the battles but losing the war. Failing to win the "confidence, respect, and admiration" of the inhabitants of the Philippines, the American general sent deceptive reports to Washington that the rebellion was all but finished. However, American correspondents told the opposite story. In the learning competition that ensued, American forces were to discover the imperative of restraint. The carrot of benevolent assimilation had to be better orchestrated with the punitive stick of aggressive military operations. Adding to the austerity, operations were often curtailed by the tropical climate of the Archipelago. For instance, actions virtually came to a halt during the monsoon season which extends from June to September.

To make matters worse, the story being told back home to the American people was much like the oppressive and often cruel measures employed by Spanish troops in Cuba. In the absence of an articulated counterinsurgency policy, many frustrated field commanders resorted to such tactics as the "water cure" to gain intelligence in the hopes of fishing out the insurgents from among the populace. This method of torture consisted of forcing water down the victim's throat until he divulged the

---

[15] Ibid., 159.
[16] *Honor in the* Dust, 140.

required information. Equally inept, the insurgents countered by terrorizing the people to ensure their continued support.

Seeking to revitalize the counterinsurgency effort, on May 6, 1900, McKinley relieved Otis, replacing him with Major General Arthur MacArthur, the father of General Douglas MacArthur. The new commander brought in a high level of competence, zeal, and a comprehensive assessment of the American COIN effort. Based on his predecessor's design, MacArthur continued the civic programs instituted by Otis. In fact, by August 1900, the Army had built no fewer than 1,000 schools, spending some $100,000.[17] Yet, facing increasing guerrilla activity, MacArthur nuanced the American strategy. He realized the imperative of isolating the guerillas from their base of support, the people, before going on to defeat them militarily.

To that end, MacArthur first announced an amnesty program for insurgents who surrendered and swore allegiance to the new Manila government. He also launched a buyback program in which he offered a monetary reward for each rifle turned in. Moreover, MacArthur encouraged the recruitment of Filipinos into police and scout forces. These formations would assume the duties of local law enforcement. By the end of 1900, Filipino policemen were taking over security duties from the Americans in provinces where the guerillas had been sufficiently beaten down. This freed up U.S. troops to operate in other areas. Despite the positive nature of these developments, to the American chagrin, they nonetheless produced meager returns. For instance, in terms of metrics, four months witnessed a mere five percent (5,022) of the insurgents surrendered and a haul of weapons that was a paltry 140 rifles.

With these initial attempts having produced poor results, MacArthur decided to limit operations until after the November 1900 elections. In the event in which McKinley remained in office, he would have a free hand to inject a new vigorous spirit into things. As he correctly judged, the Republican political victory in the 1900 election was a powerful blow to insurgent morale. In the wake of this political victory, MacArthur launched a new and vigorous COIN campaign. Albeit, what the general soon realized was that he needed to curb the excesses of his army. MacArthur and his commanders understood that success depended on

---

[17] Birtle, *U.S. Army Counterinsurgency and Contingency Operations*, 121.

the support of the Filipino people. He therefore anchored his strategy on the U.S. Army's General Order 100 (GO 100).

Among the significant by-products of the Civil War was *GO 100, Instructions for the Government of Armies of the United States in the Field*. Originally signed by President Abraham Lincoln on April 24, 1863, GO 100 was the first code of rules for land warfare ever adopted by any nation.[18] These instructions were formulated in response to the creation of the Partisan Regulars by the Confederate Army in the spring of 1862. Also known as the Lieber Code, GO 100 helped to somewhat standardize the basic conduct of the Federal Army during the American Civil War. It gave workable solutions to such questions as: How should martial law function? How should the Army treat private property and civilians in occupied territory?

A critical aspect of GO 100, emphasized by MacArthur, was the Army's responsibility to protect those natives who accepted the American occupation government. This necessitated American presence to protect Filipinos from guerrilla intimidation and demands. Pointedly, GO 100 "admonished soldiers to respect the personal and property rights of unarmed citizens, as well as their religious and social customs." Under these provisions, all forms of wanton looting, pillaging, cruel acts, torture, and revenge were strictly prohibited.[19]

Regardless of these positive measures, most U.S. field commanders viewed the Filipinos as savages, "unworthy of civilized treatment."[20] Enemy metrics backed up this dissension. By the end of 1900, insurgents racked up some 350 assassinations. Insurgents also set fire to towns accused of collaborating with the Americans. Such terror tactics received grudging support. In retaliation, ruthless measures were undertaken, including such tactics that had earlier been condemned by Americans when employed by Weyler's forces in Cuba. Such measures included the burning of villages as part of a scorched earth tactic. In a special proclamation issued on December 20, 1900, MacArthur urged his

---

[18] General Orders 100 served as the foundation for the Hague Conventions in 1899 and became part of U.S. Army doctrine in 1940 when the Army incorporated it into Field Manual 27-10, Rules of Land Warfare.
[19] Birtle, *U.S. Army Counterinsurgency and Contingency Operations Doctrine, 1860-1941*, 34.
[20] *Honor in the* Dust, 115.

subordinate commanders to enforce and comply strictly with the applicable passages of GO 100.[21] Following some initial faux pas, this old but new way forward laid the foundation for what would be a model pacification approach to counterinsurgency.

Carrot and Stick

Perhaps the greatest advantage the American generals had over the insurgents was their rich inheritance of the Army's warfighting experience in the Civil War and the Indian wars. For example, out the thirty U.S. generals who served in the Philippines, twenty-six had fought in the Indian wars. Drawing on collective experience, MacArthur innovatively adapted tactics in which his troops alternatively used a carrot and stick approach to gain the allegiance of the population. Moreover, searching for ways of making American troops look less like an occupation force, MacArthur employed some 15,000 native auxiliaries, giving efforts a more native face. Amongst the greatest achievements of the war was the U.S. Army's organizing of the Macabebe Scouts who were recruited from the village of Macabebe. The Macabebe were mortal enemies of Tagalogs, who made up the largest percentage of the insurrectionists. Recruited to be employed as scouts and guides, the Scouts providing valuable intelligence on the location of Filipino rebel forces.

Perhaps the most noteworthy episode during MacArthur's tenure was the capture of Aguinaldo. It was so audacious that it has become something of legend. Aided by the Macabebe Scouts, Colonel Frederick Funston, arguably one of the ablest and most audacious leaders in American history, infiltrated into Northern Luzon's guerrilla-infested jungles to capture the elusive rebel leader. Along with four other officers, Funston posed as a captured American prisoner of the Macabebe Scouts, who themselves were masquerading as reinforcements for Aguinaldo. On March 23, 1901, having been brought to the center of Aguinaldo's

---

[21] Kalev I. Sepp, *Resettlement, Regroupment, Reconcentration: Deliberate Government-Directed Population Relocation in Support of Counter-Insurgency Operations* (Fort Leavenworth, KS: Army Command and General Staff College, 1992), 28.

camp, Funston's force completely surprised Aguinaldo's men and captured *el presidente.*

Eager to capitalize on the nadir of insurgent morale due to Aguinaldo's capture, MacArthur treated the insurgent leader exceptionally well. He even allowed Aguinaldo's wife and family members to visit him. And in return for the promise to release 3,000 prisoners, Aguinaldo issued a proclamation to the Philippine people, accepting the sovereignty of the United States and called for an immediate end to all hostilities, saying it was "absolutely essential to the welfare of the Philippines."[22]

With Aguinaldo's capture and the reelection of McKinley, whole units of Filipino insurgents became demoralized and began to surrender. In fact, the first six months of 1900 witnessed 17,000 surrenders. In an effort to consolidate gains, on July 4, 1901, William Taft was appointed Civil Governor. Taft promptly set out to organize the civil government, writing a new legal code which gave structure to the civil service. Chairing the Second Philippine Commission, Taft sought to remedy what he determined to be the three principal grievances resulting from economic and sociopolitical issues.

One of the thorniest problems came from the archipelago's long-standing relationship between the Spanish colonial government and the Roman Catholic Church. For Taft, in order to create a new Philippine government, it was essential to end this relationship and separate church from state. For instance, friars owned some of the choicest land on Luzon. The Spanish had governed the islands through the Roman Catholic Church, particularly through the friars.

Seeking to remedy this land grievance, Taft eventually negotiated the 7.2-million-dollar sale of this acreage from the Catholic Church for redistribution to Filipino farmers. This was the first modern program of land reform in the Philippines. This innovative program sought to sell these acquired church lands to Filipinos in small parcels, but few could afford the down payment on the acreage. Moreover, those who borrowed money to finance the sale found themselves enslaved as former tenant farmers. As it turned out, the majority of land fell to *Ilustrados*, the Filipino upper class, or merchants, bankers, or politicians. Taft had

---

[22] Kenneth Ray Young, *The General's General* (Boulder, CO: Westview Press, Inc., 1994), 287-8.

touched a live nerve, but it would take more than land reform to bring hostilities to an end.

Many felt Taft to be out of touch with the realities of the war. For example, Taft called the Filipinos "our little brown brothers" and thought they welcomed U.S. rule. However, soldiers in the jungle campaigning far from Manila knew better. Sardonically, they sang: "He may be a brother of big Bill Taft, but he ain't no friend of mine."

MacArthur on the other hand believed the Philippines was not ready for civil rule. Eventually, due to his personality differences with the newly appointed Taft, MacArthur was replaced by Major General Adna R. Chaffee, who had at one time served under General Philip Sheridan during the Civil War. Chaffee, who had once helped put the torch to the Loudoun Valley in an effort to capture Mosby's Rangers, erroneously determined that the insurgents were essentially defeated. Arguably, there had been three key reversals signaling the guerrilla's demise: McKinley was reelected, General Luna was dead, having been assassinated by Aguinaldo loyalists, and Aguinaldo had been captured.

Considering the guerrillas now on the ropes, Chaffee's casual approach enabled diehard *insurrectos* to hand him several stunning military reversals. This prompted a barrage of criticism from the American media and a congressional investigation. Down but not out, in terms of the guerrilla threat that remained, there were essentially three problem areas. The Province of Batangas on Luzon, the Island of Samar, and Mindanao Island, situated some 600 miles distant and disconnected from the current struggle. With the task of bringing the war to a close, Chaffee concentrated his attention on the Batangas and Samar. Two months after he took over from MacArthur, the massacre of an American company in the village of Balangiga led Chaffee to implement the most repressive counterinsurgency policy of the war.

Situated in the Visayan group of islands, Samar had a population of over 250,000 concentrated along the coast. The island was well suited for guerrilla warfare and enabled insurgent general Vicente Lukban to establish a formidable resistance infrastructure. Earlier American efforts had driven insurgents to the interior of the island. Operating mainly on the coast, Brigadier General R. P. Hughes successfully conducted several tactics that were later used on a broader front. Working toward starving out the insurgents, Hughes blockaded the island and directed aggressive patrols he combined with extensive food and property destruction. His orders included the imperative to clear the country of all insurgents and make the region untenable for them. The success of these tactics gave the impression that Samar had been pacified. What Hughes failed to take into account was that his policy of extensive population resettlement more than tripled in size some villages. This coupled with the destruction of food and crops left the inhabitants of Samar unable to support themselves. This all played into Lukban's hands.

At a prearranged signal from the town mayor on the morning of September 28, 1901, a few dozen insurgents and residents of Balangiga, armed with bolos, beheaded the company commander, Captain Thomas Connell, and massacred another forty-seven Americans from C Company, 9th Infantry Regiment. Troops sent to Balangiga after the incident found the bodies of the victims mutilated.

The attack shocked the American public, with newspapers equating it to George Custer's last stand at the Battle of the Little Bighorn. With U.S. troop levels reaching 70,000, the highest of the war, Chaffee responded by sending Brigadier General Jacob H. Smith to Samar. Chaffee then reorganized U.S. forces in the Philippines into seven Separate Brigades, making Smith the commander of the 6th Separate Brigade, which included Major Littleton Waller's battalion of 315 Marines.

In retaliation, Smith gave Waller the following order: "I want no prisoners. I wish you to kill and burn; the more you kill and burn the better it will please me. I want all persons killed who are capable of bearing arms in actual hostilities against the United States." When Waller asked to know the limit of age to respect Smith replied, "ten years." What followed would prove to be one of the gravest blunders of the entire war.

Intent on making the island a "howling wilderness," Smith's indiscriminate retaliatory attack on the inhabitants of Samar killed some 2,500. Smith's order "kill everyone over ten" became a caption in a *New York Journal* cartoon. The caption at the bottom declared "criminals because they were born ten years before we took the Philippines." Smith's conduct eventually resulted in his court martial.

Meanwhile, the COIN tactics underway on Samar contrasted dramatically with the one conducted by Brigadier General J. Franklin Bell in the Batangas Province. In the words of Mark Moyar, "Bell's campaign in Batangas and Smith's campaign in Samar underscored one of the most important lessons of the Philippine insurrection – that drastic measures could defeat the insurgents quickly without large scale violence against the civilian populace but only if the right type of leader commanded the counter insurgence."[23]

No Neutrality

As history demonstrates, success in COIN requires the counterinsurgent to separate the insurgency from the population both physically and psychologically. Though separation of the population from the guerrillas was not a new practice in the Philippine war, Bell's employment of resettlement was unique for its unprecedented scope and success. Placing Bell in command of the 3rd Separate Brigade, Chaffee ordered him to act decisively in order to conclude the war as quickly as possible. Facing Bell and brigade commanders was one of the most capable insurgent leaders, Miguel Malvar. His provincial guerrilla movement, which was one the most organized in the archipelago, consisted of three main elements: the regulars (full-time guerrillas), the Sandahatan (village militia), and a shadow government, consisting of *principalia*, town cabezas, and other officials, who considered themselves loyal patriots to Philippine independence.

Hiding in plain sight, the Sandahatan was of immense importance to the guerrillas' survival. They were adept at doing nefarious acts and then quickly fading again into the background. This phenomenon, referred to as the "chameleon act" consisted of guerillas conducting an attack, conducting a sabotage and then hiding their weapons and transforming themselves into "amigos" in the blink of an eye. The Sandahatan's main tasks were to supply the guerrillas, conduct sabotage, and provide intelligence and fresh recruits. Additionally, they provided early warning, collected taxes, hid fighters and weapons, and intimidated collaborators. Overseeing this work was the shadow government which collaborated with Malvar to undermine American policy objectives.

---

[23] Mark Moyar, *A Question of Command: Counterinsurgency from the Civil War to Iraq* (New Haven, CT: Yale University Press, 2010), 86.

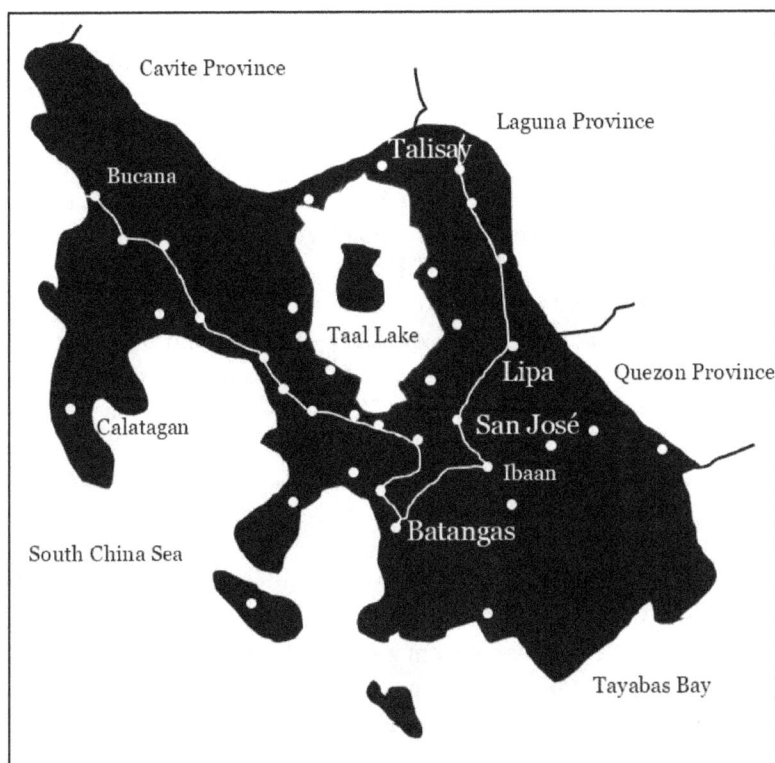

Zoned Towns of Batangas Province 1901-1902.

The regular guerrilla units operated in small highly mobile formations made up of some 30-50 riflemen and bolomen. These regulars would hole up in sanctuaries in the surrounding mountains and hills of Batangas, then, at coordinated times, they would band together with militia now turned fighters to ambush American patrols and attack *Americanista* towns and villages (those who collaborated with the Americans). Max Boot describes the carnage: "Some *Americanistas* were burned or buried alive; others had their tongues cut out, limbs hacked off or eyes gouged out. The *insurrectos* even burned down whole towns on occasion if they refused to pay taxes."[24] All told, Malvar could boast

---

[24] Max Boot, *Savage Wars of Peace: Small Wars and the Rise of American Power* (New York: Perseus Books, 2002), 113.

of some 500,000 sympathizers through Luzon. This support base
provided food and supplies, and recruits and intelligence to about 8,000
regulars under Malvar's control. With this force, he attempted to
revitalize the revolution.

With a command of about 10,000 men, Bell laid out an operational
design to stamp out what remained as the last hurdle to achieving a
better state of the peace in the archipelago. Although Bell's methods
would be severe, the actions of his troops generally fell within the
parameters permitted by the law of war.[25] In the words of Bell, "A short
and severe war creates in the aggregate less loss and suffering than
benevolent war indefinitely prolonged."[26] There would be no neutrality.

The foundation of Bell's plan called for employing the tactic of
resettlement to physically separate the insurgents from the people.
Through a series of thirty-eight circular telegrams, Bell laid out his plan.
In his Circular Order No. 2, transmitted on December 8, 1901, Bell
ordered inhabitants of outlying areas in his district to be "reconcentrated" into towns under U.S. Army control. For this purpose, Bell
committed 6,000 of his men to garrison duty. They would control the
towns and protect the people from guerrilla attack. Without a garrison,
insurgents would merely reoccupy the town after U.S. troops left and
then punish anyone suspected of having helped them.

The towns themselves constituted the center of each zone of
concentration. To clear fields of fire, commanders were authorized to
press into service villagers and their work animals. Villagers put up miles
of fence, drawing a line of demarcation around the fortified zones.
Additionally, people from the outlying rural areas were encouraged to
relocate their nipa huts (stilted homes). Those relocated were also given
land at no cost. Garrison commanders were also directed to establish
storehouses for the storage of rice and other supplies. While this was
underway, "Army garrisons also instituted sanitary reforms, and some
cleared roads and built bridges."[27] All told, there were over twenty such

---

[25] Birtle, *U.S. Army Counterinsurgency and Contingency Operations Doctrine, 1860-1921*, 131.
[26] Ramsey, *Masterpiece of Counter Guerilla Warfare*, 46.
[27] Linn, *The U.S. Army and Counterinsurgency in the Philippine War, 1899-1902*, 21.

garrisoned towns and villages in Batangas and Laguna. Eventually, over 300,000 people lived in the protected zones.[28]

Next, in his Circular Order No. 5, dated December 13, 1901, Bell announced that GO 100 would thereafter be regarded as the guide of his subordinates in the conduct of the war. Bell first ordered his commanders to concentrate the population into "protected zones" by December 20, 1901. After December 25, anyone found without a pass outside a protected zone would be considered hostile. Any food they possessed would be subject to confiscation or destruction. In this manner, Bell ensured the welfare of the villagers, while forcing the insurgents to live off the land.

Ultimately, as articulated in his Circular No. 21, what he considered his brigade's most important task was "to obtain possession of the arms now in the hands of insurgents and disloyal persons, and incidentally to capture as many insurgents as possible, especially those accused of grave offenses against the laws of war." Under the provisions of GO 100, Section 17, the starving of unarmed hostile belligerents as well as armed ones was authorized. This would provide for a speedier subjection of the enemy. In the words of Kalev Sepp, "Like Kitchener in South Africa, Bell issued directives to his subordinates to ensure adequate care for the 'reconcentrated' populace, including construction of schools and storehouses, allowance for provisions, food price controls, public works projects, and vaccination programs."[29] Contrary to anti-imperialist reports of the time, there was no starvation in those camps. All the people had to do was draw their daily rations and not cross the protective zone "dead line."

Additionally, some 6,000 Filipinos served as police officers in the newly constituted constabulary force. These constables provided invaluable assistance by identifying insurgents and their sympathizers hiding in the villages. The Philippine Constabulary (armed police force) would indeed play an integral part of this conflict, as well as future ones.

---

[28] Gates, *Schoolbooks and Krags*, 261.
[29] Kalev Sepp, *Resettlement, Regroupment, Reconcentration: Deliberate Government-Directed Population Relocation in Support of Counter-Insurgency Operations* (Fort Leavenworth, KS: Army Command and General Staff College, 1992), 28.

To root out sympathizers, Bell employed some of the best counterinsurgency officers as provost marshals, granting them authority to make mass arrests and impose hefty fines. Carrying out interrogations, the provost marshals unearthed and broke apart Malvar's shadow government.[30] In the analysis of Brian Linn, "Bell structured a pacification campaign designed to make the existing state of war and martial law so inconvenient and unprofitable to the people that they would earnestly desire and work for the reestablishment of peace and civil government, and for the purpose of throwing the burden of war upon the disloyal element."[31]

The next phase in Bell's plan called for sweeping the countryside of Malvar's regulars. To free up most of his men from garrison duty and give the COIN effort a native face, Bell authorized the arming of auxiliaries. Additionally, greatly assisting and guiding these active sweeps would be the Native Scouts, some of which were formal guerrillas. In the spirit of Sherman, Bell's troops were directed as follows: "The men will operate in columns of 50 and will thoroughly search each valley, ravine and mountain peak for insurgents and for food and destroy everything outside of towns. All able-bodied men will be killed or captured."[32]

Greatly assisting Bell's efforts at this time were the actions of the Filipino Federal Party. Created in December 1900, the Partido Federal was composed of *illustrado* and *principalia* elite. These elites called for Filipinos to work with the Americans. The Federal Party had a marked effect. It helped Bell, and the overall American COIN effort, to achieve positional advantage thereby underscoring David Galula's argument: "The counterinsurgent reaches a real position of strength when his power is embedded in a political organization issuing from, and firmly supported by, the population."[33]

Bell issued directives to his subordinates to ensure adequate care for the reconcentrated populace, including construction of schools and storehouses, allowance for provisions, food price controls, public works

---

[30] Moyar, *A Question of Command,* 84.
[31] Linn, *The U.S. Army and Counterinsurgency in the Philippine War,* 153.
[32] Vic Hurley, *Jungle Patrol, the Story of the Philippine Constabulary, 1901-1936* (London: Cerberus, 2011), 74.
[33] Galula, *Counterinsurgency Warfare,* 94.

projects, and vaccination programs. Everything of value outside the U.S. controlled "protective zones," particularly crops and farm animals, was confiscated or destroyed to deny their use to the guerrilla *insurrectos*.

As is so often the case, the scale of the "reconcentration" plan had been underestimated. Some 300,000 native reconcentrados found themselves forced to live in overcrowded and unsanitary camps. Food shortages, poor morale, bad hygiene and disease ravaged the internees, culminating in a cholera epidemic in 1902. At least 11,000 Filipino non-combatants died in Bell's "reconcentration camps," and possibly as many as 40,000 others died in like camps throughout the various military districts, primarily from disease. Notwithstanding these unfortunate events, the "reconcentration" effort immediately produced the originally intended results of halting guerrilla activities. Malvar surrendered on April 16, 1902. By July, Batangas was thoroughly purged of *insurrectos*.

Bell's reconcentration policy raised a storm of controversy in the American press who compared him to the Spanish General Valeriano Weyler. Moreover, he was scathingly criticized on the floor of the Senate, denoting his tactics as "a mixture of American ingenuity and Castilian cruelty." Nonetheless, it was clear that the population resettlement had decidedly weakened the insurgents in Batangas district. When used in combination with other tactics, resettlement significantly contributed to the insurgent defeat.

Unlike the British experience against the Boers, fortunately, the guerrillas were subjugated before the negative aspects of the camps could backfire on the Americans. Tellingly, beginning on January 1, 1902, and ending April 16, 1902, Bell's resettlement campaign took less than five months to snuff out the resistance. Kill metrics were decidedly less. In stark contrast to Smith's search and destroy campaign on Samar, during the five-month period, the total of insurgents killed was only 163 with 209 wounded. Inordinately, 3,626 surrendered. With the capture of Malvar, Bell was able to report without exaggeration that the insurgency in south-west Luzon had been crushed and Batangas district was thoroughly pacified. Then, on July 4, 1902, President Theodore Roosevelt declared the Philippine Insurrection officially over. Notwithstanding this declaration of peace, hostilities continued in some districts until 1903, and even until 1913 on the Island of Mindanao. Nevertheless, from 1903 on, the U.S. needed only a light footprint of some 15,000 troops to oversee the fledging republic. For their part, the Filipinos worked with the Americans to see that become a reality.

The lack of strong popular support in the United States for the Philippine Campaign, in large part due to the disagreeable nature of anti-guerrilla operations that included forced population resettlement, made the war, and its hard-learned lessons, best forgotten. With the outbreak of World War I in 1914, the U.S. Army's attention and energies were drawn away from the Philippines to Europe. In the end, observes Rod Paschall, "the bloody conquest produced an empty victory, as the United States quietly returned the islands to the Filipinos and—eager to bury an unpopular bout of imperialism—promptly forgot the counterinsurgency techniques that could have come in handy a century later."

Summary and Implications

As B. H. Liddell Hart has so aptly argued, "History can show us what to avoid, even if it does not teach us what to do—by showing the most common mistakes that mankind is apt to make and to repeat." In this vein, while many judge the Philippine War not a fitting model for counterinsurgency, various elements may indeed be emulated. By way of the PLAIN five laws of irregular warfare, the following is an analysis of American and Filipino efforts:

Political objective (s) – First, while the cause of the Philippine insurrection was clear, it may be said American objectives were anything but that. Notwithstanding the mission creep nature of the American effort, in the words of Brian McAllister Linn, "the United States was able to structure a coherent pacification policy that balanced conciliation with repression, winning over the Filipino population and punishing those who resisted."[34] Moreover, U.S. COIN efforts certainly had the imperatives of endurance if not restraint. At least not until MacArthur's repurposing of GO 100. The same could be said for American cultural intelligence (CQ).

---

[34] Brian McAllister Linn, *The Philippine War 1899-1902* (Lawrence: University Press of Kansas, 2000), 323.

Second, while the Army's COIN effort in the Philippines was by no means uniform, a key ingredient to success was the integration of political, military, and civilian elements. Beginning in 1902, Taft began to redress the damage of reconcentration by invigorating the native constabulary and establishing a nation-wide education system and instituted land reforms.

Legitimacy – This is the degree to which a population accepts that government actions are in its interest. In this vein, it would be a mistake to focus exclusively on misconduct by soldiers at the expense of the larger strategic picture. In spite of the excesses of some, the success of the U.S. COIN effort was due to the innovative ideas of U.S. field commanders, the perseverance of the American fighting men, and the moral integrity found in GO 100 which helped structure and guide operations.

Like other wars of this nature, the Philippine War was a contest over the soul of the Filipino people. In light of this fact, the single-point failure of the Aguinaldo-led Philippine insurgency was arguably a failure to unify a solid base of support. Aguinaldo failed to see the imperative of gaining the support of Filipinos at all levels of the society. Going along with the wealthy landowners, he effectively foreclosed the possibility of "mobilizing the mass of the people against the Americans."[35] The simplest explanation for this was Aguinaldo wanted to preserve the status quo ante. That is, it seems his desire was to maintain the traditional societal structure while taking the position of power of the outgoing Spaniards. Recognizing this, MacArthur implemented the operational employment of non-Tagalog native scouts such as the Maccabee and Ilocano Scouts to exploit this fissure, thereby further alienating Aguinaldo.

Adaptability – Bell's COIN campaign in Batangas makes for a fitting argument for employing resettlement as a viable tactic. In fact, according to some, Bell's campaign was a masterpiece of counterguerrilla warfare. His success was built upon America's previous experiences in the Mexican War, Civil War, and the War of Westward Expansion. Most laudable was the frame of reference the warfighters were provided by GO

---

[35] Max Boot, *Savage Wars of Peace* (New York: Perseus Books, 2002), 110.

100. The graph below demonstrates the relative weight of American COIN approach:

Resettlement	
Selective Strikes	Reforms
	Amnesty

As is demonstrated from history, external support is a crucial factor to winning the war of the flea. Tangible support reduction (TSR) therefore proved key to the American victory. As often forgotten, noted by Max Boot, the Army could not have won without the Navy's blockade of the archipelago. This effectively prevented Aguinaldo from "receiving foreign arms shipments or moving supplies of reinforcements."[36]

Influence – As stated before, success in COIN requires an operational design that stabilizes the affected area by identifying with the people, working with them to create a better state of the peace, winning their allegiance, and if possible, redressing grievances that led to the insurgency. At their best, Taft and his generals were able to hold out to the Filipinos the prospects for independence on the virtue of acquiescence. In practice, it proved an almost impossible task.

Moreover, from its inception, America's COIN strategy in the Philippines was anything but coherent and knowledgeable. In time, however, the information given to the Filipino people was crafted into a narrative that suited political goals. As an obvious example, Bell's circulars ensured local commanders could achieve unity of purpose while communicating a national rule of law.

---

[36] Boot, 128.

Summing up the daunting situation that faced the U.S. troops, Andrew Birtle observes, "Deployed in small, isolated detachments under the command of inexperienced junior officers, surrounded by an alien and untrustworthy population with whom they could not communicate, and frustrated by their inability to come to grips with an elusive foe, American soldiers felt the war's corrosive effects on both their morale and their morals."[37] Despite the record of some atrocities, by and large, the Americans were able to win over many Filipinos. Underscoring this truth, Manuel Quezon, guerrilla fighter and later Filipino president, once said, "Damn the Americans! Why don't they tyrannize us more?'"

Native Face – As previously stated, a successful counterinsurgency applies a discriminate use of force to separate the insurgency from the population both psychologically and physically. Early in the war, guerrilla use of terror tactics proved to be a stronger incentive to gain the support of the Filipino people than the American policy of attraction. Filipinos stopped supporting the insurgents only when the provisions of GO 100 were applied, and when they saw American troops providing them protection. Civic actions enabled American troops to identify with the people and leverage their support showing them to be the people's defenders and caretakers. Taft further fostered the support of the populace by organizing the Filipino Federal Party. This also weakened the influence of the revolutionaries over the Filipino upper classes, whose support had been so very critical to the movement. This was the crucial step toward breaking the connection Filipino insurgents had on the populace.

In his analysis of the American COIN effort, John Morgan Gates observes:

No single event or technique of pacification was responsible for the great achievement of the American army and its Philippine campaign. The Americans had found no panacea for victory. Instead, they had won with what could be termed a comprehensive approach that utilized every possible means at their disposal to bring an end to the guerrilla war. Filipinos supported the Americans

---

[37]Andrew J. Birtle, *U.S. Army Counterinsurgency and Contingency Operations Doctrine, 1860-1921* (Washington, DC: Center of Military History, 2001), 131.

for a variety of reasons, and the appeals used by the Americans to obtain support were equally varied. Filipinos were given both positive and negative incentives to accept American sovereignty to end the resistance. Benevolence, astute propaganda, the payment of money for weapons, and the promise of true civil government all represented positive incentives. American efforts in organizing municipal government, schools, public health programs, public works projects helped convince the Filipinos of the sincerity of the American claims to have come to the islands to have an abundance. At the same time, deportation, internment, imprisonment, or death awaited those who refused to end their resistance to the United States.[38]

Perhaps one of the greatest lessons from the Philippine Insurrection involves the necessity of preserving and passing on of the vast repository of knowledge that lies in the collective memory of veterans. Unfortunately, as Robert Asprey despairingly observes, "The lessons from the American experiences in the Philippines made no great impact on military thinking of the day."[39] It is now hoped that these lessons live on.

Though officially over in 1902, the war would continue unabated until 1913 on the Island of Mindanao. As they had plagued the Spanish, the Moros continued to be a vexing problem for the United States. When the resistance was finally put down, it came at a high cost in Moro blood. It was reported that as General Jack Pershing's troopers met the last bastion of holdouts, the Moros fought with stones and daggers. The United States then set up a colonial government in Manila. Under Taft, who served as the first governor of the colony, the U.S. improved public health, and established free speech and civil liberties. Filipinos began to serve in government. The end of the war also brought in a surplus of business interests. The Archipelago's economy boomed. Things were moving toward independence. In fact, on March 24, 1934, the U.S. Congress approved the Tydings-McDuffie Act which promised Filipinos self-government by 1946. Toward that end, in 1935, the Philippines became a commonwealth with its own elected government and constitution. Manuel Quezon was named the first president.

---

[38] John Morgan Gates, Schoolbooks and Krags: The United States Army in the Philippines, 1898-1902 (London: Greenwood, 1973), 270-271.
[39] Asprey, Robert B. *War in the Shadows: The Guerrilla in History.* Vol. 1. 2 vols. Garden City, NY: Doubleday, 1975), 198.

For Further Reading:

1. *Schoolbooks and Krags: The United States Army in the Philippines, 1898-1902* by John M. Gates.
2. *Honor in the Dust: Theodore Roosevelt, War in the Philippines, and the Rise and Fall of America's Imperial Dream* by Gregg Jones.
3. *The Philippine War* by Brian McAllister Linn.

5

They Remained:
Resistance in the Philippines

"Give me ten thousand Filipinos and I shall conquer the world!"
— General Douglas MacArthur

Members of Wendell Fertig's Resistance on Mindanao unload supplies.

During World War II, some of the most effective resistance movements in history took shape in the Philippine Islands to resist Japanese occupation. The aim of this chapter is to outline three resistance movements to demonstrate how they effectively waged a war within a war and set the conditions needed for the eventual return of U.S. forces to the Philippines.

Prosperity and progress toward independence in the Philippines came to a halt with the invasion of Japanese forces on December 8, 1941. Early on the day after the attack on Pearl Harbor, a Japanese juggernaut of some 100,000 completely surprised a larger combined American-Filipino force about 151,000 that parceled throughout the archipelago. American bombers and fighters were caught on the ground and destroyed. Advancing along several axes, the Oxford educated Lieutenant General Masaharu Homma, and his 14th Army, aimed to draw the Philippines into Japan's Greater Far East Co-Prosperity Sphere.[1] This pan-Asian concept strove for economic viability and freedom from "Western colonial subjugation."

The Japanese invasion came in two prongs. The northern invasion force was launched from Formosa Island and landed at Aparri and Vigan on December 10. The eastern prong launched from Palau Island and made its landing at Legaspi on December 12. Two battalions of this force also landed at Mindanao on December 20. Homma's main force then landed at Lingayen Gulf on December 22. Mindanao would be the springboard for a further thrust into Borneo. The last stage of this invasion was the landing at Lamon Bay the day before Christmas.

With his United States Army Forces in the Far East (USAFFE) retrograding southward down Luzon, General Douglas MacArthur vacated Manila, declared the "Pearl of the Orient" an open city, and invoked war plan Orange 3. This called for MacArthur's USAFFE to fight a delaying action at fixed points across the thumb-shaped Bataan peninsula. There, MacArthur expected to hold out long enough for reinforcements to arrive from the mainland United States. Despite fighting doggedly for every inch of ground, MacArthur's forces were steadily being pushed back by the weight of numbers and superior firepower of Homma's 14th Army.

---

[1] The proposed dividing line between the Nazi Germany's *Lebensraum* and Japan's Co-Prosperity Sphere was the 70th meridian east. This meridian runs from the Yamal Peninsula in Siberia to Gujarat on the western coast of India.

Imperial Japanese
14th Area Army
Lt Gen Masaharu Homma
129,435 troops
90 tanks
541 aircraft

United States Army Forces
in the Far East (USAFFE)
Gen Douglas MacArthur
151,000 troops
108 tanks
277 aircraft

10 Dec

10 Dec

Aparri

Vigan

22 Dec

Lingayen Gulf

Luzon

Baguio

Gen Wainwright
4 Inf Divs, 1 Cav

South
China
Sea

Bataan
Peninsula

Manila
Bay

Manila

24 Dec

Corregidor

Atimonan

10 Dec   Japanese Axes
Of Advance

Mindoro

Legaspi

12 Dec

Masbate

Samar

Panay

Leyte

Palawan

Cebu

Negros

Bohol

Gen Sharp
3 Inf Divisions

Mindanao

Davao

Sulu Archipelago

20 Dec

Japanese Invasion of the Philippines, December 10-24, 1941

Referring to themselves as the Battling Bastards of Bataan, despite their courage, they grew weaker by the day. Rations and ammunition were running low. Elsewhere in the Pacific, the Japanese had launched offensives. On December 10, 1941, the Royal Navy battleship *HMS Prince of Wales* and battlecruiser *HMS Repulse* were sunk in the South China Sea. Then, British Hong Kong fell on Christmas Day. This was followed by the loss of Singapore on February 15, 1942. Some 80,000 British and Commonwealth troops became prisoners of war, making it the largest surrender in British history. This conquest opened the way for the Japanese to exploit the oil-rich islands of Borneo, Java, and Sumatra.

Deep in his underground bunker of the island fortress of Corregidor, MacArthur saw the situation as bleak. A message from President Roosevelt arrived on February 20, 1942, ordering MacArthur to leave immediately for Australia, where he would assume command of all United States troops. MacArthur at first refused. He even drafted his resignation and was fully prepared to fight as a private rifleman alongside his men. His staff managed to talk him out of it.

Then on March 11, 1942, on Washington's orders, a reluctant MacArthur stole out of the island fortress on a U.S. PT boat for the Mindanao archipelago captained by Navy lieutenant John Bulkeley. After a harrowing two-day voyage, through stormy seas patrolled by scores of Japanese vessels, MacArthur and his family arrived on Mindanao.

When MacArthur arrived at General William Sharp's headquarters at the Del Monte Plantation, he was informed that the president of the Commonwealth of the Philippines, Manuel Quezon, and his vice president, Sergio Osmeña, were evading the Japanese on the island of Negros.[2] MacArthur immediately called on the expertise of Bulkeley, who managed to rescue the Philippine cabinet. Meanwhile, MacArthur and his family and staff were transported by two B-17s to the Land Down Under. This left General Jonathan Wainwright in command of USAFFE, and Major General Edward P. King in command of the forces arrayed on the Bataan perimeter.

---

[2] Kent Holmes, *Wendell Fertig and His Guerrilla Forces in the Philippines: Fighting the Japanese Occupation, 1942-1945* (Jefferson, NC: McFarland, 2015) 21.

Following three months of savage fighting, on April 9, 1942, King surrendered his command of 75,000 men. Then, within a month, following a Japanese amphibious assault of Corregidor, Wainwright surrendered on May 7. This constituted the worst defeat in U.S. history. The victorious Japanese then marched 50,000 American and Filipino prisoners on the infamous "Bataan death march" to prison camps a grueling 105 kilometers to the north.[3] An estimated 10,000 men died on the way. Determining that further resistance would be fruitless, and faced with the threat of exterminating the Bataan POWs, Wainwright ordered all USAFFE troops throughout the archipelago to surrender. In the wake of the invasion, the Filipinos civilians did what they always do. They took to the hills to fight. But for the American and Filipino soldiers, who were sworn to obey orders, the decision to surrender or not was more difficult.

The first months after the surrender of American and Filipino forces saw a total breakdown of authority throughout the islands. Armed bandits also took to the hills. Making matters worse, the Japanese demobilized the 8,000-man Philippine Constabulary (PC). In the vacuum of power, wild disorder prevailed. On Mindanao, Moros descended to plunder and pillage the lowland Christian settlements. Luzon witnessed internecine strife and the struggle for power as various factions fought each other.

After the main fighting subsided, time revealed the Japanese failure to have a doable pacification plan prepared for the islands. In a lesson on how not to do it, they tended to yield all but the coastal towns. This lack of control that ensued kept the Philippines in a state of disorder. To use guerrilla leader Russell Volckmann's three categories, those who made for hills were either fugitives, guerrillas, or criminals.[4] Outside of the Japanese controlled areas, levels of violence increased. Villagers and farmers found themselves caught in the middle. If these Filipinos kept their guns, the Japanese would learn of their possession. The penalty was beheading. If they turned in their guns, as the Japanese required

---

[3] Approximately 17,000 of these were American soldiers, sailors and marines; 12,000 were Filipino Scouts, and 21,000 were members of the Philippine Commonwealth Army.

[4] Russell W. Volckmann, *We Remained: Three Years Behind the Enemy Lines in the Philippines* (New York: W. W. Norton & Company, 1954), 69-70.

them to do, they found themselves at the mercy of marauding bandits. This caused the undecided population to either move to the coast, seeking some measure of Japanese "protection," or join the resistance, seeking protection from their own countrymen. As a result of their inability to get many off the fence, Japanese forces were never able to control more than twelve of the Archipelago's forty-eight provinces.

Japanese COIN measures were equally inept. To combat the guerrilla menace, Japanese patrols were sent to the interior hills and towns. Warned through the bamboo telegraph, the guerrillas were long gone before the Japanese arrived. To acquire intelligence regarding guerrilla whereabouts, the Japanese resorted to terror and torture. When this failed, the Japanese tried the organization of Neighborhood Associations. An appointed neighborhood leader was to inform on suspect guerrillas. As this failed, the Japanese would resort again to repressive measures which took the form of zonification and the employment of the magic eye. Using the technique of the magic eye, the Japanese would seize a barrio or town, herd the men into the town square where a hooded man with eyelets cut in the hood would be brought out to pick out spies, guerrillas, or sympathizers. These unfortunates selected by the "magic eye" would then be publicly tortured, and in most cases, beheaded.

It's fair to say that Imperial Japanese forces were ill-prepared to contain any sort of guerrilla movement within the Philippines. Equally inept was their foresight regarding prisoners of war. Some preparations were made, albeit not for the sheer numbers they encountered. The Bataan Death March POWs were initially held at Camp O'Donnell in Central Luzon. Within a few months, some 1,500 American POWs died within its cramped confines. Filipino deaths were much higher. Some estimates are 20,000. Out of clemency, Filipinos were released in July 1942 upon the promise of not taking up arms again against Japan. Lacking the necessities to house thousands of POWs, the camp was closed in January 1943. American internees were shipped out to the Central Luzon internment camp, Cabanatuan. As for the thousands of American and Allied expats, with the fall of Manila, these "enemy civilians" were interned into various camps. Initially housing some 3,000, Santo Tomas was the largest.

As the stick approach was clearly failing, the new landlords of the Philippines tried dangling the carrot. Prospects of the Filipinos having their own Republic were discussed. Then, to assist in bringing their Asian brothers in line, the Japanese rebranded the Constabulary. Pressing into service a cadre of former PC, the collaborationist version was renamed, "Bureau of Constabulary." According to guerrilla leader Ray Hunt, the Japanese rounded up and forced handfuls of former constables to work in this puppet organization, with the threat that their loved ones would be harmed if they failed to comply.[5]

Subsequent to this "liberation" of the archipelago, what followed was wholesale economic exploitation. Schools, businesses, and small shops were all closed. Services were discontinued. Radio and newspapers were banned. Within a year, some 1,000 Filipino businesses were taken over by Japanese businessmen. Mining equipment was dismantled and shipped away. Thousands of automobiles, trucks, and farm equipment were dismantled and shipped to Japan as scrap. Grain and rice harvests were seized to provide supplies for the Imperial Japanese Army. Carabaos were slaughtered for meat. This left farmers without the means to cultivate the fields, and the fields not seized were often burned. With the Co-Prosperity Sphere in full swing, and the Filipino economic infrastructure now supporting the Japanese war effort, Japan's liberation of the Philippines was proving to be anything but prosperous. It's fair to say that when it came to recruiting guerrillas, the Japanese did all the work. The Japanese expected the Filipinos to turn on their supposed American oppressors, and they were incensed when this did not materialize.

From his headquarters in Brisbane, Australia, MacArthur and his staff set about organizing the resistance against the Japanese. This unconventional warfare (UW) apparatus would become so extensive that an estimated 300,000 Filipinos organized into 277 guerrilla bands were brought into its orbit. Supporting them was an auxiliary and underground component that is estimated in the millions.

The most effective resistance movements took shape on the Islands of Luzon and Mindanao, and the Visayan Islands of Leyte, and Cebu. In

---

[5] Ray C. Hunt and Bernard Norling, *Behind Japanese Lines: An American Guerrilla in the Philippines* (Lexington: University Press of Kentucky, 2000), 107.

fact, these were among the most successful of their kind in history. The aim of this chapter is to outline the resistance movements on these islands to demonstrate how they effectively waged a war within a war, and to set the conditions needed for the eventual return of U.S. forces to the Philippines.

While fugitives chose to sit out the Japanese occupation, those who refused to surrender organized themselves into guerrilla bands. Guerrilla fighters came from all walks of life and levels of society. Most guerrilla groups were led by American or Filipino officers who refused to surrender. Some were led by Filipino soldiers who had been in Japanese internment camps, such as Luzon's Camp O'Donnell. Still, others sprang up from grassroots paramilitary and even civilian organizations. On Southern Luzon, Governor-elect of Camarines Norte, Wenceslao Vinzons, began an armed revolt against the invaders. Among the first to organize a resistance, he was nonetheless one of the first to be captured and executed. Former Manila cab driver and boxer, Marcos "Marking" Augustin, formed a group along with his mistress, Valeria Panlilio. She was the real brains of the operation.

Luzon peasant Luis Taruc formed the Communist *Hukbong Bayan Laban sa Hapon* (People's Anti-Japanese Army), or Hukbalahap, commonly called the Huks. Elsewhere on Luzon, Vincente Umali, former mayor of Tiaong in Tayabas province, formed President Quezon's Own Guerrillas. Among those to organize themselves against their new landlords were disgruntled cadets and university faculty, called the Hunters ROTC group. Then, there were of course the Moros of Mindanao who fought everyone. And in the Visayans, on Cebu, veteran soldier turned radio announcer, Harry Fenton, led guerrillas on the southern part of the island.

Initially, many of these bands exhausted themselves in uncoordinated actions. Such was the short but menacing career of Walter M. Cushing. Cushing had come to Luzon to escape the cash-strapped, hand-to-mouth days of the Great Depression. In the wake of the Japanese invasion, he persuaded a group of Filipinos and American soldiers to oppose the invaders. In a series of vicious ambushes canvassing several provinces in Northern Luzon, Cushing and his private army dispatched some 500 to 1,000 Japanese. His demise came in June 1942 as he set out to contact other resistance groups in Isabela Province. Word quickly reached the local Japanese garrison regarding Cushing's presence. When the Japanese found him holed up in a house, they riddled it with bullets. Mortally wounded, Cushing played dead. As six Japanese soldiers approached, he killed all of them, and saved one bullet for himself.[6]

Scores of other guerrillas met similar fates. As the unsurrendered struggled to survive and build their organizations, their persistent appeals for help were reaching MacArthur's new headquarters, Southwest Pacific Area (SWPA), some 3,500 miles to the south in Australia. One of the first radio transmissions to be received was that of Filipino army officer Lieutenant Colonel Guillermo Nakar. Operating out of Nueva Ecija Province on Luzon, Nakar's unit continued to report on Japanese activities until he was captured in early September 1942. He was taken to Kempeitai headquarters at Manila's Fort Santiago. The Kempeitai were secret military police, Imperial Japan's version of the Gestapo. Colonel Akira Nahaham, Kempeitai chief in the Philippines, wielded enormous power. He was by far the most formidable adversary of the resistance. Once someone came into his clutches, they were taken to Fort Santiago and never seen again. Nakar was savagely interrogated, then beheaded.

Recognizing the central importance of these guerrilla bands, MacArthur assigned Colonel Courtney Whitney the task of managing their activities under the newly formed Allied Intelligence Bureau (AIB). This was no small feat. MacArthur further dispatched U.S. Navy Commander, Charles 'Chick' Parsons, to the archipelago. To him fell the task of vetting the desperate guerrilla commands.

Parsons was a natural pick. He knew the Tagalog language. He knew the geographic and human terrain of the islands well and had lived in Manila for a number of years.[7] The plan was simple. Parsons and his team would clandestinely make contact with the various guerrilla bands throughout the archipelago, vet the leaders, then grant them authority, and provide supplies and radios. The various guerrillas could be thus networked with MacArthur and AIB in Australia. With command and

---

[6] Bernard Norling, *The Intrepid Guerrillas of North Luzon* (Lexington, University Press of Kentucky, 1999), 12.

[7] Initially detained, but convinced Parsons was a Panamanian consul, the Japanese repatriated him and his family. Following a near trans navigation of the globe, Parsons and his family reached New York on August 12, 1942.

control over such a network firmly emplaced, MacArthur gauged that victory was sure to follow.

As to where Parsons would go first was being debated, a small sailboat arrived from Mindanao. Wendell Fertig, a reserve Army officer, had sent three men to request radio equipment and supplies to further his movement. "Mindanao it would be," Parsons decided.[8]

Parsons' mission, dubbed Spy Squadron (SPYRON), undertook a sketchy four-month tour de force.[9] Launching out by the U.S. submarine *Tambor*, Parsons' odyssey began with a visit to Wendell Fertig on Mindanao. When he arrived at the port of Misamis City on March 4, 1943, he was met by an orchestra dressed in white playing "Anchors Aweigh." One sailor remarked that the submarine's skipper had taken a wrong turn and ended up in Hollywood.[10] It took only forty minutes to upload some 25 tons of ammunition, radio equipment, medical supplies, food, and $10,000 in pesos.

Fertig

Before the fall of Bataan, reserve Army Colonel Wendell Fertig made one of the last flights out of Corregidor on April 29, 1942. On orders to join the U.S. staff in Australia, he climbed aboard a Catalina amphibious aircraft. Arriving unscathed on Mindanao, he set out immediately for Australia. However, Japanese advances on Mindanao would cancel any further air operations. Having gained an initial foothold at Davao back in December 1941, the main Japanese invasion force landed at Cotabato, on Mindanao's western shore. The 5,500-man Kawaguchi Detachment made landfall the same day Fertig's plane landed.

Falling under General's Sharp's command in Mindanao, Fertig's orders were now to supervise the construction of airfields, along with the demolition of the main roads and bridges. In the chaos, Fertig did manage to meet up with his old engineer buddy Charles Hedges. As the Japanese noose tightened, Sharp surrendered on May 10, 1942, leaving

---

[8] Travis Ingham, *Rendezvous by Submarine: The Story of Charles Parsons and the Guerrilla-Soldiers in the Philippines* (New York: Double Day, 1945), 53.
[9] SPYRON fell under the U.S. Navy's 7th Fleet while SWPA's intelligence section administered the program through the AIB.
[10] Larry S. Schmidt, *Fire in the Jungle* (Fayetteville, NC: Blacksmith, 2019), 86.

his subordinate, General Guy Fort in command of the tattled remnants of the Allied forces on the island. Fertig, Hedges, and Navy chief petty officer Elwood Offret, took for the hills.

For the next few months, the three were constantly on the move as they avoided Japanese concentrations in the Province of Lanao. The local Japanese commander, Captain Yamato, learned of Fertig's presence, and sent a personal letter guaranteeing his safety if he would capitulate. No reply was sent. Then on July 4th, Fertig witnessed an "Independence Day Parade." General Fort was placed in a truck bed at the head of a column of prisoners who were mostly barefooted and shackled together with wire. Fertig's resolve to resist the Japanese at all costs began that day. [11]

Undaunted in his resolve to resist the Japanese, Fertig ventured further into the jungle. It was at this time that he made this entry in his diary:

During the months in the forest, I have become acquainted with myself and developed a feeling that I do not walk alone; a feeling that a Power greater than any human power has my destiny in hand. Like a swimmer, carried forward by a powerful current, I can direct my course as long as my way lies in the direction of the irresistible flow of events. Never have I lost the feeling that my actions have followed a course plotted by some Power, greater than any human agency.[12]

Sensing himself destined for victory as part of a master plan, Fertig waited on the hands of Fate. Opportunity knocked in the form of a boisterous young American mestizo, Luis Morgan. As a former Philippine constabulary officer, Morgan was a hard-drinking, fearless man. Of him it was said, "If a man is brave, and has a gun, he joins Morgan." However fearless, Morgan distinguished himself by systematically massacring the indigenous Islamic Moro population. In fact, he killed more Moros than Japanese.

As Fate would have it, Morgan offered Fertig command of some 600 Filipino guerrillas on the condition that he be made Fertig's chief of staff. The duplicitous Morgan did so in the hopes of using Fertig as leverage to assume power over all Mindanao. Morgan's scheme was

---

[11] Schmidt, *Fire in the Jungle*, 39.
[12] John Keats, *They Fought Alone: A True Story of a Modern American Hero* (New York: Turner, 2015), 104.

simple. Fertig would promote himself to general and unify the desperate bands into a single resistance against the Japanese. Having his general stars fashioned from coins by a Moro silversmith, Fertig promoted himself, and drafted a proclamation announcing his assumption of command. The result was immediate. Droves of unsurrendered Americans and Filipinos flocked to his banner. Among these were Army officers Captain Clyde Childress and Major Ernest McClish. Both had been battalion commanders, whose battalions were thrashed in the defense of Mindanao. Organized along divisional lines, Fertig designated the resistance the United States Forces in the Philippines (USFIP). McClish was given command of 110th Division, and Childress was appointed his deputy.

As for the hapless Morgan, Fertig leveraged the whimsical mestizo's influence, and, in time, eventually took over the command of his group, his plan all along. Fertig then sent Morgan off on a recruitment mission to bring more into the fold. While he was gone, Fertig patched things up with the Moros, who constituted a third of the Mindanao population, which was 29,000. Fertig rightly understood they were the key to an organized front against the Japanese. Moro communities were led by a *datu*, much like a chieftain. Securing the loyalty of prominent *datus* was one of Fertig's chief concerns. Following the counsel of Charles Hedges, Fertig used various strategies tailored to meet the peculiar situation of each *datu*. In this way, he was able to extract promises of support or assurances of neutrality from the Moro chieftains.[13]

Another important step in the development of Fertig's resistance was the bringing of the Filipino elite into his orbit. In his thinking, with such people back in their native governmental offices, legitimacy, not to mention financial stability, could be achieved. No doubt these landed gentry could galvanize a great deal of public support. These elites were also closely connected to the Catholic Church. While a few Filipino elites collaborated with their new Japanese landlords, as in the case of Emilio Aguinaldo, most were appalled by their senseless brutality. In Misamis Occidental, where Fertig had raised his standard, the Ozamis family had a major influence. In particular, Doña Carmen was the matriarch of the

---

[13] Schmidt, *Fire in the Jungle*, 59.

family. As the wife of former Senator Ozamis, her word virtually carried the weight of law. With her support, the province would as well.

Fertig's Resistance Movement, Mindanao Island, 1942-1945.

Calculatingly, Fertig directed his men to attend Sunday mass. The gesture was not lost. Shortly after, Fertig received a dinner invitation from Doña Carmen. Others in attendance were Carmen's priest, Father Calanan, and her close adviser, Doctor Contreras as well. With the proviso that Morgan would no longer be a menace, Doña Carmen pledged the support of Misamis Occidental, and, with Father Calahan's consent, the support of the Catholic Church as well.

Another key figure who played a prominent role in the local government was F. D. Pacana. Also connected to the Ozamis family,

Pacan was the prewar provincial treasurer. Concerned about how the guerrillas would be paid, he informed Fertig about how Quezon had appointed a Provisional Emergency Currency Board (PECB) for each province on Mindanao. Though they had no plates to print with, this board was authorized to print 3 million pesos. With this authorization by the Philippine government in exile, and the expertise of Fertig's old friend Sam Wilson, five-peso notes were soon produced. The guerrillas were paid with these notes, which, after the war, were honored by the Philippine government as legal tender.

To put Mindanao back in working order, Fertig ordered the prewar Mindanao government officials back to work in their former positions. On December 1, 1942, the Province of Free Lanao was proclaimed under Governor Marcelo Paiso. This did much to further strengthen the resistance movement as Larry Schmidt adds:

In the areas where the Japanese were strong and the establishment of a free civil government deemed impractical, Fertig established martial law under the 10th Military District. This was done in Bukidnon and a portion of Misamis Oriental. Interestingly, in some areas, government officials had to serve two masters, but Fertig considered the gains were worth the risk.[14]

Enterprisingly, similar to the methods being used by Wenceslao Vinzons in the Bicol Region of Luzon, Fertig relied upon funds gained through taxation. The people readily supported the effort.

When it came to logistics, the problem of food supply on Mindanao was less acute here than it was on the other islands. Food was in abundance. The problem was distribution. To help solve this problem, the well-to-do couple, Nick and Josefa Capistrano, created the Women's Auxiliary Service. This network of labor groups collected and helped distribute food to the guerrilla force. With control over 90% of the island, Fertig dispersed the much-needed supplies to his units through such auxiliaries by boat, carabao, and courier. Fertig also supervised the development of many farm projects, which raised chickens and hogs to support the cause. Because fuel was scarce, the guerrillas had to make their own. Fermenting a coconut sap called tuba, the gasoline substitute

---

14 Ibid., 73.

was manufactured in copper stills. This concocted fuel could run engines and machinery.

By the time Parsons arrived on March 5, 1943, Fertig's handful of unsurrendered Americans and Filipinos swelled to an impressive force of some 15,000 with 5,000 rifles. This number was complemented by a militia force and a loosely organized division of Moros.[15] He could even boast of an Officer Candidate School, a commando course, and had in his possession two airfields in the nearby Bukidnon Province.

Impressed, Parsons laid out MacArthur's ground rules. To keep the guerrillas from being wiped out and to spare the local populace the wrath of Japanese retaliatory strikes, MacArthur emphasized the necessity to lay low. Fertig's task was to set the conditions necessary for a U.S. return. To that end, he was to focus on gathering enemy intelligence. Like the rest of the guerrilla bands throughout the archipelago, Fertig's resistance was to be the eyes and ears of MacArthur who vowed to return.

Fertig did not fully agree with MacArthur's orders to avoid combat with the Japanese. The quiet gathering of intelligence would not satisfy the wrath of the Filipinos, many of which had a blood feud with the Japanese. The savage war between the guerrillas and the Japanese was played out with no quarter given by either side. The Japanese gave the guerrillas such short shrift because they perceived them to be mere bandits. Thus, counterguerrilla operations were called "punitive expeditions." So far, the Japanese attacks hadn't penetrated into the interior of the island, rather they contented themselves to looting and then destroying the fields, to starve the people.

For his part, consciously or unconsciously, Fertig followed several tenants of irregular warfare. He understood that to succeed there should be no battle lines, but rather, like a fish, his guerrillas had to swim in the sea of the people. Correlated with this, Fertig followed Mao's strategy of pitting one man against ten, and the tactic of ten men to one. Holding this dichotomous methodology in balance, Fertig's strategy was his so-called "pillow effect." This offered no defensive battle lines to be overrun. Whenever the Japanese would attack, they would be met by little to no resistance. Then, using selective strikes and ambush tactics, Fertig would

---

[15] Robert Smith, *Triumph in the Philippines* (Washington, DC: Center of Military History, United States Army, 1963), 586.

decimate the Japanese patrols brave enough to venture out of the garrisons. These patrols normally consisted of fifty to a hundred men. In one of his well-conceived schemes, the Japanese would negotiate with a farmer to buy rice. Notified of the day to have it ready, the farmer would then inform Fertig's guerrillas who would lie in wait to ambush their unsuspecting prey. As an added bonus, the guerrillas then recovered the rice from the obliterated convoy.

Fertig knew all too well that if he didn't produce the hits that would attrite the Japanese, Filipino morale, not to mention recruitment, could be a challenge. Parsons reiterated: MacArthur was serious about this lay low policy. Fertig's units were to make no more attacks on Japanese garrisons. Additionally, he was not to try to free the Allied prisoners interned in Davao Penal Colony. The question was, could Fertig accept such terms. Fertig acquiesced. Being satisfied with his ability to lead and take orders, and personally knowing McClish, Bowler, and Hedges, three of his subordinate commanders, Parsons thereby designated Fertig the Commander 10th Military District. He then presented "General" Fertig with a set of silver eagles and said, "Colonel Fertig, you're in the Army again. Or should I say still?"[16]

In addition to being a spy master, Parsons was also a skillful diplomat. He plied his skills on the last remaining hold out to joining Fertig's command, former Philippine Army Colonel Salipada Pendatun. A native of Cotabato-Maguindanao, Pendatun gathered the scattered fragments of his old unit to form the "Bolo Battalion," which was indeed a force to be reckoned with. Pendatun had racked up an impressive string of victories, including the elimination of two Japanese garrisons at Kitaotao and Malaybalay. Through sheer weight of personality and a skillful dialogue involving metrics, particularly those related to submarine-borne shipments of guns and supplies, Parsons won Pendatun over to Fertig's camp.

Commandeering one of Fertig's ships from his "navy," the sixty-foot motorized boat *Nara Maru*, Parsons hugged the Mindanao coast north to the Surigao Strait, the narrow passage between the Bohol Sea and Leyte Gulf. There, he would place the first of many "coastwatcher" teams. Consisting of two to three men, these teams would brave the elements to

---

[16] Ingham, *Rendezvous by Submarine*, 61.

send up-to-date information on weather and Japanese aerial and naval movements. Fertig would go on to emplace coastwatcher teams on the southern coast of Mindanao. These were particularly effective in notifying the U.S. wolfpack submarines that preyed on Japanese shipping. Demonstrating their lethality, within a two-year period, it is estimated that some 300 ships were sunk between Davao and Zamboanga.[17]

Before the war was over, Parsons would undergo seven more clandestine missions via submarine. Parsons' visit also had a tremendous psychological effect. It spoke volumes to the resistance on Mindanao that they had not been forgotten, and that help had finally come. While the Navy commander was emplacing coastwatcher teams, rumor had it that General Ichiro Morimoto, the Japanese commander on Mindanao, was planning a punitive expedition to wipe out Fertig's command.

Fertig's underground agents in Davao informed him the Japanese attack would come on June 19th. The force was estimated to be 20,000 strong. Morimoto had already launched a series of attacks along the northern coast of the island. Japanese radio direction finding had triangled Fertig's headquarters. As luck would have it, ten prisoners had escaped from Davao Penal Colony on April 4th and were assisted by Fertig's guerrillas after their escape.

The Japanese attack came on June 26, 1943. Supported by a destroyer and five fighter planes, 4,000 of Morimoto's troops, aided by a force of 200 pro-Japanese Bureau of Constabulary troops, struck Misamis City. The attack sent the guerrillas back into the hills, while Parsons made a forced march across Mindanao and met the submarine *Thresher* in Pagadian Bay. From there he made his escape to Australia and reported to MacArthur on Fertig's resistance movement.

After getting blown out of his g-base, Fertig had a new problem. One of the more prominent Moro chiefs informed him that Morgan had assumed command of the Mindanao resistance, replacing Fertig, who had supposedly stepped down to become Morgan's administrative officer. The next day, a letter arrived signed by "General Morgan." Fertig

---

[17] Holmes, *Wendell Fertig and His Guerrilla Forces in the Philippines*, 56.

was summoned to a conference to be held in Kolumbugan on August 10, 1943.

Fortune favored Fertig. In addition to its military supplies, a recently arrived submarine fortuitously delivered an edition of *Life* Magazine. The magazine contained a lengthy article on King Ibn Saud, regent of Saudi Arabia. Replete with pictures and expressions of friendship from the king to the United States, the article had an immense impact on the fence-sitting Moro leaders. Hundreds of Moros on Mindanao had made the hajj to Mecca. This tangible support by King Saud to the United States quickly sealed the decision for the Moro leaders in favor of Fertig. Morgan saw the writing on the wall and reluctantly boarded the submarine *Bowfin* bound for Australia. Hedges, Fertig's second in command, wanted Morgan shot.

Volckmann

Following the fall of Bataan, Army officer Russell Volckmann escaped the brutalities of the Japanese. Together with another Army officer, Donald Blackburn, Volckmann evaded capture, making his way to northern Luzon. In his evasion, which took five months, he brushed past multiple Japanese patrols. During his long trek north, his body was wracked with dysentery and malaria. Aided by local Filipinos, he and Blackburn joined a guerrilla force under Army colonels Martin Moses and Arthur Noble. Like other early guerrilla commands, this force spent itself in uncoordinated attacks on the Japanese.

After the capture and eventual execution of Moses and Nobles in October 1942, Volckmann assumed command of the guerrilla remnant. He designated his Filipino American guerrilla band the official-sounding U.S. Army Forces in the Philippines – Northern Luzon (USAFIP-NL). In June 1943, its fighting strength was 2,000. By the end of the war, it swelled to 19,660 Americans and Filipinos, roughly the size of a U.S. Army division.[18] How Volckmann brought his force to that size and how it was employed to great success is a story of his keen intellect, indomitable will, and courage.

---

[18] Volckmann, *We Remained*, 180.

Arguably, Volckmann's resistance was the most effective against the Japanese out of any other in the archipelago. This may be deduced on metrics alone. Even by the most conservative estimates, his guerrilla command of 19,000 bagged some 30,000 Japanese. As a further testament, when U.S. forces returned to clear Luzon of Japanese, Volckmann's guerrilla command fought as an independent division.

Starting pretty much from scratch, Volckmann knew the first essential element to any resistance is a cause. As he reasoned, potential for resistance existed throughout Luzon and the Islands. His major question was, could those opposed to the Japanese be organized and directed to express their opposition through subversion and guerrilla warfare? Volkmann answers this question in his autobiography:

A successful resistance movement can only be generated among people who have the courage and stamina to withstand privations, endure hardships, and face imminent death while fighting back against great odds. To make use of these characteristics, I recognized that leaders must emerge to inspire, awaken, organize, and direct this potential. From the willingness of the Filipinos to trust and be guided by American leadership, I was confident that any lack of strong native leadership could be supplemented by a few determined Americans. [19]

Volckmann may not have been familiar with Mao's three phases, but as a West Point graduate, no doubt he had read his Clausewitz. As laid down in his treatment of what he called "the people's war," Clausewitz noted that one of five conditions to creating an effective uprising was "The national character (of those fighting it) must be suited to that type of war." The Filipinos fit the Prussian's prerequisite to a "T."

That the Filipino people greatly aspired independence was self-evident. More than that, they had been promised independence. And central to the spirit that animated the Filipinos under Japanese occupation was the belief that U.S. forces would return. Such a faith was critical to the survival and success of the guerilla movement. Knowing this full-well, U.S. propaganda material and commodities were emblazoned with MacArthur's famous line, "I shall return." Recognizing the energizing force of MacArthur's slogan, in contradistinction, Volckmann entitled his autobiography *We Remained*.

---

[19] Ibid.

Taking stock of his situation in February 1943, Volckmann established his headquarters at Darigayos Cove. A nook of white sand and blue water about fifteen miles north of San Fernando. In terms of dynamics, the shell of the guerilla organization Volckmann inherited from Moses and Noble had two crucial positive attributes: A suitable organizational structure, and ideal geographic location, which at many points was and remains rugged and inhospitable.

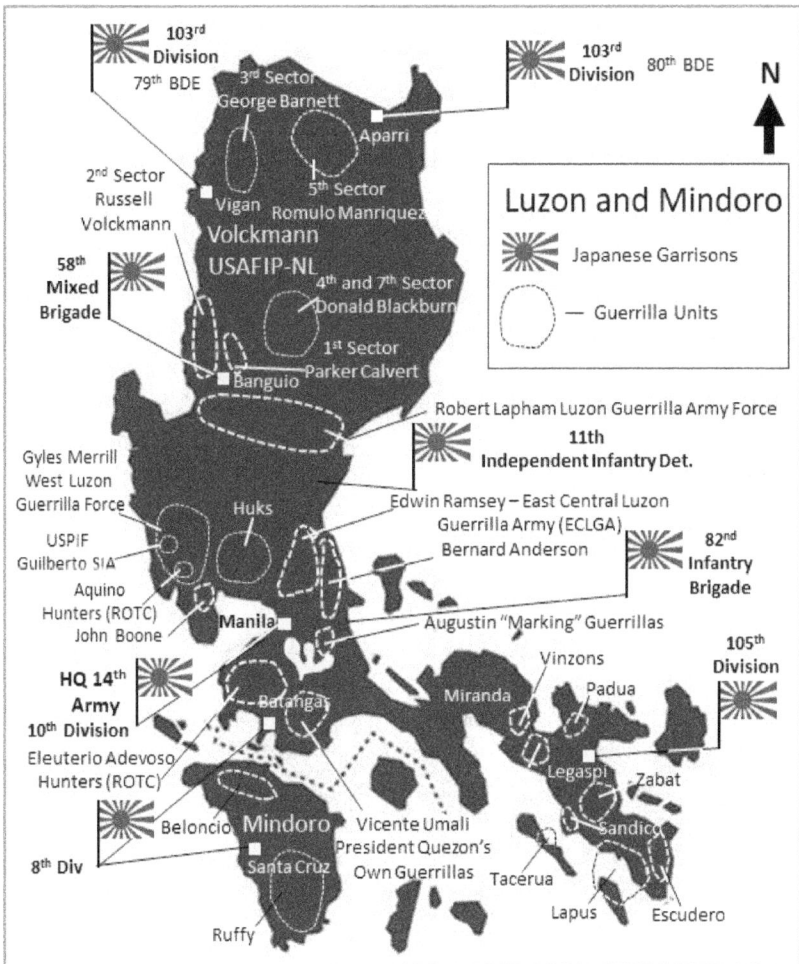

Luzon and Mindoro Guerrillas.

Shaped roughly like a football, the pine-covered mountainous extremity of Northern Luzon was barely accessible except to its indigenous population. The region proved to be highly conducive to guerrilla operations. Assessing the fifteen or so assorted guerrilla bands in Northern Luzon, Volckmann set out to organize the resistance into one cohesive unit. He was greatly assisted in his organizational efforts by shadow governor Roque Ablan. The other fragmented guerrillas in Northern Luzon began to fall in line. All except the guerrilla force of Robert Lapham on the densely populated central plains of Luzon. Lapham's organizational structure and operational design were starkly different from that of Volckmann. Volkmann wanted to bring Lapham's command under his control, but Lapham refused. At one point, Volckmann threatened Ray C. Hunt, one of Lapham's commanders, with court-martial if he didn't acquiesce. Hunt remained with Lapham.

Notwithstanding this bureaucratic tiff, Volckmann divided his organization into seven sectors. The 6th sector being reserved for Lapham, if ever he could be brought into the fold. Within this arrangement, Volckmann further arrayed his force into five regiments. These would eventually rise to a strength of about 3,000 men and include three rifle battalions of four rifle companies each. With battlefield recovery, Volckmann's resistance soon boasted of field artillery.

Having framed out a flexible formation, Volckmann also set out to build a strong intelligence network. Around this time, the Philippine Constabulary was reestablished in Bontoc, some fifty miles north of Baguio. As luck would have it, one of Volckmann's earlier acquaintances from Bataan, Mario Bansen, had been pressed into service as a clerk in this Bontoc office. Learning of Volckmann's presence and eager to help the guerrillas, Bansen sent him a message. In time, Bansen became Volckmann's most prolific spy.

It's often the simplest things that work best. As instructed, Bansen began feeding the USAFIP-NL valuable intelligence by placing an extra sheet of paper in his typed reports.[20] Through this method, Volckmann learned the names of those who were collaborating with the Japanese

---

[20] Volckmann, *We Remained*, 132.

and was kept abreast of any developing counterguerrilla operations. With his man on the inside, Volckmann was able to convey high grade intelligence to AIB, and, in the process, ferret out Northern Luzon's Japanese spies and collaborators with impunity. Once Filipinos saw that it was safe to support the guerrillas, the so-called fence-sitters began falling in the right direction.

Manila Cloak and Dagger

Notwithstanding Volckmann's counterintelligence bonanza, the Japanese had collaborators such as the talented Mr. Reyes. Having acquired the credentials of an agent for the SWPA, Franco Vera Reyes lured the unsuspecting into his web. Before spying for the Japanese, Vera Reyes was a common crook incarcerated in Manila's Old Bilibid Prison. The Japanese set him free under the condition that he would work for them. Vera Reyes' career as a deadly spy began. The crook turned spy was so effective that he was able to infiltrate Manila's underground network. Believing him to be an intelligence officer on orders from MacArthur to coordinate all guerrilla activities, Maggie Leones, a member of the guerrilla band Fil-American Irregular Troops (FAIT), was inadvertently gathering pertinent details for "Colonel Reyes."

Unwittingly, Maggie helped the talented Mr. Reyes nab 29 members of FAIT. The Kempeitai round up included some of Manila's prominent civilians, including Blanche Jurika, Commander Chick Parsons' own mother-in-law. They were hauled off to Fort Santiago, savagely interrogated, herded into Manila's North Cemetery, then bayoneted to death. Vera Reyes would continue to play a key role in the wartime drama of Manila. Astonishingly, when the Kempeitai net closed, Maggie survived the encounter. She was later brought into Volckmann's orbit and was highly useful to his intelligence network.

When it came to the Manila underground, one would like to think that it was monolithically structured. It was anything but that. As Henry Kissinger would later put it, "It is only to posterity that revolutionary movements appear to be unambiguous." That sentiment described the Manila underground to a "T." Moreover, for the resistance's sake, it was good that it was. The Manila underground was a diverse group of

American expats, priests, students, hospital staff, and shopkeepers. Many did not know each other, working diligently and covertly to aid the various resistance movements. It is believed that much of the underground's structure was set in place by Jose Ozamis, Manual Roxas, Jose Laurel, and Chick Parsons. The compartmentalized nature of the Manila network enabled it to survive the war, though many would be hauled off to Fort Santiago.

When Manila fell to the new conquerors, Jose Ozamis, Manual Roxas, and Jose Laurel all accepted posts in the Japanese occupation government. They did so with the blessings of the Filipino resistance who understood that these positions would enable them to move discreetly.[21] In May 1943, Jose Ozamis, who had been befriended by none other than Vera Reyes, traveled to his native island of Mindanao. The purpose of his visit was to meet Fertig and Parsons. After their meeting, and a visit to his sister Doña Carmen, who was by all intents the shadow governor of Misamis, he returned with some letters the Mindanao guerrillas desired to be delivered to friends and loved ones in Luzon. Somehow Vera Reyes came to learn about the guerrilla correspondence. The Kempeitai struck hard and fast, hurling Ozamis and twenty-nine others into Fort Santiago. True to form, all were savagely interrogated, then beheaded.

Claire Philipps was an American expat and out of work entertainer. As Homma's 14th Army was hammering down on Manila, Philipps followed the retreating U.S. forces and found herself serving as a nurse for the Battling Bastards of Bataan. As Japanese forces bore down on the defenders, she was persuaded by John Boone, a corporal of the U.S. 31st Infantry, to help the resistance. She then managed to stay out of Santo Tomas by adopting the alias Dorothy Clara Fuente, a Philippine-born Italian.

With her forged papers, Philipps, and a Filipina named Fely Corcuera, opened Club Tsubaki, a gentlemen's club that quickly became super popular. The alcohol-soaked atmosphere of the club became a target rich environment for Philipps and her cabaret dancing agents. Philipps was soon passing information on Japanese ship and troop deployments to Boone, who by this time was leading a guerrilla unit in Bataan. Boone

---

[21] John Keats, *They Fought Alone* (New York: Turner, 1963), 137.

in turn passed these messages on to Edwin Ramsey, another guerrilla leader on Luzon, and then on to AIB in Australia. Ramsey managed to scrape together about 13,000 guerrillas under his command he entitled East Central Luzon Guerilla Area (ECLGA). Known as "high pockets" for her penchant for carrying messages in her brassiere, Philipps collaborated with other Manila agents to smuggle in medicine, food, and supplies into Cabanatuan prison. In her endeavors to assist the POWs in Cabanatuan, Philipps worked hand-in-hand with Margaret Utinsky and her spy ring.

Due to the large numbers of mosquitoes, malaria was a constant threat. Medicines such as quinine tablets were crucial to keeping the guerrillas alive. These and other medical supplies were acquired primarily through various methods, including sympathetic local populace, via American submarines, or from the underground. During his numerous submarine voyages, Parsons brought the guerrillas various medical supplies and vaccines. Such items were highly prized. In fact, Volckmann's guerrillas raided Japanese installations with the intent of capturing medical instruments and supplies.[22]

For their part, the Japanese had their puppets. When the invasion came, former president Emilio Aguinaldo was quick to side with the Japanese. On February 1, 1942, he even made a radio address calling for the Battling Bastards of Bataan to surrender. Following the capitulation of Allied forces, he was tasked by his Japanese handlers with creating a new constitution for a Japanese puppet state. When U.S. forces returned to Luzon, guerrilla forces under Marcos "Marking" Augustin and "Mammy" Valeria Panlilio captured him. By the Spring of 1943, Santo Tomas' internment population crested to 7,000. Some 2,000 were transferred to the newly constructed Los Baños internment camp. Built on the south shore of Laguna de Bay, the backyard of Hunters ROTC resistance, the first group of 800 arrived in May 1943.

Lapham and Ramsey

Led by Robert Lapham, the Luzon Guerrilla Armed Forces (LGAF) was positioned in Luzon's main agricultural area between Baler Bay and

---

[22] *We Remained*, 128.

Lingayen Gulf. Due to him being on a densely populated lush plain, Lapham had to operate differently from other guerrilla units. Like John Singleton Mosby, Lapham's 13,000 men lived at home, then converged at points to do kinetics. According to Lapham, his biggest problem at first were the turf disputes, complicated by quarreling over jurisdiction, about who would rule who, all of which was exacerbated by personal grudges. As mentioned, his biggest tiff was with Volckmann. This issue was causing problems outside of Northern Luzon. In December 1944, SWPA sent a memorandum to American commanders telling them to stop quarreling. This was backed up with another message with the announcement that MacArthur did not want an overall guerrilla commander on Luzon.[23] Lapham's strategy was to steer clear of such rivalries and operate autonomously.

Like Volckmann, one of Lapham's major hurdles was communication. In fact, he did not have a radio until 1944. Prior to radio contact, like other guerrilla groups, Lapham used couriers to communicate with other groups, and with SWPA. This was time consuming. Not to mention the fact that the sender could never be certain that a message got through, and couriers were at risk of capture. This detriment was soon remedied after Lapham's group succeeded in setting up its own radio and transmitted it to Colonel Charles Smith on Samar on July 1, 1944. The radio used was delivered from Fertig via Robert Ball. Fully aware of their needs, on August 31, 1944, SWPA sent Lapham's force weapons, equipment, and medicine via the submarine *USS Narwhal*. Notwithstanding the friction, Ray Hunt, one of Lapham's commanders, gave a radio to Volckmann.

Like Robert Lapham, Major Edwin Ramsey also hailed from the 26th United States Cavalry Regiment. In fact, before he organized the East Central Luzon Guerilla Area (ECLGA), he took part in the last cavalry charge by the U.S. Army on January 16, 1942, on Bataan. For Ramsey, the Huks represented an irreconcilable enemy. Ramsey noted that he fought the Huks almost as much as the Japanese. In fact, the leader of the Huks, Luis Taruc, put a price on Ramsey's head, referring to Ramsey as a "Japanese sympathizer and traitor." In one incident, the Huks

---

[23] Robert Lapham and Bernard Norling, *Lapham's Raiders: Guerrillas in the Philippines, 1942-1945* (Lexington, KY: University Press of Kentucky, 1996), 92.

feigned a parlay with one of Ramsey's units. The Huks then ambushed and killed them all. Ramsey ordered for Huks to be shot on sight. Despite this war within a war (within a war), Ramsey's ECLGA would eventually swell to some 45,000 men, with 7,000 arms.

Ramsey's fight against the Japanese was primarily through means of sabotage. Ramsey was in regular contact with the Manila underground, and John Boone's Bataan guerrillas. Tapping into his extensive intel network in Manila, his operatives conducted limited low-risk sabotage operations. These actions included setting fire to supply depots and pouring sugar into vehicle gas tanks. One such sabotage mission included the destruction of a Japanese oil tanker in the port of Manila. Though dramatic, its net effect was to raise the morale of the resistance.

Visayan Island Guerrillas

During the initial invasion, the Japanese bypassed the Visayan Islands of Panay, Negros, Cebu, Bohol, Leyte, and Samar. When they finally arrived, these islands were only partially garrisoned throughout the occupation period. For this reason, the guerrilla movements on these islands were decidedly different than that of Luzon and Mindanao. Prior to Wainwright's surrender, Colonel Albert Christie was the commander of the USAFFE forces on Panay.[24] Obeying orders, he promptly surrendered on May 19, 1942. Christie's operations officer, Lieutenant Colonel Macario Peralta, led 5,000 troops from the 61st Division as they fled to the hills and formed the core of the guerrilla group on the island. Peralta was able to secure most of the division's weapons and supplies.

Philippine Army Colonel Ruperto Kangleon commanded the guerrilla movement on Leyte. He did so after a bitter struggle with other contenders.[25] Comprising the islands of Leyte and Samar, Kangleon's command was designated as Ninth Military District. Kangleon escaped Japanese captivity and was in hiding with his family on the island when he was asked by Parsons to assume command of the guerrilla effort. It was through his leadership on the island that he unified the various

---

[24] Kent Holmes, *Wendell Fertig and His Guerrilla Forces in the Philippines: Fighting the Japanese Occupation, 1942-1945* (Jefferson, NC: McFarland, 2015) 44.
[25] Ibid., 3.

guerrilla groups and restored discipline to their ranks. Kangleon commanded over 3,000 guerrillas on Leyte and his forces helped provide intelligence used in the liberation of the island in October 1944.

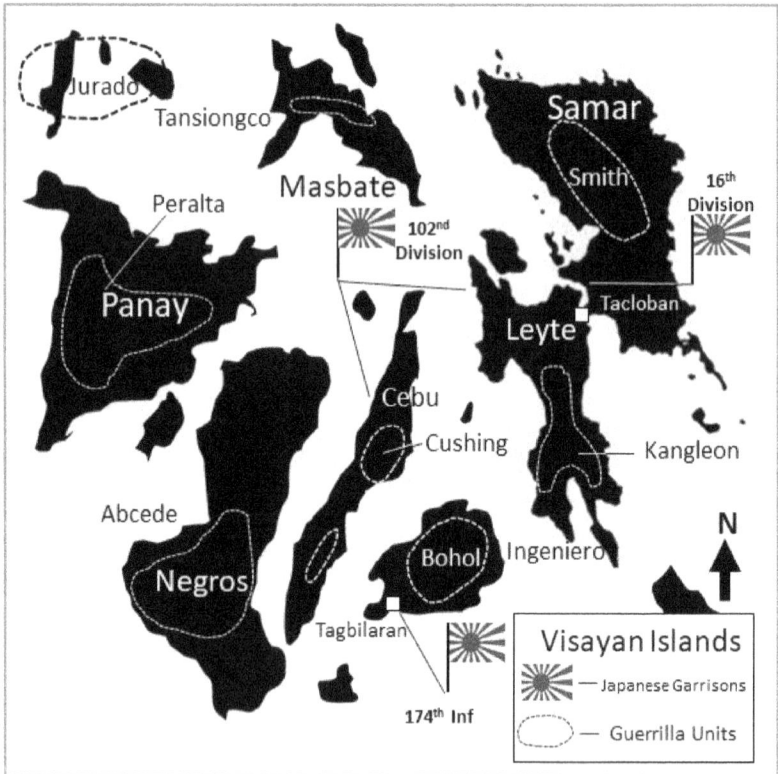

Visayan Island Guerrillas

As the Japanese were able to occupy much of the island of Cebu with little effort, they soon sent most of their troops elsewhere, co-opting pre-war mayors and using local constabulary troops to maintain their new order. As to resistance, there was no organized effort against the Japanese until September 1942. Two major guerrilla groups emerged. Those in southern and central Cebu were under James M. Cushing, an American-born mining engineer, and brother of Walt Cushing. Cushing was commissioned as a U.S. Army captain by Brigadier General

106

Bradford Chynoweth, commander of units in the Visayas before the surrender of USAFFE forces. Those in northern Cebu were led by radio announcer Harry Fenton, who was an enlisted American soldier in Manila before his discharge from the Army. Since the war began, Fenton had been transmitting anti-Japanese messages. The two groups merged in the fall of 1942 with a shared command: Fenton assumed administrative duties and Cushing served as the group's combat leader.

Initially this arrangement worked well, resulting in many successful attacks against Japanese troops. However, in time, conflict between Fenton and other guerrilla leaders eventually split the organization. Disguised as a priest, Cushing ventured to neighboring Negros Island to meet with Major Jesus Villamor, a Philippine Air Force officer sent by General MacArthur to assess the resistance movement in the Visayas, establishing this vital contact with SWPA. While Cushing was away, Fenton's erratic and violent behavior led his men to execute him. When he returned, Cushing discovered the fate of his hapless former co-commander. Furious and shaken, Cushing retreated to his hut for several weeks. Officially recognized by MacArthur on January 22, 1944, Cushing was promoted to lieutenant colonel and appointed sole commander of 8th Military District. A month later, a Navy submarine delivered the first load of guns, ammunition, medicine, and a long-range radio capable of reaching SWPA HQ in Brisbane. By the time MacArthur returned in October 1944, Cushing's force had swelled to some 25,000. About half were auxiliaries.

One of the most important contributions of Cushing's guerrillas, and indeed of the resistance in the Philippines, was the capture of the so-called Koga Papers. On March 31, 1944, Admiral Mineichi Koga, commander-in-chief of the Imperial Japanese Combined Fleet, and his staff, were enroute to Davo. When their two Japanese flying boats encountered a typhoon, they crashed into the Cebu Straight. On board was a briefcase containing documents that detailed the top-secret Japanese "Plan Z." In light of the impending U.S. invasion of the Mariana Islands, Plan Z was Koga's concept for the defense of an American attack. Koga and some of his staff were killed in the crash. Rear Admiral Shigeru Fukudome and eight other staff officers survived. Not wanting the plans to fall into wrong hands, Fukudome swam toward shore and tried to send the briefcase to the bottom of the Cebu Straight.

Investigating the wreckage, a local fisherman recovered the briefcase and turned it and the Japanese survivors over to Cushing's guerrillas.

Cushing radioed SWPA: "Believe Japanese party came from Palau. One thought Admiral Koga. Recovered case of important documents. Some look like cipher system." The news threw SWPA into a frenzy. AIB immediately radioed Cushing to transport the 10 Japanese survivors to Negros Island. There they would be picked up by submarine and taken back to Australia for interrogation. Cushing knew such a mission was impossible without detection as large Japanese sweeps were being carried out in an all-out search for the survivors. The Japanese commander of Cebu, Lieutenant Colonel Onishi, also delivered a threatening message. Unless Cushing turned over his captives within thirty-six hours, scores of villages would be burned to the ground.

Forced to walk a fine line, to prevent unnecessary bloodshed, Cushing arranged for the prisoners to be released to Onishi, and the documents to be picked up via submarine. MacArthur sent a direct order, "Prisoners are to be held by you at all costs." Cushing radioed back, "Impossible to comply." According to historian William Breuer, MacArthur relieved Colonel Cushing of command on Cebu and reduced him to the rank of private.[26] For the time being, three of Cushing's guerrillas huffed the briefcase to Negros, and turned it over to the awaiting submarine.

Once in Australia, AIB quickly translated the plans and forwarded them to Admiral Chester Nimitz, Commander in Chief of the U.S. Pacific Fleet. Knowing what the Japanese would do beforehand, the U.S. Navy's Fifth Fleet sank three Japanese carriers, damaged six others, and shot down about 600 enemy aircraft. The Navy shot down so many Japanese planes it was called the "Great Marianas Turkey Shoot." In the follow up Battle of Leyte Gulf, Japanese carriers had so few aircraft that they were essentially used as decoys. As for Cushing, he and his men managed to survive Onishi's counter guerrilla sweeps. And despite not knowing if he would be court martialed when the dust settled, Cushing continued to send a steady stream of intelligence to SWPA.

A similar intelligence bonanza was uncovered in Luzon. Volckmann's guerrillas were able to reach the site of a Japanese plane crash. Inside

---

[26] William Breuer, *MacArthur's Undercover War: Spies, Saboteurs, Guerrillas, and Secret Missions* (Hoboken, NJ: Castle Books, 1995), 154.

the wreckage were documents detailing the planned defense of Luzon. In September 1944, General Tomoyuki Yamashita, known as the Tiger of Malay for his conquest of Singapore, assumed command of the Imperial Japanese 14th Area Army, which numbered some 262,000 troops. With the inevitable return of U.S. forces at hand and assessing a Japanese defense of the Philippines as untenable, the Tiger of Malay opted for the strategy of resistance in depth. He would not resist beach landings but would draw his forces into the mountainous recesses of Northern Luzon. There he hoped to buy time for his nation's forces much as MacArthur had at Bataan. Discovering Yamashita's plans, Volckmann had the documents translated and radioed to SWPA.

Over the Threshold

It has been said that by the time MacArthur returned to the Philippines, he knew what every Japanese lieutenant ate for breakfast, and where he had his hair cut. The guerrilla units of the archipelago certainly set the conditions necessary for the U.S. return. By the fall of 1944 there were 126 radio stations and 27 weather stations reporting regularly back to SWPA.[27] Providing detailed reconnaissance, SWPA knew the composition, disposition, and strength of every major Japanese command in the Islands. The guerrillas also worked with AIB agents to identify suitable landing sites.

Eager to make good on his promise and determined to recover from the ignominy of having been driven from the Philippines, MacArthur's return would commence with Leyte Island. Kangleon's guerrillas provided Kruger's 130,000-man Sixth Army with an accurate picture of Japanese dispositions. Prior to the landings, the elite Alamo Scouts of the 6th Army Special Reconnaissance Unit arrived to coordinate guerrilla actions with the upcoming Allied operations. Commander Parsons also snuck onto the island to warn villagers. They had ringside seats for what followed. The virtually unopposed landings began on October 20, 1944, with an eruption of naval gunfire and air bombardment.

---

[27] Robert Lapham and Bernard Norling, *Lapham's Raiders: Guerrillas in the Philippines, 1942-1945* (Lexington, KY: University Press of Kentucky, 1996), 209.

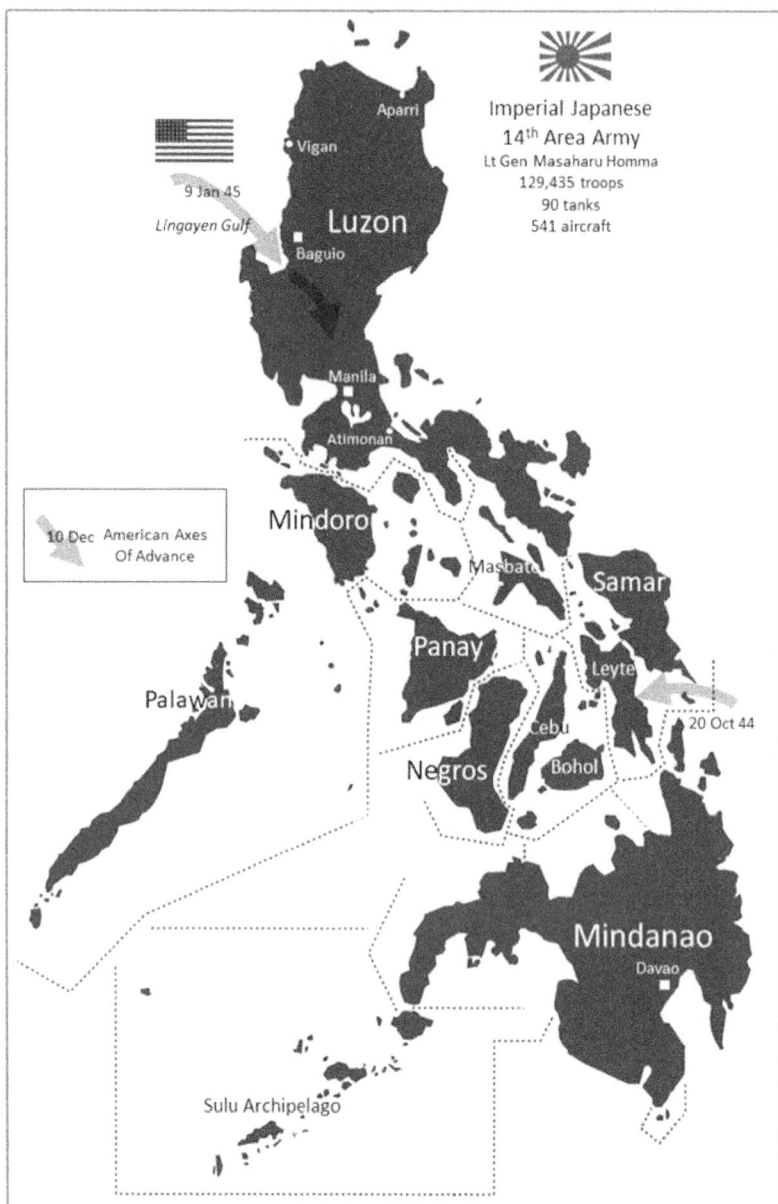

MacArthur's Return to the Philippines, 1944-1945.

In support of the landings, Kangleon's guerrillas sabotaged key bridges which blocked Japanese reinforcements. Seeking to blunt the U.S. invasion, for three days, from October 23 through October 26, the Imperial Navy fought the U.S. Navy's Third and Seventh Fleets. Involving over 300 warships and over 1500 aircraft, what followed was the Battle of Leyte Gulf, the largest naval battle in history.

This engagement was remarkable for several reasons. It included the last battleship-to-battleship action in history, and the first time Americans encountered the kamikaze attack. In the lopsided U.S. victory, Japanese losses were appalling; 28 ships, including three battleships, three carriers, 10 cruisers, and 11 destroyers were sunk. By Christmas of 1944, the 20,000-man Japanese 16th Division, the perpetrators of the Bataan Death March, were laid to rest on Leyte.

The U.S. victory on Leyte sent a warm glow across the archipelago. The dark days of occupation were nearly over. The next landing in the liberation of the Philippines was planned for Mindoro, an island south of Luzon. It would be used as a springboard onto the archipelago's largest island. The landings came on December 15, 1944. Barely occupied by the Japanese, and with much of it held by Filipino guerrillas, Mindoro was quickly overrun. In the wake of the American juggernaut, the Japanese Imperial General Staff decided to make the Philippines their final line of defense. To halt the American advance toward Japan, every available Japanese soldier, airplane, and naval vessel was sent to the Philippines.

For the Luzon guerrillas, the shift from intelligence gathering to kinetic operations occurred with a warning order. Commencing January 4, 1945, guerrillas were to sabotage enemy supply dumps, rolling stock, aircraft, and telegraph lines, to generally "unleash maximum violence against the enemy." Ray Hunt promptly raided the Japanese garrison in San Quenton. Volckmann's guerrillas made a night raid on the Japanese naval base at San Fernando. Sixty-thousand liters of fuel were destroyed.[28] In preparation for the coming U.S. landing at Lingayen, Volckmann's men cleared the gulf of 350 sea mines, sabotaged numerous rail lines (with the explosives from the sea mines), and put the Agno River Power Plant out of commission. [29]

---

[28] *We Remained*, 176.
[29] Ibid., 176.

On the morning of January 6, 1945, an Allied force of 875 ships commanded by Admiral Jesse B. Oldendorf steamed into Lingayen Gulf. Unopposed by Japanese naval forces, the Allied ships came under aerial and Kamikaze attack. In a rain of steel, of the 600 Japanese aircraft that took part in the attempt to destroy the Allied fleet, an estimated 200 were used as kamikazes.[30] All told, some 24 Allied ships were sunk, and scores of others were damaged. Then, following a deafening barrage of naval gunfire by roughly 70 battleships and cruisers, the U.S. Sixth Army under General Kruger landed on a 25-mile beachhead between the towns of San Fabian and Lingayen. Within days, 203,000 U.S. troops were on Luzon.

As Sixth Army steamrolled toward Manila, Lapham was concerned about the POWs at Cabanatuan. The previous month on Palawan Island, the Japanese had executed 150 POWs. Traveling thirty miles on horseback, on January 26, 1945, Lapham presented himself to General Kruger. He proposed that a rescue attempt be made to liberate the estimated 500 POWs at Cabanatuan before the Japanese could dispose of them as well. Due to his proxemics, Lapham's force was intimately familiar with the internment camp at Cabanatuan. He estimated the Japanese had some 200–300 troops within the camp, and around 5,000 within the town of Cabanatuan. In Kruger's estimation, his forces would need another week before they could reach the camp. A week might be too late. Assuming command of all Luzon's guerrillas, Kruger selected the 6th Ranger Battalion, commanded by Lieutenant Colonel Henry A. Mucci, to conduct the raid.

One of the most daring in history, the Cabanatuan raid force consisted of 121 Rangers, 13 Alamo Scouts, and 286 of Lapham's guerrillas. Lapham's men guided the force to the camp, then provided security and laid an ambush along the main road coming into camp. Accompanying the force, Mucci chose Captain Robert Prince to lead the operation. Launched on the night of January 30, the Rangers had to crawl across nearly a kilometer of open field to get into position. Mucci also arranged for a P-61 Black Widow to buzz the camp to distract Japanese guards. The guerrillas also cut the camp's telephone lines.

---

[30] Japan would launch some 1,900 Kamikaze aircraft during the Battle of Okinawa, April 1, 1945.

The raid commenced at 19:40. Within the first minute, all the camp's guard towers and pillboxes were targeted and destroyed. Breaching the main gate, the Rangers poured fire into the officer quarters and guard barracks. Having received accurate intelligence from the Alamo Scouts, the Rangers eliminated the enemy positions while withholding fire to the prisoners' huts. Within thirty minutes, the Rangers wiped out the entire Japanese garrison and evacuated 522 POWs. The only casualties occurred in the withdrawal. A lone Japanese soldier managed to fire three mortar rounds, hitting several Rangers, Scouts, and POWs with shrapnel. The battalion surgeon Captain James Fisher was mortally wounded.

Meanwhile, the litter patients from the camp were transported by guerrilla-organized carabao carts. Many were later evacuated to a hospital at Guimba. To cover the Rangers' withdrawal, a guerrilla delaying action successfully fought off Japanese counterattacks. What makes the raid at Cabanatuan so impressive is the overwhelming success despite the lack of rehearsals.

Following the success of Cabanatuan, another raid liberated more than 2,100 prisoners from the Los Baños prison camp on February 23, 1945. Situated east of Manila at the southern end of Laguna de Bay, the camp was constructed by the first group of 800 men who arrived in May 1943. AIB had long known of its whereabouts through the reports from guerrillas. Closest in proximity was the guerrilla unit Hunters ROTC, led by former Philippine Army Cadet Eleuterio Adevoso. Having intimate details of the area, Hunters ROTC guerrillas would serve as the reconnaissance element for the raid. This intel, which came by way of an escaped prisoner, included sketches of the interior layout, as well as patterns of life. The actionable intel included the fascinating fact that the Japanese guards conducted physical training without equipment or weapons.

Apparently, the enemy commander, being one more inclined for ceremony than readiness, directed the encumbrances to be neatly collected in one location. On the morning of the 23rd, one element of the 11th Airborne (511 Parachute Infantry Regiment) crossed Laguna de Bay in amphibious craft while another element took off by plane for a daylight parachute drop. All forces converged in a swift and coordinated attack which caught the Japanese guards of the camp in the middle of their morning calisthenics. The entire garrison was annihilated with practically no loss to the Allies, and the Los Baños prisoners were evacuated across the Bay.

In one of his major engagements of the war, Volckmann's guerrilla army laid siege to the Japanese garrison at San Fernando. Volckmann embedded his troops with the 24th Marine Air Group. From their forward position, his troops called in precision-dropped 500lb. bombs from the 24th's F6F Hellcats which leveled the playing field. The Japanese released an official communique which declared the Americans had "perfected a new aerial bomb which was attracted by concentrations of ammunition and fuel."[31] San Fernando was then liberated on March 14, 1945. Next, in support of the Allied push northward, Volckmann's 20,000-strong USAFIP-NL operated as an independent division of Kruger's Sixth Army. Their task: dislodge the Japanese regiments at Bessang Pass who were blocking the way to Yamashita's stronghold.

Following the liberation of La Union on March 23, Volckmann's forces commenced the assault on Bessang Pass. Counterattacking on May 17, the Japanese 73rd and 76th Regiments pushed them back. In a renewed offensive, supported by the U.S. Army's 122nd Field Artillery Battalion, USAFIP-NL cleared the ridges, and on June 15, 1945, liberated the town of Cervantes.[32]

As Kruger advanced toward Manila, Lapham's and Ramsey's guerrillas were formed into regiments and attached themselves to Sixth Army units. These elements continued to provide enemy intelligence, serve as porters to move equipment and supplies, and as guides, bypassing Japanese minefields. In many ways, guerrillas freed up American troops for use elsewhere. When the battle for Manila came, it would be the scene of the worst urban fighting fought by American forces in the Pacific theater.

Not intending to defend Manila, Yamashita did not declare Manila an open city, as MacArthur had done. Instead, he ordered the destruction of

---

[31] Allison Ind, *Allied Intelligence Bureau: Our Secret Weapon in the War Against Japan* (New York: David McKay, 1958), 156.

[32] From Jan 9 to June 15, 1945, the USAFIP-NL sustained 3,375 casualties, including 900 killed.

all bridges and installations and then evacuated the city. In defiance of Yamashita's orders, and with no real fleet to command, Japanese Rear Admiral Sanji Iwabuchi decided to entrench his force of 19,000 within the city, the wellbeing of Manila's one million inhabitants not being a factor. On February 3, lead elements of Kruger's Sixth Army encountered dug-in Japanese defenders. That night, soldiers of the U.S. Army's 1st Cavalry Division liberated the internees at Santo Tomas, the hell hole where 3,785 captives had spent 37 months.

Some of the darkest days of the war then came to Manila. Determined to fight to the last man, Japanese defenders fought doggedly for every block. Civilians were used as human shields. During lulls in the fighting, Japanese soldiers committed acts akin to the Rape of Nanking. Girls between the age of 12 and 14 were gang-raped to death. Then, their mutilated bodies were doused in gasoline and set on fire. Pregnant women had their bellies ripped open. Nothing was sacred. Nuns were raped in convents, and nurses in hospitals. When the dust settled on March 3, about 100,000 Filipinos civilians were killed, both in the massacre and from artillery and aerial bombardment by Japanese and U.S. forces.[33] On the morning of February 26, Iwabuchi and his officers committed suicide.

In the fall of 1944, more than half of the Japanese garrison on Panay Island left to defend Leyte. Taking full advantage, Peralta's guerrillas seized all available airstrips and took on the Japanese force of some 2,750 men. By the time of the liberation of Panay, the American 40th Infantry Division faced only a token resistance from the Japanese.

On Cebu, Cushing's 8,500-man guerrilla force joined Major General William H. Arnold's Americal Division, as they sent a force of 12,000 Japanese into the hills. The combined assault commenced on March 26, 1945. So effective were the guerrillas in combating the Japanese, that on June 20 the Americal Division left Cushing to clear the island of any remaining Japanese.

D-Day for Mindanao and Fertig came later than most of the Islands. It was now March 1945, Fertig's guerrillas had killed some 7,000 Japanese, while tying up another 60,000 still intent on breaking the

---

[33] The Battle for Manila claimed 1,010 dead and 5,565 wounded for the Americans and some 16,000 Japanese killed in action.

resistance. Early on, the Joint Chiefs of Staff believed that Mindanao had the most to offer an invasion force. The U.S. invasion of Mindanao was set for November 15, 1944, with a follow-on strike at Leyte Gulf on December 20, 1944.[34] Admiral Halsey opted for a Leyte first approach, having assessed the overall Japanese defense of the island to be minimal. Fertig didn't know it at the time, but Mindanao would be one of the last major islands to be invaded.

Fertig's guerrilla units conducted pre-assault shaping missions, severing communications lines, clearing beachheads, sabotaging bridges, and supply depots, and attacking Japanese garrisons. In support of the invasion, Eighth Army had intended to use an airstrip on Palawan, but it wouldn't be ready in time. This problem was solved by the guerrilla's covert airstrip at Dipolog. On March 8, 1945, an airbridge was established as C-47s landed two reinforced companies of the U.S. Army's 24th Division onto the airfield. The same day, sixteen Marine Corsairs arrived. The airfield's proximity would later enable the Marine pilots to carry out bombing runs. The invasion of Mindanao began on March 10 with a naval and air bombardment along the beach defenses east of Zamboanga City. As the U.S. Army's 81st Division made their landings, the Japanese fled into the hills in disorder.

Launched on March 20, 1945, the raid on Talisayan was one of the most unusual guerrilla operations of the war. Gaining support of the U.S. Navy, a force of 350 guerrillas, of the 110th Division, raided the Japanese garrison of about 200. Supported by the U.S. Navy's Task Force 70.4, created to aid the guerrillas, Fertig's men on four landing crafts captured the coastal fort, and annihilated the garrison. Following up on this success, Fertig's guerrillas supported General Eichelberger's 53,000-strong Eighth Army landings at Parang on April 17, 1945. After screening Eighth Army's drive to Davao, Fertig's guerrillas independently cleared the eastern end of Mindanao of Japanese stragglers until the cessation of hostilities. Although the formal end of hostilities in the Pacific Theater occurred on August 15, 1945, Yamashita did not surrender until

---

[34] Joint Chiefs of Staff, Future Operations in the Pacific: Report by the Joint Staff Planners, March 10, 1944, pp. 10, 21; See also General Headquarters, Southwest Pacific Area, "Basic Plan for Montclair Operations," February 25, 1945, pp. 3-4 for the concept.

September 2, the day of Japan's official surrender. When Yamashita came out of the hills to surrender, it was to the guerrilla commander Volckmann. Yamashita was executed a year later for war crimes.

Disarmament, Demobilization, and Reintegration

Upon establishing the beachhead on Leyte Island, MacArthur immediately transferred authority to Sergio Osmeña, the successor of Manuel Quezon, as Philippine Commonwealth president. This reverted the Philippine Islands back to its pre-war U.S. commonwealth status. Then, with the liberation of Manila, a ceremony was held announcing the restoration of the Government of the Commonwealth of the Philippines. Osmeña's cabinet was formed. Slowly but steadily, the Philippine government was gradually reestablished. As the archipelago was freed, military authorities transferred power back to the municipal and provisional governments. The Philippine National Bank and the Philippine Congress were both reorganized. Approved by the U.S. Congress, the Philippine Commonwealth received $900 million for the payment of war damages, of which $1 million was earmarked to compensate for church losses.

There were many pressing matters for the first Commonwealth Congress to solve. Among these pressing matters were reconstruction, and the disarmament, demobilization, and reintegration of the guerrillas. By way of what is now referred to as the DDR process, the 200 or so guerrilla bands throughout the archipelago would be stood down, disarmed, and reintegrated into civilian life. For the most part this process was uneventful, except in the case of the Huks. By the end of the war the Huks could boast some 100,000 guerrillas and supporters. Their Communist ideals for a different post-war Philippines placed them at loggerheads with the Commonwealth Government. In the end, they were disarmed at gunpoint and their movement went back underground for the time being.

A few other details pertaining to Volckmann deserve attention. During the war Volckmann gave about 10 million pesos worth of IOUs. These were all honored by the government. Amongst issues to settle, the family members of the collaborators who had been executed under Volckmann's orders came forward. Another thorny issue had to do with the Philippine collaborators who remained alive. Many of whom were on trial or awaiting to be tried. All these issues were settled when President Roxas granted full amnesty to all. This amnesty act exonerated Volckmann, and many others, of war crimes. Then on July 4th, 1946, as promised, the Commonwealth of the Philippines became the Republic of the Philippines. Manuel Roxas was its first president. Many of the old guerrilla formations later become Philippine Army units. Volckmann's USAFIP-NL formed what became the Second Division of the Philippine Army.

Summary and Implications

It's no overstatement that within the annals of history one would be hard-pressed to find a better example of a more effective UW campaign than the one organized by General Doulgas MacArthur. A very close second would be that of the one prosecuted by the Office of Strategic Services in various other theaters of the war, particularly Operation Jedburgh in Europe (see chapter six).

Political objective (s) – As has been argued, for an insurgency to win, it must have a cause. For the Filipinos, the communist Huks to a lesser degree, the cause was of course independence. For their part, the Japanese were plagued by a lack of coherent policy directives with a few meaningful exceptions.

Legitimacy – Legitimacy is the degree to which a population accepts that government actions are in its interest. Perhaps one of their worst decisions, the Japanese disbanded the Philippine Constabulary Forces. This contributed to the chaotic atmosphere and left the population without a layer of protection from banditry and lawlessness. To what extent would the PC have assisted their new masters is anyone's guess. In its place, the Bureau of Constabulary did little to bring legitimacy to the occupational government. In the last analysis, the Japanese failed to pacify the Philippines because they failed to gain widespread support of the Filipino people. They saw the Japanese for who they were, rapacious oppressors who came to plunder them for all they had.

<u>Adaptability</u> – In irregular warfare, who adapts wins. The Japanese were also plagued by their inability to modify their approach. Conversely, the myriad of guerrilla units did everything to stay alive and relevant to the cause. Based on the autobiographical accounts of many guerrilla leaders, perhaps their greatest adaptation came in the form of the Lay Low order which curtailed their actions against the Japanese, at least until it was time to unleash hell. As an example, in addition to their primary mission of intelligence gathering, the relative weight of the various tactics employed by Volckmann and Fertig could be surmised as such: 60% guerrilla warfare, i.e., ambushes and raids, 30% sabotage, that is the wrecking of rail lines, and destruction of bridges and supply dumps, with the remaining 10% accounting for the subversion conducted through various messaging, and propaganda leaflets and posters.

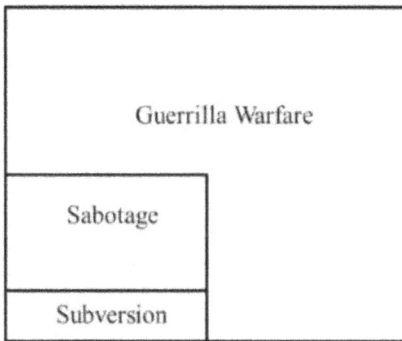

Relative Weight of Volckmann and Fertig's Resistance Movements.

<u>Influence</u> – If the two prime movers of IW are legitimacy and influence, the Japanese had nothing to stand on. With rare exceptions, the Japanese had no capacity to influence save that of fear and loathing.

During the war, the Japanese maintained control of the eastern third of Mindanao. In this coastal area, the Japanese treated the population with uncharacteristic respect. As a result, the Filipinos there became more fence-sitters. Unbeknownst to the Japanese, this conciliatory approach to the people could have wrecked the resistance movement on Mindanao and elsewhere. As a principle there could be no neutral, middle ground. To remedy this potentially disastrous situation, Fertig's guerrillas had to conduct sabotage and assassinate Japanese leaders in this area in order to provoke Japanese reprisals against various eastern towns. The guerrilla calculation proved effective. Reprisals came, and in so doing, the Japanese were seen again as the real enemy.

Throughout the archipelago, in implementing the more brutal aspects of their occupation policy, the Japanese effectively undermined any political objectives they had. This includes the rather patronizing gesture of occupational "independence" under the puppet President Laurel. In the words of historian Robert Ross Smith, "One phenomenon of the reconquest of the Philippines was certainly far different from any other experience of the war in the Pacific. That was the presence of a large, organized guerrilla force backed by a generally loyal population waiting only for the chance to make its contribution to the defeat of Japan."[35]

Native Face – Despite the great sounding slogan, "Asia for the Asians," and its overarching program, "The Greater East Asia Co-Prosperity Sphere," it all had a hollow ring. Actions speak louder than words. The Filipino people saw through the scam. Conversely, buttressing cause of the resistance movement in the Philippines was MacArthur's immortal slogan "I shall return."

Regarding guerrilla warfare, there are many takeaways, namely these:

1. Guerrillas require a sanctuary in terrain that is considered inaccessible in which to organize and train.
2. A guerrilla force requires at least the passive support of the affected population and external support to achieve success.
3. The enemy army of occupation must be sufficiently weak in numbers so as to be unable to occupy the disputed territory in depth.
4. While guerrilla forces are incapable of gaining a military decision, they nonetheless set the conditions needed for the forcible entry of future conventional forces into a denied area.

---

[35] Robert Ross Smith, *United States Army in World War II: The War in The Pacific- Triumph in the Philippines* (Washington, D.C.: Center of Military History, U.S Army, 1984), 657.

5. An irregular force can operate in close collaboration with conventional forces. The success of the Philippine resistance underscores this principle.

In the last analysis, the resistance movements of Volckmann and Fertig remain archetypal of insurgency par excellence. What their resistance movements, along with many other guerrilla formations, demonstrate is UW is more like alchemy than science. It bends leadership, psychology, money, guns, and guts. If war costs blood and treasure to prosecute, the Philippine resistance saved a lot of both.

For Further Reading:

1. *We Remained* by Russell Volckmann.
2. *Fire in the Jungle* by Larry S. Schmidt
3. *MacArthur's Undercover War* by William Breuer.

6

War in the Shadows:
Resistance in Europe

"To a very high degree, the measure of success in battle leadership is the ability to profit by the lessons of battle experience." – General Truscott

Parachute resupply of Jedburgh team during World War II (photo CIA archives).

No less successful than the U.S.-led unconventional warfare campaign in the Philippines, were the operations conducted by the Special Operations Executive (SOE) from Britain and the Office of Strategic Services (OSS) from the United States. This combined force played a crucial role in liberating Europe from Nazi occupation during World War II, most notably through the D-Day invasion of Normandy in June 1944, which marked a turning point in the war against Germany. As the primary American intelligence and special operations unit, the OSS supported resistance groups, gained crucial intelligence, conducted sabotage operations against Axis powers, and greatly contributed to the Allied victory in Europe. The aim of this chapter will be to highlight the UW effectiveness of two Allied missions in occupied France.

On September 1, 1939, Hitler invaded Poland. In under thirty days, the Wehrmacht Blitzkrieg of modern tanks, fighter planes, and infantry overran Poland's obsolete forces. Czechoslovakia and Austria had already been forcibly drawn into the Reich. Then, in a series of lightning thrusts beginning in May 1940, Denmark, Norway, Holland, Belgium, Luxembourg, and France all fell to Nazi domination. Evacuating its forces from the continent, Britain now faced the Nazi onslaught alone. A severe test of British resolve came in the Battle of Britain in which the Royal Air Force withstood the numerically superior Luftwaffe. By the fall of 1941, Hitler had invaded Yugoslavia, Hungary, Bulgaria, Romania, Greece, and Russia, including the Russian-occupied Baltic States of Lithuania, Latvia, and Estonia. This left an estimated 250 million people to fall under Nazi occupation (see below).

Ensconced on the island fortress of Britain, Prime Minister Winston Churchill called for the creation of a secret army for the purpose of aiding local resistance movements, conducting espionage, sabotage, and reconnaissance in occupied Europe. Formed in July 1940, to "set Europe ablaze," this secret army was designated the Special Operation Executive or SOE.

Watching the western democracies in Europe fall like dominoes to the Nazi juggernaut, President Franklin D. Roosevelt was concerned about American strategic intelligence deficiencies. He therefore appointed retired Major General William Joseph "Wild Bill" Donovan to draft a plan for an intelligence service that could fill this void. Donovan, a World War I veteran, and Medal of Honor recipient, was sent to Britain in 1941 where he drew inspiration from two British organizations: the SOE, and MI6 (Secret Intelligence Services). Impressed with the SOE, Donovan was determined to establish an American counterpart based upon it. Roosevelt promptly founded the Office of the Coordinator of Information (COI) and named Donovan its coordinator.

Following the Japanese attack on Pearl Harbor, Roosevelt placed COI under the Joint Chiefs of Staff and renamed it the Office of Strategic Services or OSS. The OSS was then tasked to collect and analyze strategic information and conduct special operations in Nazi and Japanese occupied countries. Donovan began by dividing the OSS into two functional groups: intelligence and special operations (SO).

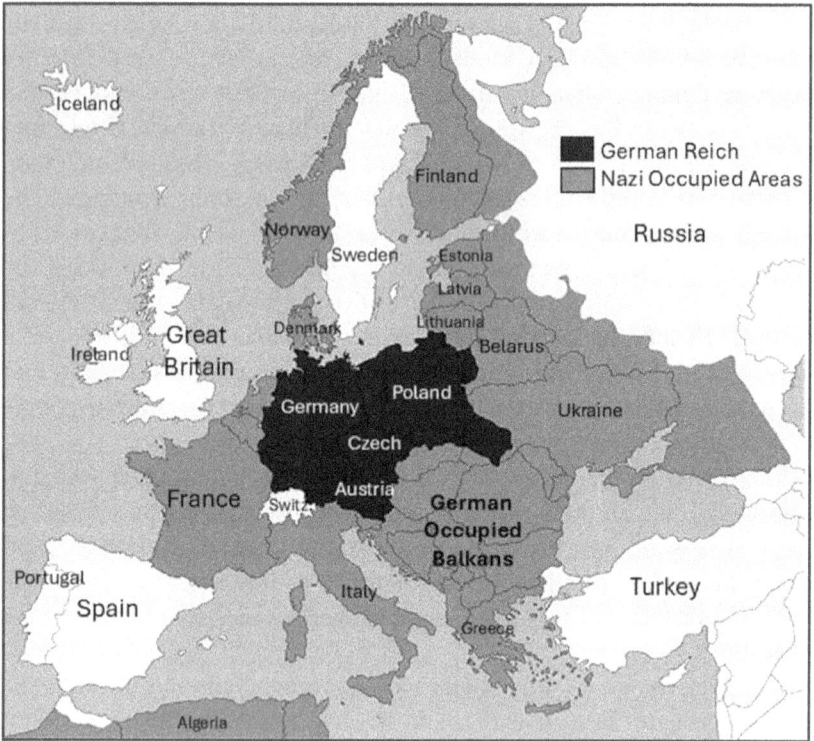

Nazi Occupied Areas, 1944.

The Special Operations (SO) arm was tasked to organize and conduct sabotage operations behind enemy lines. SO furnished agents, communications, and supplies to underground and guerrilla groups in Norway, France, Denmark, northern Italy, and China. SO also organized special teams sent behind enemy lines for the destruction of specific targets, securing intelligence, and waging guerrilla warfare, all of which included: maritime operations, special projects, research and development section, and morale operations, which was the deceptive title for psychological operations (PWO).

In 1943, Donovan added the Operational Groups (OG). Trained in parachuting, guerrilla warfare, infantry tactics, foreign weapons, and having attached radio, weapons, medical, and demolitions experts, the mission of the OGs was to organize, train, and equip local resistance organizations, to conduct guerrilla operations. The OGs, consisting mainly of U.S. Army officers and enlisted, were recruited chiefly from line units and service schools. The organizational structure of the OG units commanded by a captain consisted of two OG sections, with one officer, a first lieutenant, and 15 non-commissioned officers each. The function of the OGs was twofold: serve as the operational nuclei of guerrilla organizations, which were formed from resistance groups in enemy territory, and execute independent operations on enemy targets as directed by the theater commander.

Operation Jedburgh

In May 1944, SOE and OSS established the Special Forces Headquarters (SFHQ) in London. By the time the OSS became joined at the hip to their British counterparts, there were 50 SOE clandestine networks operating throughout occupied France. Operating in various geographic areas, these "circuits" were tasked with aiding resistance movements in occupied Europe by intelligence gathering, espionage, sabotage, assassinations, propaganda, and guerrilla operations against Axis forces. Supplying weapons and training, and coordinating attacks on key targets, all while blending into the local population and operating behind enemy lines, these agents aimed to "set Europe ablaze." Coordinating activities, SFHQ infiltrated agents by parachuting them behind enemy lines. Once embedded with the French Resistance, OSS agents communicated with SFHQ via coded radio messages and dead drops, enabling them to deliver critical intelligence, and to receive weapons and supplies by way of clandestine parachute drop.

By the spring of 1944, Germany was on the verge of defeat due to major Allied advances on multiple fronts. In preparation for the invasion of Fortress Europe, a new UW operating concept was imagined – Operation Jedburgh. Created to serve as the vital link between Allied actions and the resistance, three-man Jed Teams would parachute into France, then coordinate actions with the invading force.

Jedburgh Teams in France. Courtesy AP.

Ninety-three Jed teams were dropped into France and the Low Countries in support of Operation Overlord, the code name for the Allied operation to liberate German-occupied Western Europe. The Jed mission was to link up with French resistance groups, provide leadership, and coordinate sabotage activities. Parachuting at night, through the belly of B-24 bombers, Jed teams linked up with the resistance groups, called Maquis, short for maquisards.

In 1944, the Germans had some 60 Wehrmacht and SS combat divisions in France. Including foreign conscripts and auxiliaries, German strength numbered about a million men. In support, the Germans had about 5,000 fighter planes to protect occupied French airspace. As D-Day approached, London ordered an increase in sabotage. In the previous year, between June 1943 and May 1944, the resistance sabotaged thousands of trains, rendering some 200 locomotives and 2,000 freight and passenger cars out of commission. The resistance sabotage efforts caused so much trouble in the Axis rear area, that the Germans had to divert a great number of forces away from the front.

In preparation for the invasion, the Allies developed four plans for the French resistance to execute starting at D-Day and continuing in support of the beachhead at Normandy. Plan VERT called for wreaking havoc on the rail system for 15 days; the time that would be required to establish the beachhead at Normandy. Plan BLEU dealt with destroying electrical facilities. Plan TORTUE planned for the delaying of enemy troops that would naturally be coming to reinforce the axis forces at Normandy. Finally, Plan VIOLET issued instructions for the cutting of underground cables. Each of these plans focused on the use of sabotage.

Team GEORGE

The first Jed team, code named Hugh parachuted into Chateauroux on June 5, 1944, the night before the Allied invasion of Normandy. Three days after the D-Day landings, Team George, consisting of American Captain Paul Cyr, French Captain Philippe Ragueneau, and French radioman Sergeant Pierre Gay, jumped into the vicinity of Redon, France. Their mission was to assist the SAS in training the resistance, to destroy bridges, and cut rail roads. These actions would tie down as many German troops as possible, keeping them from reaching the D-Day beaches. The three-man team was met on their drop zone by some 20 Frenchmen, who pounced on them with joy at their arrival. Coming to a large barn on the outskirts of the town of Redon, they had no sooner dropped their gear when girls came running in kissing them, giving them flowers and wine. The Maquis were ready to attack the German garrison in Redon immediately. The Resistance made so much noise, Cyr thought they would certainly get compromised.

The next morning, radioman Pierre Gay started sending broadcasts to London. Cyr estimated the size of the Maquis force to be about 5,000. He therefore asked for the needed weapons and ammunition to be airdropped. The Germans however immediately jammed the frequency, forcing Sergeant Gay to broadcast on their secondary channel. Soon the Germans jammed this one as well, forcing the team to broadcast on their emergency channel, asking SFHQ for a new primary frequency. Needless to say, the team's communication with London remained shoddy at best. Adding to the difficulty, London believed that the Gestapo had compromised the team's mission.

London had good reasons for concern. The Gestapo was in fact able to significantly infiltrate the French Resistance. The notorious Nazi secret police had managed to penetrate several networks and captured key leaders. This was done through a combination of tactics including double agents, extensive surveillance, and brutal interrogation techniques, which allowed them to disrupt resistance operations and arrest numerous members. Despite these setbacks, the Resistance still managed to carry out significant acts of sabotage and intelligence gathering.

For the next several days, Cyr and his team busied themselves to organize and arm the Maquis. The team also reestablished contacts between their Maquis and other Resistance groups, keeping London informed as to their situation. Within a few nights, more planes were dispatched to Brittany, dropping arms and supplies. As men and equipment poured into the camp, Frenchmen were coming from hundreds of kilometers away to receive weapons, then returned home talking about it. The continuous parachute drops and the poorly led Maquis attacks on several German garrisons and depots began to attract considerable attention from the Wehrmacht. Cyr soon realized his team had stayed at the farmhouse too long.

On the morning of June 18, Cyr and his team were awakened by machine gun fire and exploding grenades. The Germans had hastily organized a task force of several thousand paratroopers and infantry. Before long, the team found itself in a desperate fight alongside the SAS and about 1,000 Maquis.

As the Germans pressed their attack, Cyr ordered for the team's documents and code book to be burned. The SAS called in a RAF air strike which managed to hold off the Wehrmacht's attack until about 4:00 pm. In an effort to force the Germans back, Cyr and Ragueneau led company sized groups of Maquis in counterattacks. As the German noose was tightening, Team George, and what was left of the SAS, loaded their wounded on trucks, and stacked what they couldn't carry in the barn then lit it. A great deal of arms and supplies meant for the Resistance went up in smoke. After dodging several enemy patrols, Team George made a week-long, 35-mile trek inland to the village of La Roche-Blanche. There, partisans provided the team with a safe house and a place to hide their wireless radio. Sergeant Gay radioed to London "Arrived safely." Having made link up with more Maquis, the team set out at once to start over. However, believing the team to be compromised, and in light of the numerous other circuits that had been rolled up, SFHQ refused to drop any more supplies. Waiting out the situation, and on the verge of being demoralized, the team sent in its status report, "one radio, one battery, a few weapons and our clothes."

Finally, in late June, with London fully assured of Team George's implacable status, they became the de facto Jed Team in the French Département of Loire-Inferieure. The team would train the local Maquis. Setting out at once to create a guerrilla force capable of operating against the Germans, George spent July and August organizing and training the Maquis to conduct sabotage and raids.

Team HAMISH

A few hundred miles to the east of Team George, another Jed team was infiltrated into occupied France. On the night of June 12, just six days after the Normandy invasion, Team Hamish jumped out of the belly of a B-24 that had earlier braved a barrage of anti-aircraft fire. Hamish was composed of American Lieutenant Robert Anstett, French Lieutenant

René Schmitt, using the nom de guerre Lucien Blacere, and American radioman, Sergeant Lee Watters. As Lieutenant Anstett recounts, "All of us were worried about German night fighters, because ours was a solo flight. I remember just sitting there with that stuff bursting all around us. My greatest fear was that we might all be killed before we had a chance to get started."

Dropped into the Departement of Indre, near the town of Chateauroux, 190 miles south of Paris, and roughly 350 miles behind the German lines, Hamish's mission was to recruit, organize, train, and lead Maquis in harassing attacks on all roads, bridges, and communications lines. In particular, the Paris-Limoges railway ran through the Team's assigned area. Their principal mission was to cut it and keep it cut. Additionally, the team was to keep London furnished with all possible intelligence on German troop movements. Scores of other Jed Teams were detailed for similar missions across occupied France. The Jed teams were to function as the vital link between the Allied command and the resistance forces in the effort to slow the Nazi response to Operation Overlord. This would be accomplished by sabotaging enemy supply lines, blowing bridges, and derailed trains.

Parachuting in civilian clothes at 300 feet above the ground, the team barely had enough time to get their feet settled before they flopped into a wheat field. With great elation, the team was now in France. They had barely gotten out of their harnesses when they heard shouting in French. Anstett shouted the password in French, "Does Grandfather like milk?" Back came the reply, "Yes, only when he's in the woods." To the team's surprise, within minutes, their accompanying parachute-dropped containers were rapidly packed into one of the underground's trucks. Hamish was soon brought to Team Hugh's secret lair and brought up to snuff. After a meal and loud toasts of "Vive la France; vive l'Amerique; vive la victoire!" Team Hamish was taken in an old car to a farmhouse they would use as headquarters. Describing the ride there, Lieutenant Anstett recounts, "I was amazed at the casualness with which the Maquis seemed to cruise along over these roads in broad daylight. I was slightly apprehensive after we ran straight through a German convoy at a crossroads. Our French escorts were so calm about this meeting that I was not aware of it until we suddenly burst into high speed. The Germans evidently were as surprised as I, because they did not follow us."

When the team arrived at the farmhouse, they met Robert, the guerrilla chief, a tall, clean-cut Frenchman of twenty-four years-old who had been in the underground for four years. He had about 300 men, but only 50 of these were armed. Their spirits were high, and our arrival seemed to give them a tremendous lift. After pleasantries, Anstett learned about their guerrilla activities. Just the week before, they had celebrated D-Day by attacking the local German garrison. The battle lasted all day, and they used every weapon they had, along with most of their available ammunition. Anstett immediately wired London for a supply drop.

While they waited, Anstett organized a five-man team and led them out to sabotage the Paris-Limoges line. Finding a stretch of line with a road running along it, the team drove down the road and stopped every few hundred yards and blew the tracks. They made 24 cuts that night and the same number the next two nights. The Germans would come out and fix it in the daytime, and the team would blow it again at night. After leading the first three nights, Anstett didn't bother going out anymore, but specially tasked one group with this assignment. From the time that Team Hamish entered the region until they left, not a single German train got through on that railroad. And though there were plenty of close calls, they never lost a man while sabotaging it.

With his primary mission firmly in hand. Anstett set out to harass the local German occupiers. A half-hour drive to the north at Chateauroux, was an airfield for training Stuka pilots. A large German force serviced the field and manned the anti-aircraft guns around it. There were an additional 3,000 German troops who were billeted in town. Unsurprisingly, the town of Chateauroux hated their German occupiers. Nonetheless, the town was full of Gestapo agents and Vichy sympathizers whose presence made the existence of the underground extremely perilous. Ever the clever propagandists, the Nazis had effectively split the townspeople from the local resistance, labelling the Maquis and their "puppet master" British and American helpers as bandits and terrorists.

Meanwhile, the first airdrop had arrived a few days after the team started its sabotage mission on the Paris-Limoges railway. When the guerrilla chief Robert saw the tons of weapons and ammo that were in the containers, he wanted to take on the whole German army. Within 48 hours, new Maquis recruits joined up by the hundreds. With the farmhouse already overcrowded, the team began billeting the new recruits on several nearby farms. Whenever German patrols were reported in the area, the whole guerrilla force took to the fields, where they spent hours crouching in the bushes with weapons at the ready.

With his Maquis force now swelling into several hundreds, Lucien developed the team's training program, teaching the French how to use the newly acquitted weapons. He likewise devoted most of his time to building out the organization, arranging for the reception of more supply drops, and planning additional sabotage missions. The railroad-cutting was proceeding so well that Anstett turned his attention to blowing bridges and interrupting road traffic. Among the first targets were a couple of small culverts which the team destroyed without interference. The ease with which these operations were accomplished made a great impression on the French. The team soon had a large group learning how to handle explosives and undertake operations on their own.

The reaction of the Germans to the team's new attacks was prompt and emphatic. Seven hundred specially trained anti-Maquis Austrian troops moved into La Chatre. A large enemy force was now less than five miles from the team's farmhouse guerrilla base. Following the usual procedure of an area infested with Maquis, these troops started terrorizing the local population. The Maquis had to be continually on the alert as heavily armed counterinsurgent patrols were constantly sweeping the countryside. To the consternation of the Germans, the underground intelligence network managed to keep the Maquis and the team abreast of planned German movements, warning them ahead of time.

While Anstett considered his next move, guerrilla chief Robert brought him two men who proposed an answer to the new German menace. They were both telephone workers. To stop the German communications system in DépartementIndre, they proposed one would cut and the other would repair. Their system was as simple as it was effective. Their telephone company was of course controlled by the Germans and used by them for all their local communications. One of them took the job of traveling around in his official car. He would cut the lines then report the breaks, which he had "discovered." His partner in turn would then take his time repairing the broken lines. The two kept up their unique campaign under the very noses of their German supervisors without ever arousing suspicion.

While these activities were going on, the team kept receiving additional drops. Then, on the Fourth of July, the team learned that a large German truck loaded with butter had broken down in La Chatre. The team jumped at the opportunity. A relief truck had already been dispatched by the Germans to pick up the dairy product. The team sent in a squad-size element to ambush it. The ambush was a complete success. The team killed the driver and captured the truck along with the butter. They even repaired the first truck and stole it too. The Germans were incensed.

Team GEORGE II

By this time, the Allied invasion of Normandy on June 6, 1944, succeeded at all points. While Allied control of the sea and air ensured the rapid buildup of follow-on forces, sabotage continued to play an important role. After D-Day, the French resistance was responsible for preventing, or at least delaying, the flow of German men and supplies to the new front. In a famous example, a Panzer Corps was on its way to join the fight at Normandy. Its presence might have spelled defeat for the Allies. The Panzers were going to cross the Eure River via the sole bridge not knocked out by RAF planes. Saboteurs proceeded to demolish the bridge only three hours before the Panzers arrived.

Following the success of operation Cobra, the offensive that took place from July 25–31, Allied forces were able to break out of Normandy. Then, by the first week of August, the U.S. Third Army began fighting its way into Brittany. In all, 12 Jed teams would be deployed to the Brittany peninsula to employ some 19,000 Maquis alongside a further 35,000 Free French infantry (FFI) in support of 3rd Army's campaign.

Team George's new mission was to organize, train, equip, and lead the Maquis for action against the Germans at St. Nazaire and to protect the 3rd Army's southern flank. Under the command of General George S. Patton, the U.S. Third Army's campaign commenced on August 3. Opposing Patton's advance was German General Fahrmbacher's well-entrenched 50,000-man XXV Corps. Fahrmbacher's ruthless policy was to execute anyone suspected of supporting the Resistance. In one of the more remarkable exploits of the war, Captain Paul Cyr had managed to gather a wealth of information on the German fortification plans on the

Brittany coast. This intelligence trove included data regarding the German U-boat base at St. Nazaire. Navigating through the seams of German units, with the help of resistance elements, Cyr made his way to Patton's Third Army headquarters where he passed on this valuable information. Armed with this intelligence, Patton's 3rd Army managed to bottle up some 20,000 Germans in St. Nazaire, removing them from the battlefield.

Team HAMISH II

Back in the Département of Indre, Team Hamish continued its sabotage mission. Then, on August 5, a courier brought the team news. A detachment of French Melice, Vichy militia whom the Maquis hated even more than the Germans, were coming to Chateauroux to pick up some gasoline. Anstett quickly dispatched ambush parties to points along every possible road that the Melice party could take then waited. In due time they entered one of the Maquis' thirty-man ambushes. But being a sizable force, of about 80, they managed to blast their way out of certain death. As the Melice sped away, the Maquis blasted them with bazookas, killing 19 more. The Melice would prove harder to avoid than the Germans.

About a week later, the Maquis captured a suspicious Frenchman. The twenty-one-year-old Frenchman was interrogated and admitted that he was a Melice agent sent to capture or assassinate Lieutenant Anstett. He implicated nine others in his confession, who were all quickly rounded up, tried before the Maquis military court, and sentenced to be shot for treason. Anstett was able to talk to the young Frenchman before he went to his death. He said he was sorry and realized that he had made a mistake. He was particularly sorry that he was going to be shot on the charge of treason and asked Anstett if he might command his own firing-squad. He desired to show the Maquis that he was a real Frenchman and could die bravely. Anstett asked the Maquis leaders what they thought. They readily approved. The team then witnessed one of the most extraordinary spectacles in war.

Being a mild midsummer evening, the prisoner was stripped to the waist. He was led to the site of his execution, then given the traditional glass of wine and last cigarette, the custom of French justice decreed for doomed men. He calmly drank his wine and smoked his last cigarette. Then, flicking the spent cigarette away, he called the firing-squad to attention. "Aim at the heart,'" he said, pointing to the left side of his chest. Then he made a little speech in which he admitted his crime and asked the forgiveness of his people. Then, taking up his position against a wall, he looked straight at his 12 executioners, and without a tremor in his voice, slowly gave the commands: "Ready! Aim! Fire!" The bullets ripped into his body, and he fell dead; a brave but misguided man, who had fallen prey to Nazi propaganda.

On August 14th, Team Hamish received word of the Allied invasion of southern France, which was to take place the next morning. The team immediately intensified its rail sabotage activities and set up roadblocks all over the area. The Maquis under their tutelage had by now ballooned up to well over 2,000 well-armed guerrillas. The team's main target was the Chateauroux airfield. While the team made plans for its capture, they were greatly assisted by two German deserters who had worked on the installation's defenses.

Another character who was very helpful in this regard was a fabulous Maquisard named Jacques, a member of the local underground. Jacques outfitted himself with false papers attesting to the fact that he was a member of the French Melice. He had a high-speed car with two Bren guns kept constantly loaded and at the ready. Jacques readily performed some truly incredible exploits. On one such occasion, following an ambush in which a Maquisard was injured, Jacques raced into Chateauroux, forced his way into the hospital, and spirited away the wounded guerrilla before anyone knew what had happened.

Jacques was also known to have a fierce hatred of collaborators. On another occasion, he drove into Chateauroux in broad daylight, then boldly sought out the leading French collaborator and shot him down in the street. His real tour de force, however, was the assassination of the Gestapo chief at Chateauroux. Having learned that this German was dining at a certain place, he walked into the restaurant, inquired as to the Gestapo chief's whereabouts, then walked over to the man's table and calmly shot him four times. Later that day he learned that the man had not died but was in the town hospital. Undeterred, Jacques obtained entry to the patient's room with his false papers. He then marched up to the bed, promptly stabbed the man to death, then calmly walked out of the hospital. Needless to say, Jacques and the underground kept the team pretty well posted on German moves in Chateauroux.

Anstett was just about ready to attack the airfield when he received word on August 20th that the German garrison was starting to move out. It was now or never. Anstett gave orders for an all-out attack. Only, instead of attacking the airfield, the Maquis would ambush the German troops as they fled. Ambush parties of 30 men armed with bazookas and machine guns were deployed on every possible roadway out of the airfield.

The first German column of 2,000 men started moving out at about two o'clock in the morning on August 21, 1944. Unfortunately, they headed north, away from the team's main strength, but were attacked nonetheless with every available man. Under harassing fire, it took the German column twenty-four hours to move less than 50 miles. As the Maquis pressed the attack, the German's suffered the loss of over 300 killed before they managed to evacuate the Département. The following day, the team received word that a second column was heading south in the direction of La Chatre. This would take them within a few miles of the team's main camp. As the German column rolled south, they stopped short of the ambush to bring heavy fire down on the Maquis' position. The Germans had been tipped off by a man who was later apprehended and shot. Overall, 19 Maquis were killed in the failed ambush.

The next day, however, the Maquis managed to get even, attacking another German column of about a thousand men. Almost continuously for three days, over numerous roads, the Maquis chased them relentlessly. An estimated 300 made it back to Germany. Over a six-day period, which culminated in the liberation of Chateauroux on August 27th, though outnumbered three to one, the team killed over 700 Germans and destroyed more than 100 vehicles. Maquis losses were only 30 killed and a few wounded.

Team Hamish and the Maquis entered Chateauroux to the jubilant cries of the population who were wild with joy at their liberation. Wine and champagne flowed freely as the liberators were hailed as conquering heroes. While the celebration was underway, the Germans were streaming away as fast as they could for the Rhine.

Regarding the menacing activities of the OSS, Eisenhower wrote, "They surrounded the Germans with a terrible atmosphere of danger and hatred which ate into the confidence of the leaders and the courage of the soldiers."

Summary and Implications

By way of the five PLAIN laws of irregular warfare, the following is an analysis of the OSS UW effort:

Political objective (s) – The objective of the Allies was the liberation of Europe and the end of the wicked and oppressive regime of Hitler. The mission of the OSS was to gather intelligence, conduct espionage and sabotage operations behind enemy lines, while supporting resistance movements to aid the Allied cause; essentially acting as a clandestine intelligence gathering and unconventional warfare unit.

Operating hundreds of miles behind enemy lines, the Jeds facilitated subversion and sabotage actions against the Nazis by providing not only leadership and logistics but a cohesive bond with the Allied command. Coordinating through SFHQ in London, Jed Teams dropped tons of supplies, guns, and ammunition to outfit the Resistance. All told, the combined efforts of SOE / OSS in France from June to August 1944 tallied the following: 75 roadways cut, 44 factories sabotaged, 140 telecommunications lines cut, 885 rail lines cut, 322 locomotives destroyed, 22 convoys attacked, accounting for tens of thousands of German deaths, and 7 aircraft shot down.[1]

On the day after D-Day, twenty-six major rail lines were rendered unusable. As a direct result, the 2nd Panzer Division's movement from Toulouse to Normandy was stymied, taking the Panzers twelve days to cross the 400 miles, thereby nullifying their potential contribution at Normandy. Eisenhower's headquarters estimated that the value of the Resistance to Operation Overlord amounted to the equivalent of 15 military divisions.

Legitimacy – Legitimacy is the degree to which a population accepts that government actions are in its interest. This was a zero-sum game which the German occupiers lost completely. Committing numerous atrocities against civilians including mass executions, hostage-taking, forced labor,

---

[1] Brown, *The Secret War Report of the OSS*, 460.

and the deportation of Jews to concentration camps. German actions significantly fueled the growth of the French Resistance.

Adaptability – Throughout the war, the OSS modified their tactics, particularly adapting their methods based on the theater of operations. Tactics therefore depended on the region, with more reliance on guerrilla warfare and resistance movements in areas like occupied Europe compared to the Pacific theater where different tactics were needed to combat the Japanese. In the European Theater, the OSS focused primarily on supporting established resistance groups behind enemy lines through sabotage, intelligence gathering, and coordinating attacks with local partisan forces. Learning from experiences in the field, the force incorporated new intelligence gathering techniques, including significantly developing their counterintelligence operations through the X-2 branch. The overall relative weight of the various tactics employed by the OSS effort could be surmised by the following graph:

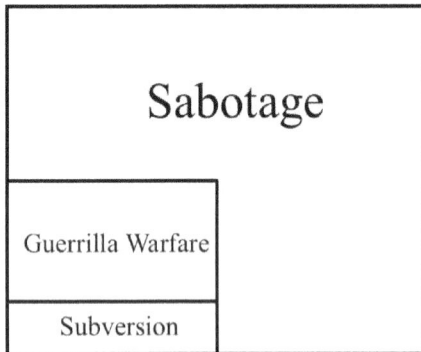

```
+-----------------------------------+
|                                   |
|                                   |
|            Sabotage               |
|                                   |
|                                   |
+---------------------------+       |
|                           |       |
|     Guerrilla Warfare     |       |
|                           |       |
+---------------------------+       |
|      Subversion           |       |
+---------------------------+-------+
```

The Germans were likewise highly adaptive, employing the Propaganda Staffel, a unit of the Luftwaffe dedicated to psychological warfare. Despite their best efforts, the unit had limited success in occupied France. This was largely due to the widespread resistance sentiment and the French population's strong anti-German feelings. Most French disregarded or actively resisted such propaganda messages.

<u>Influence</u> – Never underestimate the power of a small group of innovative people to change the world. The OSS employed nearly 13,000 men and women at its peak and operated for a little more than three years, from 1942 to 1945. In that short time, it helped shorten the war and save lives in Europe, North Africa, and Asia.

<u>Nat</u>ive Face – The efficiency of a unit conducting IW depends largely on its knowledge of the people and the terrain. The threat of infiltration forced the Resistance to become more secretive and cautious in their activities. The Gestapo often tried to turn captured resistance members into double agents to infiltrate networks further. They used extensive spying and monitoring techniques to identify resistance members and their activities. The Gestapo was notorious for using torture and harsh interrogation tactics to extract information from captured resisters. Notable examples like Jacques Desoubrie, who infiltrated resistance groups and provided crucial information to the Gestapo, lead to numerous arrests.

From its inception in WWII, the United States unconventional warfare capability has since evolved from an undeveloped, undesirable approach into a coherent intellectual framework. This form of irregular warfare is recognized today as the "soul" of the U.S. Army Special Forces.

For Further Reading:

1. *The Jedburghs: The Secret History of the Allied Special Forces, France 1944* by Will Irwin.
2. *Operation Jedburgh: D-Day and America's First Shadow War* by Colin Beavan.
3. *Eisenhower's Guerrillas: The Jedburghs, the Maquis, and the Liberation of France* by Bejamin F. Jones.

Part Three

The Cold War (1947-1991)

7

The Hukbalahap Rebellion

"If we understand that this is war over people, then we can start understanding the real human values in it – which go far beyond sizes of forces, reports of battles, statistics on casualties, and differences in quality of weapons." – Edward Lansdale

Huk leader Luis Taruc reads the newspaper while his partisans look on.

The Cold War was a time of ideological and geopolitical rivalry between the Soviet Union and the United States. Remaining under the constant threat of nuclear war, this period of competition between the two superpowers expressed itself through various proxy wars including those in Korea, Vietnam, and Central America. While the Hukbalahap Rebellion failed to gain external support from either China or Russia, it is nonetheless included because it was a conflict against monolithic communism, nonetheless.

Despite the American-Filipino forces' defeat of the Japanese in 1945, several weighty grievances remained to despoil the victory. The war devastated the islands: Over a million were killed, hundreds of thousands were wounded, villages were destroyed, agricultural capacity was in ruins, schools and hospitals were in shambles, services were nonexistent, millions were homeless and hungry, and the political apparatus was fractured. As for the Philippine capital, according to historian Ray Hunt, Manila was "laid waste more thoroughly than any cities on earth save Hiroshima, Nagasaki, and Warsaw."[1] In addition to the paucity of the state, below a thin veneer of nationalism lie a deep-seated discontentment that in time would rear its ugly face. In short, the Philippines after WWII had all the necessary ingredients for insurgency: Land grievances, lack of state services, and monetary disputes.

Regarding land grievances, landowners, newly enlightened by capitalism, tried to squeeze more and more of the crop out of their tenant farmers. As the status quo returned following liberation, landlords built up private armed groups that came to be known as Civilian Guards. These paramilitary forces were employed to reassert control over Central Luzon. As Huk revolutionary leader Luis Taruc describes, "when the Americans came, they made boasts about having brought democracy to the Philippines, but the feudal agrarian system was preserved intact."[2] This deep-seated peasant grievance had been the cause of Filipino dissent before World War II.

According to historian Benedict Kerkvliet, under U.S. rule, the plight of many Filipinos worsened. For example, the traditional landowner-peasant relationship during Spanish rule meant the landowners were responsible for the well-being of their tenants, which included free loans of rice and other assistance during the hard years. In return, the tenants were expected to reciprocate as the landowner might request. However, this relationship began to deteriorate as landowners got a taste of U.S. capitalism and became more concerned with profit-making and less concerned with the welfare of their tenants.[3] This system exploited the

[1] Ray C. Hunt and Bernard Norling, *Behind Japanese Lines: An American Guerrilla in the Philippines* (Lexington: University Press of Kentucky, 1986), 71.
[2] Luis Taruc, *Born of the People*, (New York: International Publishers, 1953), 26.
[3] Benedict J. Kerkvliet, *The Huk Rebellion: A Study of Peasant Revolt in the Philippines* (Lanham, MD: Rowman & Littlefield, 2002), 8.

peasantry while Filipino elites grew rich. Land grievance being the prime mover of discontent, other grievances stemmed from guerrilla recognition along with the benefits of backpay.

Assuming the presidency upon the death of Manuel Quezon, Philippine's wartime president, Sergio Osmena was in every respect the wrong leader at this precarious time.[4] As the Philippines moved toward full independence by July 4, 1946, to revitalize its ailing ally, Washington pumped in millions of dollars. Albeit, in bed with corrupt politicians in Manila, the feeble Osmena regime failed to address any of these grievances, and to a very large degree, maintained the status quo ante. Seizing on this ancient deep-seated land grievance were the Hukbalahap.

In their quest for a better state of the peace, the Huks had a cause that centered on the age-old cry of "land for the landless." In 1945, Filipino peasant farmers still only owned about 10% of the land they worked. This disparity stemmed from the four-hundred-year-old Spanish colonial encomienda system. By this system, Filipino elites gained 50% of the profit from the land-tending farmers, who owned only 10% of the land they tilled. Genuine land reform could have settled this. Such socio-economic conditions have long been fertile breeding ground for Communist ideology. Some positive steps might have ameliorated the situation. However, the tone-deaf nature of the post-WWII Filipino administrations (Osmena-Roxas-Quirino) failed to address these agrarian based grievances, giving ample fuel to the Communist-inspired Hukbalahap insurgency.

Taking the Tagalog name, Hukbo *Na Bayan Laban Sa Hapon*, literally, the People's Army [To Fight] Against Japan, the Japanese invasion enabled the Huks, pronounced "Hooks," to become an organized and motivated fighting force. Filipino Communists established the Huks as their armed resistance wing and named Luis Taruc its commander. During World War II, the Huks repeatedly rebuffed attempts by American guerrilla leaders to fall under the unified U.S. Forces Far East (USAFFE) banner. As a result of their refusal to join or operate with Allied forces, and because of their Communist sympathies, the Huks received no U.S. equipment.

---

[4] Mark Moyar, *A Question of Command: Counterinsurgency from the Civil War to Iraq* (New Haven: Yale University Press, 2009), 92.

Beginning in 1942 with about 300 fighters, the Communist guerrilla force reached 10,000 by September 1944. Like many other Filipino guerrilla bands, the Huks proved highly successful. However, as is widely reported, they killed more Filipinos than Japanese. According to historian Mark Moyar, the Huks killed over 20,000 Filipinos, as opposed to 4,000 Japanese.[5] Capitalizing on the war state of the Philippines, they seized many of the large estates in central Luzon, and established a regional government, collected taxes, and administered their own laws.

When the war ended, the Huks had to be demobilized at gunpoint. To add further indignity, unlike other guerrilla bands that fought the Japanese, they were granted no occupation-period "back pay." Nonetheless, the Huks made seemingly honest attempts at affecting the political process legally by participating in the April 1946 national elections, even winning six seats in the House of Representatives. However, the new president, Manuel Roxas, refused to allow the Huk candidates to take their seats.[6] One of these ousted Huk congressional candidates was Luis Taruc. He and his deputy were arrested and jailed on trumped-up charges of wartime crimes. Weeks later, Taruc was released. He and his Huk followers promptly took to the hills and armed themselves with the weapons they had cached. Reverting to their wartime organization, the Huks established an underground "politburo" in Manila, and reactivated their 500,000-man auxiliary force, the *Kaisahan Ng Mga Magbubukid* (National Peasant Union, or PKM).

Reverting to their tried-and-true methods, much like they did during the Japanese occupation, the Huks moved in small groups in and out of the peasant communities, relying on "barrio organizers" for political and material support. The Huk's base, referred to as the Barrio United Defense Corps (BUDC), carried out the functions of intelligence gathering, recruitment, logistics, as well as various civil affairs. Each BUDC included five to twelve people in each barrio.[7] Serving as shadow government representatives, Huk organizers performed such magistrate functions as conflict resolution.

---

[5] Moyar, *A Question of Command*, 91.
[6] Douglas S. Blaufarb, *The Counterinsurgency Era: U.S. Doctrine and Performance, 1950 to the Present* (New York: The Free Press, 1977), 25; Smith, *The Hukbalahap Insurgency*, 67.
[7] Lansdale, *In the Midst of Wars*, 7.

The guerrillas were organized into Field Commands (FCs). Ranging anywhere from a hundred to seven hundred guerrillas, Huk FCs carried out ambushes, raided outposts, rendered roads impassable, and "confiscated" funds and property to sustain the movement.[8] Notwithstanding these kinetic activities, the Huks also performed civil tasks, the likes of which formally fell to the privileged landowners. These activities included officiating at weddings, baptisms, funerals, and the issuing of marriage licenses and birth certificates.[9] In time, Huk popularity and efficiency transcended the government, establishing them as the law and order in Luzon.

Iron Fist

In response, President Roxas vowed to destroy the Huks within 60 days. He ordered the Philippine armed forces and Constabulary to eliminate the threat. However, he did so without regard to the security or rights of the people. The peasant population was caught between the government's inept and brutal repression and Huk intimidation-through terror. Central Luzon, an area 150 miles by 50 miles, fell under the Huks control. This heartland of the Huk movement, consisting of the four central provinces of Nueva Ecija, Pampanga, Tarlac, and Bulacan, became known as "Huklandia." At the outset, it is estimated that of the people of Luzon, some 10% actively supported them, another 10% opposed them and 80% was undecided (see figure 7-1). It would be this undecided 80% that would constitute Mao's famous water in which the guerrilla fish would swim.[10]

In ways reminiscent of the Japanese in WWII, the government's initial approach was indiscriminate, brutal, and ineffective. Galvanizing the Constabulary and the army, barrios (villages) were shelled and burned. Suspected Huks were rounded up and shot. In fact, the Roxas Administration even resorted to such Japanese tactics as the "magic eye." Unsurprisingly, in less than two years, the Roxas' "iron fist" campaign led to the Huk's doubling in size. However, on April 14, 1948,

---

[8] Kerkvliet, *The Huk Rebellion*, 210.
[9] Ibid.
[10] Asprey, 749.

the death of President Roxas and the accession of Vice President Elpidio Quirino led to the prospects for a peaceful solution to the conflict.[11] Quirino employed every COIN approach known to man. Beginning with amnesty, he offered those who surrendered and gave their arms a twenty-day grace period. Incredibly, Huk Supremo Luis Taruc came down from the hills. He was even permitted to take his seat in the House of Representatives, which had been denied him for his Communist sensibilities.[12] Maintaining his lenient policy, Quirino extended the grace period to 50 days.

In all fairness, Taruc did his part to persuade his fellowmen to surrender. But after a mere 100 Huks acted on Quirino's terms, the Supremo once again took to the hills, blaming the failure on the government. By this time, Huk forces swelled to some 12,000 guerrillas. These in turn were supported by some 150,000 sympathizers.

Notwithstanding this support, lacking a coherent strategy, the Huks were guilty of many improprieties and blunders. On April 28, 1949, Huks ambushed a thirteen-vehicle convoy, killing Doña Aurora, widow of the Philippines' second president, Manuel Quezon. The attack brought worldwide condemnation. Public sympathy for the Huks also waned due to their predilection for banditry, rape, and murder. As the Huk movement grew in strength, American policymakers became alarmed. Not to be undone, on Good Friday, 1950, Philippine army troops massacred a 100 men, women, and children in Bacalor, Pampanga, and burned over one hundred homes in retaliation for the killing of one of their officers.[13] By 1950, the situation got so bad that the U.S. decided it needed to get involved.

---

[11] Kerkvliet, *The Huk Rebellion*, 200.

[12] Uldarico S. Baclagon, *Lessons from the Huk Campaign in the Philippines* (Manila: M. Colcol, 1960), 4.

[13] Lawrence M. Greenberg, *Analysis Branch, U.S. Army Center of Military History, The Hukbalahap Insurrection: A Case Study of a Successful Anti-Insurgency Operation in the Philippines, 1946-1955* (Washington, DC: Government Printing Office, 1995), 76.

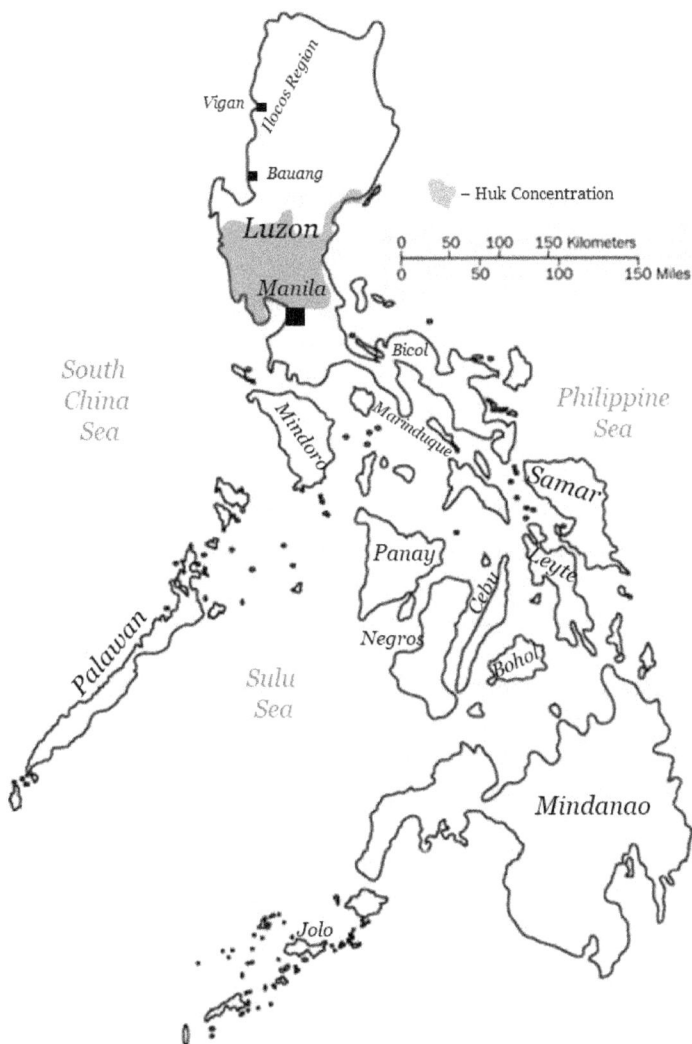

Figure 7-1. Huklandia.

Mink-Glove Covered Iron Fist

The fortunes of the Philippines changed for the better in 1950 with three important developments. First was a huge influx in U.S. aid to end

147

the conflict. As Michael McClintock observes, "American alarm over the Huk Rebellion prompted a presidential order for a program for the rapid reorganization and expansion of Philippine combat forces." With U.S. support to the Philippines tripling, as to American strategy, the Joint Chiefs of Staff (JCS), chaired by General Omar Bradley, believed that U.S. troops were not called for. Instead, seeing the big picture, remedial political and economic measures were settled on to remove the causes of the insurrection. Such measures were to be funded by the diversion of 9.3 million from other Cold War aid allocations.[14] An additional 10 million would be approved by President Truman in May 1951.[15] These funds would create a new 33,000-man Philippine Army, and bolster the Constabulary, which in time would be nearly twice as large as the army. Together, these forces would carry out the lion share of government actions against the Huks.

In a second positive step for the government, the USAFFE guerrilla leader Ramon Magsaysay was chosen to be the new Philippine Secretary of National Defense. Previously, he was appointed military governor of Zambales by the U.S. Army in 1945. After the war, he was elected to Congress in 1946, then became head of the House Defense Committee in 1948. As a third component to what would be the ingredients of success, U.S. Air Force Colonel Edward G. Lansdale was sent to advise and assist Ramon Magsaysay. Commissioned in the U.S. Army, Lansdale served in the Office of Strategic Services during WWII, served a three-year tour in the Philippines as an intelligence officer, then transitioned to the Air Force in 1947, and began working for the CIA in 1949.

Magsaysay and Lansdale made an excellent team. Due to the close nature of their association, it is difficult to decide who should receive the credit for the many positive and effective measures the Magsaysay-Lansdale team made. Suffice it to say, in his efforts to reform the military and the government, Magsaysay greatly benefited from his relationship with Colonel Edward Lansdale, who became the virtual face of U.S. advisory efforts against the Huks.

---

[14] Michael McClintock, *Instruments of Statecraft: U.S. Guerilla Warfare, Counter-Insurgency, Counter-Terrorism, 1940-1990* (New York: Pantheon, 1992), 398.
[15] Donald Robinson, *The Dirty Wars: Guerrilla Actions and Other Forms of Unconventional Warfare* (New York: Delacorte, 1968), 187.

Concerned for his family's safety, Magsaysay sent them to Bataan, then moved into Lansdale's bungalow. In his memoir, *In the Midst of Wars*, Lansdale describes his mission: "My orders were plain. The United States government wanted me to give all help feasible to the Philippine government in stopping the attempt by the Communist-led Huks to overthrow that government by force. My help was to consist mainly of advice where needed and desired. It was up to me to figure out how best to do this." Unsatisfied with the briefings he had received, Lansdale understood that Philippine and American leaders facing down the Huk problem barely concerned themselves with the political and social factors. In fact, he states, for them that was the only factor. "They dwelt almost exclusively on the military situation."[16] To get a grass roots assessment, Lansdale ventured out on a personal fact-finding tour. In his private talks with Filipinos in Manila and the countryside barrios, he pieced together an altogether different narrative: Soldiers weren't defending the people.

Diminishing the Motives

Discussing the poor socio-economic situation that the Huks were exploiting, the new Magsaysay-Lansdale team developed a package of countermeasures. These would in time diminish the motives that had fueled the rebellion. It began with military reform. Like Roxas, Quirino's relationships with his military cronies neutered the Minister of Defense's authority to decide on officer promotions and reliefs.[17] Recognizing this negative interference, Magsaysay demanded a free hand to do what was necessary to address the Huk threat effectively. Coupled with U.S. pressure, Quirino granted Magsaysay full authority to fire ineffective commanders and promote others based on merit.[18] Magsaysay wasted no time reforming the military. Within days of taking office, he recommended the retirement of 13 older and underperforming officers, including the Philippine Air Force commander. Then, in terms of force structure, about half of the money he received from the United States

---

[16] Lansdale, *In the Midst of Wars*, 19.
[17] Ibid., 45.
[18] Smith, *Philippine Freedom*, 155. Lansdale, *In the Midst of Wars*, 434.

went to raising the army's strength to almost 30,000 troops. This nearly doubled its size from a year before.

Additionally, Battalion Combat Teams (BCT) were created. Containing 1,100 men, each included a K-9 Corps to track Huks, a battalion of Scout-Rangers, and a cavalry squadron. BCT structure reflected the change in tactics from traditional to irregular. Twenty-six such teams were created. Magsaysay also began to clean up the military's reputation for abuse of the peasantry. Napoleon Valeriano, one of Magsaysay's best BCT commanders, observes, "Magsaysay was new, dramatic, infinitely energetic, determined to overcome, by any means necessary, the obstacles to effective action against the Huk."[19]

For his part, Lansdale was a committed anti-communist who believed in the American ideals of democracy. As such, perhaps his greatest contribution he made upon Magsaysay was the emphasis he placed on people. As he presciently observed, "If we understand that this is war over people, then we can start understanding the real human values in it – which go far beyond sizes of forces, reports of battles, statistics on casualties, and differences in quality of weapons." For Lansdale, the people were the actual "land."

With this new concerted effort underway, the people of Luzon saw in these endeavors of the government an effort made to defend the people against the Huks instead of the government defending themselves. Moreover, Magsaysay instilled the imperative of escalation of force, and empathy toward the population. He even implemented a low-cost telegram complaint system by which the people could field their grievances. With these reforms, the people began to see clear evidence of military and police professionalism. Following this professionalization of the constabulary and military, the Magsaysay-Lansdale team developed a new but old COIN strategy. Known as the carrot and stick approach, their two-pronged combination of psywar designed to win the support of the population (carrot), with selective strike measures to annihilate Huk fighting potential (stick).

---

[19] Valeriano and Bohannan, *Counterguerrilla Operations*, 139.

The Department of Dirty Tricks

As any carpenter knows, a sharp nail splits the wood, a blunted tip keeps it whole. The sharp nail acts like a wedge almost like a tiny splitting axe. Yet, blunting the nail acts as a whole puncher. The blunted nail, serving to keep the wood intact, is a fitting analogy to Lansdale's media management and psychological warfare campaign. Hidden behind the benign title "Civil Affairs Office" (CAO), this carrot effort aimed to influence the enemy and the public. Particularly, the target audience were the Huks' part-time supporters, who could be converted to the government side. Through the CAO, student organizations were funded which disseminated anti-Huk and anticommunist materials in schools. A carefully articulated narrative was also passed through newspapers, leaflets, and public radio. Funds were readily available to "employ" journalists and radio announcers. Over a two-year period, millions of Huk villainizing propaganda leaflets were distributed.

According to Seneca, "Religion is regarded by the common people as true, by the wise as false, and by the rulers as useful." Lansdale's psywar campaign also turned to organized religion as a medium. With the assistance of the Far Eastern Broadcasting Company, run by American evangelical missionaries, free radios were distributed which offered CAO ideological messages. This medium gave the government a means of strategic communication (SC). By way of this strategic narrative, bounties were placed on Huk leaders, and rewards were given for firearms surrendered. In the first year, this cash for guns netted an estimated half of all available weapons in Huklandia.

Moreover, the cash for guerrilla scheme equally had measured success, identifying some 1,175 Huks. Posters announced that all Huk "field commanders" were wanted "dead or alive" and had a price on their heads. Bounties were paid to civilians who brought in the bodies of Huks. Such measures gave the people of Luzon hope that the government could do good things for them, which the Huks were trying to get by violence.

Drawing on his own ingenuity, some of Lansdale's psywar methods nefariously sowed discord among the Huk ranks. Revitalizing some old OSS pranks, these black operations included exploding radios, flashlights, and doctored weapons in the ranks of the Huks who used them to their detriment. Describing these tactics, Lansdale recounts,

I took up the problem of Huks buying war materials from corrupt soldiers with the Philippine Army's intelligence and research chiefs. I asked them if contaminated ammunition could be made and inserted into the stocks being delivered secretly to the Huks. They agreed. Soldiers started reporting they had heard grenades exploding right in the hands of Huk ambushers.[20]

Under the efforts of Magsaysay, the Philippine Air Force was organized and implemented into the psywar fight. Having identified enemy base camps, and particular names of Huk personnel, such aircraft would fly over a village, then by loudspeaker, these "informers" would be thanked for their contribution by the so-called "eye of God."[21]

The Civil Affairs Office also provided a mechanism through which Magsaysay could influence his COIN force: Civil Affairs officers were attached to most units and held both advisory and supervisory functions. Magsaysay's slogan was "All out friendship or all-out force." Reminding his troops that their duties were first, to act as an ambassador of goodwill from the government to the people; second, to kill or capture Huk. Magsaysay rapidly gained a national reputation as a hands-on defense chief who made surprise inspections of troops in the field and took a personal interest in enforcing discipline and punishing the security forces for casual brutality which had been the norm in central Luzon since the war.

Perhaps the most imaginative of Lansdale's tactics was the one that played upon the popular dread of an *asuang*, or vampire. As it's regaled, one night as a Huk patrol came along a trail, ambushers silently snatched and killed the last man of the patrol. Then they punctured his neck with two vampire-like teeth holes, held the body up by the heels, drained it of blood, then left the corpse back on the trail. When the Huks returned to look for the missing man and found their bloodless comrade, every member of the patrol believed that an asuang had got him. Fearing that they would be next, when daylight came, the whole Huk formation moved out of the vicinity. Another of Lansdale's psywar tactics was devised to terrify and clear out an entire community. As recounted by Napoleon Valeriano,

---

[20] Lansdale, *In the Midst of Wars*, 45.
[21] Ibid., 73.

The army unit captured a Huk courier descending from the mountain stronghold to the village. After questioning, the courier, who was a native of the village, woefully confessed his errors in helping the Huks. His testimony was tape-recorded and made to sound as if his voice emanated from a tomb. The courier was killed. His body was left on the Huk-village line of communications. Soldiers in civilian clothes then dropped rumors in the village to the effect that the Huks had killed the courier. The villagers recovered the body and buried the Huk. That night army patrols infiltrated the cemetery and set up audio-equipment which began broadcasting the dead Huk's confession. By dawn, the entire village of terror-stricken peasantry had evacuated! In a few days, the Huks were forced to descend the mountain in search of food. They were quickly captured and/or killed by the army unit.[22]

Force X and BCT Sweeps

As for stick approaches, these consisted of strikes, sweeps, and stay behind missions. As a special strike force, the Sixteenth Constabulary Company was designated "Force X" and given the mission of penetrating Huklandia on "Huk-hunting" forays. Dressed and equipped like Huks, Force X ventured into remote areas and conducted counterguerrilla operations. Under the command of Lieutenant Colonel Laureño Maraña, Force X even deputized civilian gunmen, and paid bounties for successful hunters. Charles Bohannan, an American survivor of the Bataan Death March, recalled, "The killing of leading enemy personalities may be far more important than the destruction of a certain army unit." This was the aegis behind Force X's many decapitation strikes.

An important innovation in the campaign against the Huks was the deployment of hunter-killer counterguerrilla teams. One such was the elite 7th Battalion Combat Team (BCT). Lessons learned from Force X and other strike units were incorporated into the 7th BCTs tactics. Its mission, as described by Major Valeriano, was to find and finish Huk leaders, including the Supreme Commander, Luis Taruc. CAO posters offered rewards for information leading to the arrest of Luis Taruc, Jesus Lava, and Guillermo Capadocia. Major sweeps employing a clear-hold-build methodology were then carried out. These operations resulted in the Huks losing control of the populated areas, leaving without reliable

---

[22] Bohannan and Valeriano, *Counterguerrilla Operations*, 198.

sources of food or supplies. Trapped as they were in the swamps and forests of Luzon, the Huks could not retreat to an area that offered them a long-term ability to sustain themselves. It was only a matter of time before they starved to death, fought a pitched battle out of desperation, or surrendered. By 1952, BCT sweeps accounted for 13,000 Huk casualties.

Other operations appealed to the Huks' sense of humanity. For example, a Filipino government sector commander would discover that a Huk commander's wife was expecting a baby. Upon delivery, the sector commander sent a doctor and nurse to the village, offering their help and congratulations. Such acts of thoughtfulness contributed to the Huk commander's decision to surrender himself and the rest of his men. Additionally, Philippine Army Judge Advocate lawyers were assigned to provide free counsel for peasants in land court cases against wealthy landlords. The Magsaysay-Lansdale team made huge gains with the affected population in a short amount of time. Despite these positive measures, according to estimates, based simply on their proximity to Manila, the Huks might have been nonetheless capable of overthrowing the government. Then, a remarkable turn of fate, a devasting blow befell the Huks.

Charismatic, and personally brave, Magsaysay publicly offered to speak with any Filipino who wanted to help fight the insurgents. The Huks sent a young officer posing as an ordinary civilian to accept the offer, and then assassinate him. Intent on killing Magsaysay, when the young man showed up at the meeting, he first let the defense minister speak. The would-be assassin was so impressed that he decided to switch allegiances. Then, in a stroke of remarkable good fortune, the convert offered Magsaysay the Manila-based Huk politburo on a silver platter. The top Huk leadership could be scarfed up by following a 12-year-old girl in Manila who passed messages while peddling a meat and vegetable cart. As observed by historian Mark Moyar, placing the girl under surveillance, "Magsaysay apprehended 150 Communist Party members, including the top twelve members of the politburo, along with the complete records of the Communist Party."[23] Among those captured was the general secretary, Jose Lava.

---

[23] Moyar, *A Question of Command*, 101.

The capture of the Politburo prompted the Huks to vent a fresh torrent of murders, kidnappings, and robberies, all of which served to further alienate them from the affected populace. As recounted by Lansdale, one had a Jesse James flavor to it, "when the Huks held up the Manila Railroad at Binan and took $76,000 worth of loot."[24] As a follow up, on the night of November 25, 1950, a Huk squadron of about a 100 raided Magsaysay's hometown of Aglao. Armed with information gathered in the Politburo raid, the government responded with a new offensive. In January 1951, Philippine Army units thrashed Huk camps near Mount Arayat. The Huk massacre at Aglao would haunt Magsaysay the rest of his days.

EDCOR

The most remarkable of the novel concepts implemented by the Magsaysay-Lansdale team was the Economic Development Corps (EDCOR). Designed to address the age-old land grievance, in direct competition with the Huk political agenda, EDCOR served to undermine the Huk cause, and usurped their slogan, "land for the landless." In Mao's analogy, the guerrilla fish swims in the water of the population. EDCOR dealt with the fish. Formally instituted on 15 December 1950, EDCOR offered Huk guerrillas an incentive to surrender: 15 to 25 acres of free land on the far away island of Mindanao, a house, a carabao (water buffalo), seed, farm implements, police protection, education, medical aid, electricity, and free transportation to the site.

Arguably, while as a resettlement program EDCOR did not net as many Huks as would be desired, its psychological payoff was enormous. At its peak, EDCOR boasted a population of only 5,175. This number constituted some 950 to 1,200 families.[25] Notwithstanding the paucity of these numbers in respect to the affected area, EDCOR completely discombobulated the Huk propaganda claim of government resistance to land reform and demonstrated to the peasantry that the government was responsive to their foremost concern. Unable to stay ahead of the news cycle, Huk counterpropaganda did its best claiming that the EDCOR

[24] Lansdale, *In the Midst of Wars*, 65.
[25] Kerkvliet, *The Huk Rebellion*, 239.

resettlement was a modern-day "reconcentration camp." However, Huk fake news failed in the face of media reports and persistent rumors to the contrary.

The significance of the carrot aspect of the Magsaysay-Lansdale COIN methodology is seen when comparing it to the stick. As borne out by the metrics, between 1950-1955, some 10,000 Huks were killed with 1,600 wounded and 4,200 captured but 15,800 surrendered. Consequently, as recounted by Kalev Sepp, "as late as 1959, nearly five years after serious combat had died down, the original EDCOR project was still running, as the re-settled farmers worked off their small debts to the government for the materials they had received."[26]

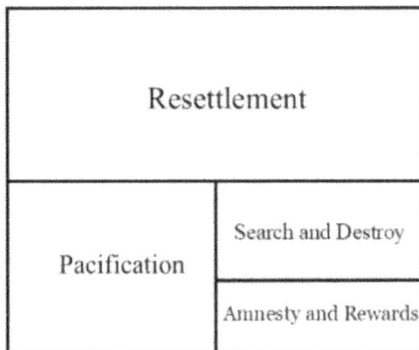

```
+-------------------------------------+
|                                     |
|           Resettlement              |
|                                     |
+-------------------+-----------------+
|                   | Search and Destroy |
|   Pacification    +-----------------+
|                   | Amnesty and Rewards |
+-------------------+-----------------+
```

Relative Weight of the Magsaysay-Lansdale COIN Strategy.

As Kalev Sepp observed, "EDCOR was so strongly touted as a success throughout Asia that British officials visiting from Malaya in 1951 expected to see a massive, nation-wide program instead of a single village populated by a few dozen families. Eventually, two more EDCOR camps were built on Mindanao, and in 1953, in an intentionally dramatic move, an EDCOR resettlement camp was established on Luzon in San Luis, the hometown of Taruc, the Huk Supremo (supreme commander)."[27] For their part, the only adjustment the Philippine Communist Party made

[26] Ibid.
[27] Sepp, *Resettlement, Regroupment, Reconcentration*, 47.

was to change their name to Hukbong Magpapalaya ng Bayan, or People's Liberation Army (HMB).[28]

By the fall of 1952, with the relentless pursuit of Magsaysay's BCTs, only a small remnant of Huk forces remained in the field. Realizing the loss of the Huk "mass base," Taruc surrendered a year later in May 1954, effectively ending the Huk rebellion.

Summary and Implications

By way of the five laws of irregular warfare, the following is an analysis of the Magsaysay-Lansdale COIN effort:

Political objective (s) – Success in COIN requires an operational design that identifies the root cause of an insurgency and its leadership. For Lansdale, solving the problem of the Huks was elementary. He recounts, it "was a classic example of counterinsurgency, with lessons all too often overlooked or misunderstood by those who face the problems of insurgency in other countries of the world."[29]

Legitimacy – In the last analysis, the genius of the Magsaysay/Lansdale COIN design integrated military activities, civic action projects and psychological warfare into one holistic endeavor. As an established rule, a COIN force must not be perceived as occupiers. On June 25, 1950, North Korea invaded South Korea. Having been fully immersed in the war on the Korean Peninsula, the U.S. was hesitant to intervene directly in the Philippines. Instead, recognizing the Filipino soldier's potential and contributing adequate military aid to outfit the BCTs formed under U.S. advisors' guidance, the United States formed a doable strategy that maintained a national Filipino face.

Adaptability – The essence of COIN involves identifying and separating insurgents (physically and psychologically) from the population and

---

[28] Lansdale, *In the Midst of Wars*, 12.
[29] Lansdale to Lavinia Hanson (Valeriano), personal correspondence, 5 March 1984, Lansdale Papers, Box 15, Hoover Institution Archives, Stanford University, Palo Alto, CA.

influencing them to renounce violence and accept the sovereignty of the host nation. Doing this makes counterinsurgency prove to be more alchemy than science. The efficiency of a unit conducting COIN depends largely on its knowledge of the people and the terrain. This was a strong suit of the Magsaysay-Lansdale team who brimmed over with both cultural intelligence and competence. Moreover, as IW is a learning competition, flexibility and adaptability are key ingredients to success. It may be said that the American experience in the Philippines represents the gold standard in COIN. Through EDCOR, the root cause of the insurgency was compromised. The Huk experience teaches that COIN strategies should be designed to not only defeat an insurgency but to also address the root cause(s) that led to the outbreak.

Influence – As stated before, success in COIN requires an operational design that stabilizes the affected area by identifying with the people, while working with them to create a better state of the peace, winning their allegiance, and if possible, redress grievances that led to the insurgency. From an advisory standpoint, the implementation of Lansdale's advice to the end of major resistance took a mere 18 months, with only scattered resistance lasting until 1956. This alone makes Lansdale's advisory role one of the most successful in history, rivaling that of T. E. Lawrence.

Native Face – Lansdale's winning formula was every bit a demonstration of the axion, finding local solutions to local problems. Under Lansdale's careful tutelage, the Huks were defeated, and the United States was able to realize its foreign policy goals in the Philippines with a handful of advisors, with minimal U.S. troop presence, and the expenditure of less than $2 billion.

The Lansdale methodology may be summed up as he recalled saying, "our efforts involved teaching and employing paramilitary forces in the tactics of political warfare."[30] Sadly, again, these lessons learned would be squandered. Lansdale's achievement was largely overshadowed by the Korean War. In 1954, he was sent by the CIA to Vietnam, where he would

---

[30] Bohannan and Valeriano, *Counterguerrilla Operations*, 236.

try to replicate the success he enjoyed in the Philippines. As borne out by history, Lansdale's success would be the road not taken (see chapter 8).

For Further Reading:

1. *In the Midst of Wars* by Edward Lansdale.
2. *Counterguerrilla Operations* by Charles Bohanan and Napoleon Valeriano.
3. *The Huk Rebellion: A Study of Peasant Revolt in the Philippines* by Benedict J. Kerkvliet.

8

A War Without Fronts: The Vietnam War

"Thank God we are leaving Vietnam. Now we can get back to the type of warfare we know how to fight." – Anonymous

Detachment A-113 (1st SFG) and the CIDG Program at Buon Enao, South Vietnam.
Courtesy of Associated Press.

The Vietnam War was one of the most complex and challenging conflicts in history. It began as an anticolonial war against the French and evolved into a Cold War confrontation between democracy and communism. As it progressed, the conflict matured into a full-blown proxy war pitting communist North Vietnam, supported by the Soviet Union and China, against the US-backed South Vietnam, which was attempting to maintain a democratic system. Before the Afghanistan War, the Vietnam War was America's longest military conflict.

As part of the larger Cold War rivalry between the United States and the Soviet Union, America's main goal in the war was to prevent the takeover of South Vietnam by communist North Vietnam. Lasting from 1945-1975, the conflict can generally be divided into two phases: the First Indochina War (1946-1954), and the Second Indochina War (1954-1975). During the second phase, North Vietnam effectively controlled and directed the Viet Cong insurgency in South Vietnam, utilizing them as a proxy force to fight the South Vietnamese government with the goal of unifying the nation under communist rule. America's counterinsurgency efforts to prevent this may be seen to play out in three stages: The Advisory Years (1954-1965), Intervention (1965-1967), and Withdrawal (1968-1973). This chapter will capture the salient aspects of these stages with the aim of understanding the evolution of the U.S. COIN effort.

The First Indochina War (1946-1954)

World War II deprived France of her colonial possessions of Laos, Cambodia, and Vietnam. Following Japan's defeat, on 2 September 1945, in hopes of finally achieving Vietnamese independence, Ho Chi Mihn, who had long fought the Japanese, established the Democratic Republic of Vietnam (DRV). However, unbeknownst to him, Truman, Churchill, and Stalin had already decided Vietnam's future at the Potsdam Conference. They agreed that the southern half would be under British control with the northern half under the Nationalist Chinese. These forces encountered a well-established Viet Minh force under Ho Chi Minh's general Vo Nguyen Giap who already moved his 50,000 strong force to control much of the country.[1]

At the conference table, the French and Viet Minh were at an impasse. While the French recognized Vietnam as a "free state" within the French Union and agreed to a referendum toward the reunification of the three states within Vietnam, Tonkin, Annam, and Cochin China, they nonetheless reneged and proclaimed a separate government for Cochin China. For their part, the British actively assisted the French in regaining

---

[1] Fredrik Logevall, *Embers of War: The Fall of an Empire and the Making of America's Vietnam* (New York: Random House, 2012), 127.

control in the south. The Chinese however, demurred, allowing the Viet Minh to expand its support base in the north.

The Two Vietnams. 1954. Courtesy AP.

By mid-1946 these occupation troops departed leaving the French to face the growing Viet Minh nationalist movement. A fragile peace continued for much of 1946. The French and Ho Chi Minh signed several agreements, granting a vague independence to Vietnam. France's objective, hegemony over Vietnam, ran headlong into burgeoning Vietnamese nationalism. The deadlock fractured on November 23, 1946, when the French cruiser *Suffren* bombarded Haiphong harbor, massacring some 6,000 Vietnamese, and injuring an estimated 15,000 more. This proved to be the first salvos of what would be called the First Indochina War. Ho Chi Minh and his Communist shadow government then went underground.

During World War II, Ho Chi Mihn's anti-Japanese resistance fighters helped to rescue downed American pilots and furnished information on Japanese forces in Indochina. In July 1945, an OSS team code-named Deer, parachuted into northern Vietnam to assist Ho's insurgents. This team trained, advised, and assisted the Viet Minh to drive out the Japanese. In the wake of the Japanese departure, Ho hastily organized the Viet Minh as a nationalist front, masking its Communist leadership in order to appeal to broad patriotic sentiment. He then moved on Hanoi, which fell to him in September 1945, while French and various factions in the south struggled for power in Saigon. According to the Pentagon Papers, in late 1945 and early 1946, Ho Chi Minh wrote at least eight letters to President Truman and the State Department requesting American support for Vietnamese independence and even said that Vietnam would adopt the U.S. constitution and the Bill of Rights. There is no record of the United States ever answering.

As the French would discover, the Indochina War was "la guerre sans fronts," a war without fronts. By early 1947 the French had reestablished control over the large urban centers, an important series of fortifications along the Chinese border, and the main roads. They could generally seize and hold any terrain they wished. Nevertheless, the Viet Minh quickly eroded whatever French support existed in the countryside and built up its forces. For their part, France completely misunderstood the nature of the conflict in which they were engaged. What they failed to realize initially was that, although they controlled the roads, they were fighting an enemy that had no need of roads. They seized strong points, but these strong points commanded nothing, since the enemy was not stationary but fluid. Most importantly, the French sought to defeat the elusive Viet Minh in set-piece battles but neglected to cultivate the loyalty and support of the Vietnamese people.

If there is one thing we learned from the Vietnam War, as with all wars, it is that military actions only have value if they contribute to the political goal of the government that prosecutes them. When Giap met with Mao Zedong in China in 1948, Mao informed him that Vietnam's situation was like that of China's in 1936, and that his principles of a protracted war should form the basis for Giap's strategy. Clearly using Mao's model of a three-phased insurgency, history shows that the Vietnam War was an insurgency brought to its planned conclusion.[2]

As for the French, for the next nine years they would achieve tactical successes, albeit they were all strategically inconclusive. Following the Maoist principle, where the French generals massed forces, the Viet Minh withdrew. Where the French held fast, the Viet Minh harassed. Where the French economized forces, the Viet Minh attacked. Where French forces tried to withdraw, the Viet Minh pursued ruthlessly. Then something happened in 1949 that strengthened the Viet Mihn position considerably. Mao's Chinese People's Liberation Army defeated the Nationalist Army led by Chiang Kai-shek. This development gave the Viet Minh a cross-border sanctuary and a steady stream of arms and logistical support.[3] The French realized they were now facing a long war.

In an effort to interdict this external support, the French constructed a series of fortifications. Then, also in the South, on 14 June 1949, the French established the State of Vietnam, a rival anti-Communist government, with Emperor Bao Dai as head of government. The United States recognized Bao Dai's South Vietnamese government and began sending military and economic aid. Despite these developments, against the Viet Minh's cause and articulated conclusion to the war, the Bao Dai regime lacked any perceptible alternatives to the Communists. Echoing this sentiment, John F. Kennedy, then congressman, made a visit in 1951 and criticized the situation as follows:

---

[2] Thomas X. Hammes, *The Sling and the Stone: On War in the 21st Century* (St. Paul, MN: Zenith Press, 2004), 58.

[3] In the early 1950s, its estimated that Mao supplied the Viet Minh with some 150,000 rifles and 4,600 artillery pieces with millions of rounds of ammunition and shells.

In Indochina we have allied ourselves to the desperate effort of the French regime to hang on to the remnants of an empire. There is no broad general support of the native Vietnam Government among the people of that area.[4]

France continued to petition for American aid, warning of the danger of the spread of communism throughout Southeast Asia if Vietnam fell to the Viet Minh. Sharing this concern, but preoccupied with the Korean War, and increasingly frustrated with the French conduct of the war, in September 1950, President Truman nonetheless established the Military Assistance Advisory Group (MAAG). MAAG oversaw the delivery of transport planes, equipment, and trucks. Following Truman, President Eisenhower deepened America's commitment, allocating burgeoning U.S. support. Early on, U.S. leaders realized that defeating the Viet Minh would require the formation of a sizable Vietnamese Army. Something the French staunchly resisted. The pro-French Vietnamese National Army (VNA) was created on January 1, 1949, and fought in a few major campaigns.

May 1953 witnessed the selection of French general Henri Navarre. Like his predecessors, Navarre did not understand the Maoist tactic of retreating before pressure to ensure survival of the insurgent forces. Using his smaller force as bait, Giap moved the war into neighboring Laos and lured the French into a trap. The French would soon have the set-piece battle it was looking for. In the massive airborne operation, a French force of some 4,200 paratroopers captured the Japanese-constructed airstrip near the town of Dien Bien Phu. What followed was a fifty-seven-day siege resulting in a crushing defeat for the French.

Following the French defeat, the on-going Geneva Accords hastened France's disengagement from Indochina. On 20 July 1954, France and the Viet Minh agreed to end hostilities. The Accords further concluded an agreement that Vietnam was to be divided at the 17th parallel until elections, scheduled for 1956, after which the Vietnamese would establish a unified government. Accordingly, Vietnam was to be temporarily divided into two zones at the 17th parallel: the north fell under Ho Chi Minh, and the south under Bao Dai, the former emperor

---

[4] George M. Kahin and John W. Lewis, *The United States in Vietnam* (New York: Dial Press, 1967), 33.

of Vietnam. President Eisenhower pledged economic aid to assist the government of South Vietnam.

The question may be asked, why did America get involved in Vietnam? The short answer is, following our Cold War doctrine of containment, America feared that if South Vietnam fell to communism, other countries in the region would follow suit in a domino-like fashion. In the words of CIA historian Thomas L. Ahern,

The final Communist victory in China in 1949 and Pyongyang's invasion of South Korea in 1950 reinforced the American view of Communism as an implacably expansionist monolith. Indochina came to be seen as critical to the defense of the Asian littoral.[5]

From the beginning, America's strategy was one of containment, to prevent the spread of communism in Southeast Asia. The mission creep nature of America's strategy to defeat the NVA and their Viet Cong insurgency, the sustained political goal was to maintain an independent non-communist South Vietnam.

The Geneva Accords marked the beginning of the transition from France to the United States being the dominant external influence on Vietnam. In terms of blood and treasure, the human cost was steep. Sources estimate over 100,000 French and 500,000 Viet Minh troops died during the eight-year conflict. In materiel, the United States spent nearly $3 billion in aid to France.[6]

The Advisory Years (1954-1965)

While the great powers were debating Vietnam's future in Geneva, the most controversial Vietnamese figure to feature in the U.S. effort in Vietnam stepped off a plane in Saigon in June 1954. Ngo Dinh Diem would serve as the newly appointed prime minister to Bao Dai. In a letter to Diem, the Eisenhower administration promised to maintain a "strong, viable state capable of resisting outside aggression." Arriving before

---

[5] Thomas L. Ahern, *CIA and the Hose of Ngo: Covert Action in Vietnam: 1954-1963* (Langley, VA: CIA Staff History, 2009), 3.
[6] Support included 2,000 tanks and combat vehicles, over 30,000 transport vehicles, 360,000 small arms and machine guns, 5,000 pieces of artillery, 500 million rounds of ammunition, 10 million artillery shells, over 400 vessels, and almost 400 planes.

Diem was Colonel Edward Lansdale, fresh from the Philippines. Having trounced the Huks, Secretary of State John Foster Dulles sent him to Vietnam to repeat his success. However genius he was, compared to Lansdale's experience in the Philippines, Vietnam was far different. The Philippines had a stable government with regular elections, an established army that outnumbered its enemy, was geographically isolated from outside communist support, and had no external threat. Vietnam had none of these attributes. Moreover, the situation Lansdale entered was beset with intrigue. Of all of America's conflicts, Vietnam presents the most enigmatic of puzzles. Thus, deploying Lansdale as a sort of mythical cleanup man was a testament to his talent but was in many ways naïve. Lansdale nevertheless quickly gained the ear of Diem and became one of his biggest supporters and most frequent American confidants.

As demonstrated against the Huks, Lansdale was a pioneer in psychological warfare. In his attempts to prop up Diem, Lansdale pulled out all the stops. The Geneva Accords provided for a three-hundred-day grace period before the partition in order to allow Vietnamese to move from North to South or vice versa. Realizing that Diem's political base was minimal, his first order of business was to conjure the idea of using the U.S. Navy to transport willing Vietnamese refugees south. "Operation Passage to Freedom," would not only bolster Diem's political base, but it would also prove useful in the media war against communism. Between August 1954 and May 1955, the U.S. Navy relocated some 900,000 refugees, most of whom were Catholics, and equipment from North to South Vietnam.

Also following the Geneva Accords, the Viet Minh removed nearly all their troops from the South but left a residual cadre of approximately 10,000. In time, this cadre would indoctrinate and build a base of support that would become the armed communist revolutionary organization in South Vietnam, Laos, and Cambodia known as the Viet Cong (VC).

Lansdale next turned to solidify Diem's position in Saigon. The French defeat had left a power vacuum, and groups besides the Vietminh were jockeying for turf. In 1955, three groups united in opposition to Diem: the Cao Dai and the Hoa Hao, religious sects, and the Binh Xuyen, an organized-crime society with a private army of ten thousand men. Diem neutralized the religious sects by the expedient of having Lansdale use CIA funds to buy them off. As Max Boot relates, the amount may have been as high as twelve million dollars, which would be a hundred million dollars today.[7] But the Binh Xuyen, which controlled the Saigon police, remained a threat.

As the U.S. supplanted France in its role in Indochina, the foremost question resolved two central issues: how Vietnam's armed forces were to be trained, and whether Diem was to remain at the head of Vietnam's government. Worried that Diem was not strong enough to hold the country together, Dulles sent cables to the American embassies in Saigon and Paris authorizing officials to find a replacement. Lansdale, who had foiled a military coup in 1954, warned Diem that U.S. support was waning, prompting him to launch an attack on the Binh Xuyen. After the Binh Xuyen gang was routed, Dulles countermanded his order.

As his powerbase solidified, Diem was emboldened to undermine Bao Dai's political standing. Motivated by this and his refusal to countenance a future loss to Ho Chi Mihn in the coming 1956 election, Diem called for a referendum to determine whether he or Bao Dai, the former Emperor, should be head of state. Diem won. His brother, Ngo Dinh Nhu, rigged the election, giving Diem 98.2% of the vote. Lansdale's main contribution to the campaign was to suggest that the ballots for Diem be printed in red (considered a lucky color) and the ballots for Bao Dai in green (a color associated with cuckolds). With Diem's consolidation of authority, he achieved the height of his power and influence.

As in the Philippines, Lansdale advocated for civic action programs to bring governance to the political vacuum in the remote areas where the Communists were most likely to re-emerge. Additionally, in August 1954, the number of advisors in MAAG rose to 342 with the task of training the nationalist army. This force would eventually become the Army of the Republic of Vietnam (ARVN), which was established on December 30, 1955. Likewise, Diem created the Civil Guard (CG) out of provincial militias that had existed during the French war. In conjunction with a village-level Self-Defense Corps (SDC) of armed villagers, this force was Diem's first line of defense against the communist guerrillas. With many

---

[7] Max Boot, *The Road Not Taken: Edward Lansdale and the American Tragedy in Vietnam* (New York: Liveright, 2018), 62.

of these projects still in their embryonic form, Lansdale left Vietnam in 1956 and was soon assigned to head Operation Mongoose, charged with devising methods for overthrowing Fidel Castro.

While it may be said that Lansdale's efforts did much to lay the groundwork of solidifying the fledgling democracy in the South, without Lansdale at his side, Diem was not up to the task. According to the Pentagon Papers, Diem was "authoritarian, moralistic, inflexible, bureaucratic, and suspicious." His mentality was much like that of a Spanish inquisitor. Complicating matters, his political machine was organized over a centralized family oligarchy. Trusting only his family members, particularly his brother Ngu, Diem was convinced that Vietnam was not ready for a representative government and democracy. As such, no organized opposition to his regime was tolerated, and his critics were eventually repressed.

In an effort to address peasant grievances, in competition with what was happening in the North, Diem initiated land reform in the South. He lowered rent payments in 1955, and in 1956 he limited landowners to owning no more than 300 acres and allowed peasants to buy land on credit.[8] However, the effect of this reform was minimal. Moreover, in many areas, the Viet Minh had seized land during the French War and signed it over to the local southern peasants who now considered it theirs. In many cases, Diem simply reclaimed such land and offered it for sale to the same peasants.[9] From the peasants' perspective, such treatment was not land reform, but merely corruption in a new form.

In the 1950s, the population of South Vietnam was about 12 million divided into several ethnic groups. The largest group was the ethnic Vietnamese who comprised 80 % of the population. This population resided in South Vietnam's 16,000 villages. Fearing that communists might gain power in them, Diem abolished the traditionally elected village councils. He then replaced them with northern refugees and Catholics that were loyal to him. Diem's political party, the National Revolutionary Movement, was controlled by his brother Nhu. Nhu also

---

[8] CIA, "Memo for Director of Central Intelligence, Subject: Agrarian Reform and Internal Security in South Vietnam," CIA (April 30, 1957), 3.
[9] Webb, *The Joint Chiefs of Staff and the Prelude to the War in Vietnam, 1954-1959*, 141–42.

led the Secret Revolutionary Personalist Labor Party, commonly known as Can Loa, consisting of 20,000 ardent Diem loyalists. Carefully chosen and placed in key positions within the civil bureaucracy and military, the Can Loa provided intelligence and worked to manipulate the civil bureaucracy and the military.

Diem's motivation to solidify his position led him to repressive measures. He curbed freedom of speech and jailed dissidents. He additionally alienated large segments of the armed forces by promoting officers on the basis of loyalty to his family rather than on the basis of ability. Diem then refused to hold unification elections as required by the Geneva Accords. To prevent what it believed would be a communist electoral victory, the Eisenhower Administration backed Diem. Alienating the people he sought to control, in a further consolidation of power, Diem detained some 50,000 to 100,000 suspected communists who were placed in detention camps. While this effort did net a large number of proto-Viet Cong cadre, not surprisingly, these "programs" designed to increase security in the countryside were carried out so badly that they drove a wedge between the farmers and the government and led to less rather than more security.

The Road to the South

As opposition to Diem's rule in South Vietnam grew, an incipient insurgency began to take shape. Viet Minh cadre, who had remained in the South, had hidden caches of weapons, and began to organize against Diem's puppet government. According to the Pentagon Papers, the Saigon-Cholon Peace Committee, the first Viet Cong front, was founded in 1954 to provide leadership for the insurgency. Initial attacks began with bombings and assassinations. A clandestine radio station also began broadcasting support for armed opposition to Diem's regime. After neutralizing the military and religious sects, Diem turned his attention to the Viet Cong and drove them into the swamps. Then, in March 1956, Le Duan, a South Vietnamese communist leader, presented his plan to Hanoi entitled "The Road to the South." In this plan to revive the insurgency, Le Duan argued:

In the past two years, everywhere in the countryside, the sound of the gunfire of U.S.-Diem repression never ceased; not a day went by when they did not kill patriots, but the revolutionary spirit is still firm, and the revolutionary base of the people still has not been shaken. Once the entire people have become determined to protect the revolution, there is no cruel force that can shake it.

As things took shape, Le Duan was directed by the Hanoi Politburo to lead the nascent insurgency in the South.

By the end of 1957, some 400 South Vietnamese officials were killed or kidnapped. During 1958, Le Duan solidified the insurgencies sanctuary by establishing safe areas in the Mekong Delta. Then, in May 1959, encouraged by the increasing dissatisfaction with the Diem government, Hanoi initiated the revolution, calling for the overthrow of the puppet regime through guerrilla warfare. Hanoi would support guerilla operations within South Vietnam to create conditions necessary for the North's eventual invasion. Pursuant to this plan, the National Liberation Front (NLF), what would later be called the Viet Cong, was formed. A contraction of Việt Nam cộng sản, Vietnamese communist, the Viet Cong were referred to as VC. General Giap's strategy was to use the Viet Cong for incessant guerrilla warfare to wear down Saigon and its American ally, while the People's Army of Vietnam (PAVN), the official title for the North Vietnamese Army, would be sent in only to fight conventional set-piece battles at times and places of its choosing. The PAVN would send troops and supplies through a network of clandestine paths known as the "Ho Chi Minh Trail." The first shipment of a few dozen rifles was sent via the trail in 1959. By 1964, the trail's supply capacity would reach 20–30 tons per day. By 1966, 300 tons of supplies moved through the trail per day. As to troops, in 1964, the NVA managed to use it to transport a regiment down it. By the end of 1968, over 300,000 enemy troops had used it to enter South Vietnam.

As taught by Mao Zedong, peasants were the key to success. The guerrillas had to move among the peasantry "like the fish swims in the sea." Like Mao's instructions to his Chinese soldiers, all Viet Cong soldiers were issued a series of directives. They were to avoid damage to the land, crops, houses, and belongings of the people. They were not to insist on buying or borrowing what the people are not willing to sell or lend. They were never to break their word, and they were to help the people in their daily work. This suited the interests of the peasants.

However, the NLF's gentility and magnanimity could be tempered by ruthlessness. During the Tet offensive, for example, 'unfriendly' people were dragged out of their houses in Hue and were shot, clubbed to death or buried alive. Nonetheless, Communists worked hard to win over the peasantry, offering them a fairer distribution of land and urging Communist soldiers to avoid the rape and pillage that was characteristic of the South Vietnamese ARVN behavior. The disorderliness and thievery of Diem's forces induced innumerable villages to join an open revolt against the government.

Reciprocating in turn, villagers often gave the Viet Cong food, shelter and a place to hide. This placed both ARVN and American troops in an invidious position – they had no way of knowing who they could trust in the villages, so that the very people they were allegedly there to protect became indistinguishable from the enemy. Even in areas supposedly controlled by the Saigon government, there was a web of informants and a host of various social organizations managed by the Viet Cong. Collectively, this became known as the Viet Cong Infrastructure (VCI).

One of the principal tasks in counterinsurgency involves separating the insurgency from the affected populace. This must be done to remove the insurgents' base of support as well as to secure and protect the population. This key principle in counterinsurgency strategy isolates the insurgents and denies them the support they need to operate effectively. This is achieved by securing the population and providing them with essential services, thereby undermining the insurgents' ability to gain influence and recruit within the community.

By 1959, realizing that he was losing South Vietnam to the Viet Cong village by village, Diem and his brother, Nhu settled on a nuanced resettlement approach. Modelled after the success of a similar program used by the British in Malaya, Diem's forces relocated some 210,000 villagers into 147 centers which had been carved out of about 220,000 acres of wilderness. These rural communities, called "agrovilles" physically separated the people from the Viet Cong's contact and influence. Diem hoped to create a "human wall in the depopulated area near the three frontiers of Cambodia, Laos, and North Vietnam."[10]

---

[10] "Memorandum of a Conversation," May 9, 1957, U.S. Department of State, *FRUS, 1955-1957, Vietnam*, vol. 1 (Washington, DC: GPO, 1985), 801.

Drawing from the historic precedent, Diem's agrovilles were basically miniature cities that required the free labor of villagers and forced relocation. Despite some successes, various factors doomed the program. Relocations necessitated long transits between their fields and their new Agroville. Complicating matters further, Diem's relocation force, consisting of a mishmash of police, self-defense corps, civil guard, were poorly trained, equipped, and miserably led. Resentful villagers, who were driven out of their ancestral homes and fields, often made eager recruits by the Viet Cong. These factors resulted in the program to be finally abandoned, but it was not the end of Diem's resettlement plan.

According to the Pentagon Papers, a United states embassy special report concluded: "The people cannot identify themselves with the government." As most historians agree, the Vietcong's rapid success after 1959 is largely attributed to the Diem regime's mistakes. According to the Pentagon Papers, on the interrogation of 23 Viet Cong cadre members regarding the communist success, the South Vietnamese people were ready for rebellion. "The people were like a mound of straw, ready to be ignited. If at that time the government in the South had been a good one, if it had not been dictatorial, then launching the movement would have been difficult." It may be said, the growth of apathy and considerable dissatisfaction among the rural populace was a major cause of the insurgency.

Following Mao's philosophy of engaging through surprise, mobility, and hit-and-run tactics, to wear it down, VC drew the ARVN and American forces into a long-drawn-out war. Basing themselves in the thick forests of South Vietnam, VC began by taking control of the villages in the rural areas. As their strength grew and the enemy retreated, they began to take the smaller towns. In the villages they controlled, the VC often built underground tunnels which led out of the villages into the jungle. This underground system also contained caverns where they stored everything from their printing presses, surgical instruments, and the equipment for making booby traps and landmines. In their hit and run forays, VC were sent out in small units of between three to ten soldiers in attacks on small patrols or poorly guarded government positions. VC were given limited knowledge of other units. If captured, they would be unable to give the enemy information. Following another one of Mao's dictums, they rarely attacked unless they outnumbered the enemy and were certain of victory.

Fully committed to the cause of reunification, the VC engaged in total war whereby every man, woman, and child was mobilized in either fighting, nursing, or supplying in order to defeat what it saw as the foreign enemy and its puppets. To gain psychological advantage, the VC employed Punji traps. These sharpened bamboo sticks lined the floor of a pit covered by a lattice of twigs and leaves. American troops would unwittingly step into such traps. As a form of germ warfare, Punji stakes were filled with excrement to maximum infection and long-term incapacitation or death.

In a May 1959 assessment, the CIA reported that the Communists numbered only 2,000 active guerrillas in the South. This same year, South Vietnamese Civil Guard and Self-Defense Corps, who bore the brunt of anti-guerrilla activities, along with the South Vietnamese armed forces (ARVN), had a combined total of 250,000 personnel. To train them and the ARVN, Eisenhower sent additional advisors, bringing the U.S. commitment to 685 by 1960. The Viet Cong followed suit, bolstering their ranks to about 3,000 at the end of 1960. However, by 1961, their numbers bolstered to 12,600 with an estimated support base in the hundreds of thousands.

Strategic Hamlet Program

The fall of 1961 brought a glimmer of hope as the aspiring South Vietnamese government was graced with another veteran COIN practitioner. Sir Robert Thompson, a veteran of the Malayan counterinsurgency effort, was appointed head of the British Advisory Mission to South Vietnam. Thompson proposed what would later be known as the Strategic Hamlet Program. His plan called for the securing of the rural populated areas from which the VC were drawing support. The plan called for armed villages to be surrounded by barbed wire or bamboo fences. The peasants were armed and trained to defend themselves. If attacked, the South Vietnamese Army, who was stationed regionally, would come to their aid. Following the "oil spot" strategy, hamlets would be constructed as a system in which each one was connected to each other. As the first hamlets were secured from the enemy, the next could be added and secured shortly after, expanding like an ink blot.

More than just providing physical security, the aim was to also improve the peasants' lives. Inside the hamlets, schools and hospitals, with electricity, were to be constructed. Along with some modern conveniences, it was hoped that improvements would encourage the peasants to move in. Thompson's plan called for the resettlement of most of the rural population, and to enhance the likelihood of success, the program was to start small in areas considered to be low threat from the Viet Cong. MAAG agreed with the plan as the national counterinsurgency strategy.

The program's goals were ambitious and historically unprecedented: In fourteen months, 14,000 of South Vietnam's estimated 16,000 hamlets were to be rebuilt into fortified Strategic Hamlets. This equated to 10 million people to be resettled in 14 months. The program got underway on March 22, 1962. Contrary to Thompson, however, Diem and Nhu decided to relocate most of the villages rather than hardening them according to plan. Additionally, restrictions and prerequisites recommended by Thompson, such as limits on hamlet distances from family lands and avoidance of Viet Cong strongholds, were disregarded by Nhu in his rush to stay on his impossible schedule.[11] Much like earlier models, the Strategic Hamlet Program involved the forced removal of peasants from their villages, most often at gunpoint. By September 1962, six months after the first resettlement effort, 3,225 hamlets had already been completed which relocated 4.3 million people. By July 1963, 8.5 million people had been settled in 7,205 hamlets. However, with this incredible rate of construction, the South Vietnamese government was unable to provide adequate and necessary support for the hamlets and its residents. There were other ominous signs. A minimally low number of military age males moved into the hamlets. This indicated that a substantial number of young men had joined the insurgency.

---

[11] Kalev I. Sepp, *Resettlement, Regroupment, Reconcentration: Deliberate Government-Directed Population Relocation in Support of Counter-Insurgency Operations* (Fort Leavenworth, KS: Army Command and General Staff College, 1992),81

VÙNG ĐỒNG BÀO ĐANG Ở DO VIỆT CỘNG KIỂM SOÁT SẼ
BỊ TIỂU HỦY.

ĐỒNG BÀO HÃY MAU DỜI CƯ VÀO CÁC ẤP TÂN SINH, NỚI
ĐÂY SẼ ĐƯỢC CHÍNH QUYỀN GIÚP ĐỠ.

Leaflet used to encourage Vietnamese civilians to move to the strategic hamlets.

Like the earlier Agrovilles venture, several problems plagued the Strategic Hamlet program, all of which contributed to the program's eventual failure. Most peasants were Buddhists who practiced ancestor worship, and their relocation prevented this. This alone made it extremely unpopular with the people. Displaced villagers lacked adequate security from the VC. Complicating matters, the forces Diem tasked to organize the program, the Civil Guard and Self Defense Corps, were still under-trained, and horribly corrupt. They were known to require the peasants to purchase the supplies necessary for building hamlets which had already been provided by the US. Further dooming the program was the fact that the Civil Guard and Self Defense Corps had been infiltrated by Viet Cong. In fact, the Viet Cong frequently moved back into the areas pacified by government troops. It was later discovered that the man chosen by Nhu to oversee the program was a communist agent named Albert Pham Ngoc Thao.

Security shortcomings became another major problem. Ignoring the "oil spot" principle, Nhu built strategic hamlets at reckless speed without much consideration. It is estimated that fewer than 20% of the 8600 hamlets met with defensive readiness and conditions worthy of such funding to the project. Many hamlets were isolated, not mutually supporting and became easy targets for the Viet Cong. Military support for the hamlets was another weakness. Armed peasants could only hold out for so long. Summing up the essence of the failure, Kalev Sepp recounts:

Adapted from the Malayan model to a significantly dissimilar terrain, population, leadership, and enemy, the concept was inappropriate to the situation in South Vietnam.[12]

Given the VC penchant for night attacks, and the unreliability of promised reinforcement, in time the Viet Cong overran the poorly defended hamlets. By early 1963, it was quite obvious that the program was failing as the Viet Cong had already controlled about a fifth of all South Vietnamese villages.

Due to its proximity, Laos became involved in the war because its eastern region, known as the "Panhandle," served as a crucial supply route for North Vietnam to send troops and supplies to the South through the Ho Chi Minh Trail. The Panhandle was also controlled by the communist Pathet Lao. In 1961, William Colby directed the training of a Hmong army to interdict this Panhandle sanctuary. Colby hoped that this secret war would help to prevent the spread of communism in Laos and help to siphon the ever-flowing stream of man and material to the insurgency. Frequently attacked and abducted by the Pathet Lao, the communist political movement and organization in Laos, the Hmong

---

[12] Sepp, *Resettlement, Regroupment, Reconcentration: Deliberate Government-Directed Population Relocation in Support of Counter-Insurgency Operations*, 84.

made a great ally of the United States. Rising to a force of some 50,000, Hmong attacked communist supply lines, provided valuable intelligence, and suffered immense casualties. It is estimated that Hmong soldiers, some as young as ten years old, died at a rate ten times as high as that of American soldiers in Vietnam.

Also operating out of Laos was Air America, a covert airline owned and operated by the CIA. Air America's operations in Laos included dropping supplies, conducting photo reconnaissance, and medevac missions, transporting troops and refugees, and inserting-extracting agents. All told, during its years of operation from 1950-1974, Air America transported thousands of troops and refugees, and rescued downed airmen throughout Laos.

Civilian Irregular Defense Group

Under Diem, South Vietnam's counterinsurgency effort included three basic elements: pacification, resettlement, and reforms. The second major resettlement effort was the Civilian Irregular Defense Group (CIDG) program. Launched in late 1961 to counter the influence of the Viet Cong in the Central Highlands, CIDG was the brainchild of CIA station chief, William Colby. Colby designed CIDG to develop irregular military units from ethnic minorities, like the Montagnards and Rhade. It was hoped that the program would also develop a sense of national loyalty among the participating ethnic groups. In time, CIDG proved to be one of the chief works of the Special Forces in Vietnam who seemed to be tailor-made for the program.

Although the program was an astounding success and greatly increased the presence of Special Forces in Vietnam, it was nonetheless opposed by the Diem government, particularly because the South Vietnamese government feared the training and arming of its indigenous population. Unlike the Strategic Hamlets, the CIDG program called for the Rhade and Montagnards to remain in their own villages and receive modern weaponry and defensive tactics to defend their homes and families.

By early 1962, an additional eight A-Teams were committed to Colby's project. By this time, Military Assistance Command Vietnam (MACV) was also created, with General Paul D. Harkins as its first commander.

To control all in-country Special Forces teams, MACV created the U.S. Army Special Forces (Provisional) Vietnam. By August of 1962, CIDG boasted of 200 Montagnard villages. To assist with the CIDG strike camps, MACV enlisted U.S. Navy Seabees and U.S. Army civil affairs advisors. These teams provided invaluable services and resources, and greatly contributed to the success of the CIDG strike camps. CIDG camps were also positioned to detect and impede the Viet Cong. Camps were established astride infiltration corridors and near enemy base areas, especially along the Cambodian and Laotian borders. But the camps themselves were vulnerable to enemy attack, and, despite their presence, infiltration continued. At times, border control diverted tribal units from village defense, the original heart of the CIDG program. In all, the CIDG program organized and led an army of over 50,000 hamlet militia, 10,000 strike forces soldiers, as well as thousands serving as medical and recon scouts.

However outstanding these positive moves toward security the population were, the overall effects of the CIDG program were overshadowed by the corrupt and inept Diem government. There were other opponents to the program. The Army brass did not like the Special Forces operating under the CIA. It all came to a head following Operation Switchback in 1963, when operational control of the program was transferred from the CIA to Military Assistance Command Vietnam (MACV).[13]

Meanwhile, Viet Cong actions against Diem's government increased significantly from 1959 to 1963. To guide the insurgency, a North Vietnamese political and military headquarters called the Central Office for South Vietnam (COSVN) was established. Politburo member, Nguyen Chi Thanh, the leading strategist and military commander of the Viet Cong, became its dominant figure.

Newly trained and equipped Viet Cong units began to engage Civil Guard and ARVN units. This increase in insurgent capability and the inability of Diem's forces to control the situation was highlighted by the battle of Ap Bac on January 2, 1963, when 300 Viet Cong held their

---

[13] Operation Switchback was the operational name given to the transference of responsibility for overt UW paramilitary activities from the CIA to MACV, as specified by NSA Memorandum 57.

positions in the face of seemingly overwhelming ARVN assaults and escaped intact.

In a further demonstration of the South's ineptitude, by 1963, the Viet Cong obtained some 200,000 U.S. weapons. The Viet Cong had a penchant for kidnappings and political assassinations. In terms of metrics, between 1957 and late 1963, the VC kidnapped some 23,000 and assassinated over 5,000. As Viet Cong activity increased, so did the repressive nature of Diem's regime.

Another crisis occurred on May 8, 1963, in the City of Hue. Diem's officials banned the display of Buddhist flags on Vesak, the birthday of Gautama Buddha. When Buddhists protested, the police and army opened fire, gunning down eight children and one woman. What followed was a three-month period of civil disobedience and violence. Buddhist monks and nuns held hunger strikes. The crisis took a significant turn when Buddhist monk Thich Quang Duc set himself on fire in front of the Saigon Central Market. The photograph of this event went viral around the world. The American people were shocked and began to question American support of the Diem government. Diem's sister-in-law, Madam Nhu, added insult to injury by referring to the incident as a "barbeque."

Kenedy's new U.S. ambassador, Henry Cabot Lodge, reminded Diem that continued U.S. support would be tied to governmental reform. Nevertheless, Nhu's secret police went on to arrest over 1000 Buddhists in Saigon. Lodge warned Diem that his brother's presence would no longer be tolerated in government. Arguably, this public relations disaster set the stage for Diem's eventual removal. Speaking from his own nation's long experience in Indochina, Charles de Gaulle told President Kennedy, if the United States intervened deeply in Vietnam, it would become entangled in a "bottomless military and political swamp."

In a September 2, 1963, interview, President Kennedy issued a stark warning to Deim, saying there could not be a successful outcome to the counterinsurgency unless there were fundamental changes in the South Vietnamese government. Change would come within two months when Diem and his brother Nhu were assassinated in a coup on November 2, 1963. Evidence suggests that Kennedy gave tacit approval to the military junta. Then, Vice President, Lyndon B. Johnson, later said that it was the worst mistake we ever made. A fundamental change occurred in the

American government as well when Kennedy was assassinated less than a month later on November 22, 1963.

Overall, it was the inability of the U.S. to induce needed reform and ensure the correct execution of the program that led to failure. The strategic hamlet program died along with Diem and Nhu. The legacy of the failed Strategic Hamlet Program was 4,000,000 refugees, and the economic, military, and political problems they presented. Short of withdrawal, the United States now had no choice but to take over the war. The absence of a national strategy to stem the tide of the Viet Cong insurgency, coupled with the obvious deficiencies with the South Vietnamese armed forces, set the stage for the expansion of the U.S. role in Vietnam.

Intervention (1965-1967)

By the time Lansdale arrived for his second tour in 1965 he found that the political landscape had dramatically changed. With Diem's removal, the American military was fully in charge, and it had little interest in the sort of covert operations Lansdale specialized in. Following the Gulf of Tonkin incident, Congress passed a resolution giving President Lyndon B. Johnson the authority to escalate U.S. military involvement. Johnson sent an initial wave of about 100,000 troops in March 1965, with a gradual escalation during his tenure, topping out at 500,000 by the end of his presidency.

Shifting from an advisory role to direct combat involvement, throughout 1965, troop strength increased unabated, ballooning up from 24,000 in 1964 to 185,000 in 1965. Under the war's new manager, General Westmoreland, the strategy now was attrition: kill as many of the enemy as possible. In short, Westmoreland's strategy was to chase the enemy into the jungle to save South Vietnam. To be fair, by 1965, the insurgency's strength had grown exponentially, forcing the U.S. four-star general to deal with this concentrated threat first and make pacification secondary. By draining the enemy of supplies and manpower, Westmoreland believed North Vietnam would have to give up in the face of their enormous casualties. The offensive strategy he employed was search and destroy. U.S. troops would actively comb through suspected

VC areas to locate and engage enemy forces in combat. Such missions required house-to-house and village to village sweeps.

The first large-scale search and destroy campaign was the 42-day Binh Dinh offensive which resulted in the deaths of about 2,000 enemy soldiers. However, Viet Cong guerrillas used underground bases and tunnels to hide from U.S. bombs, and most of the NVA regulars were able to slip away. The net enemy body count notwithstanding, over 100,000 civilians were forced to leave their homes. Like the Binh Dinh Offensive, search and destroy missions created many refugees. If evidence of VC collaboration was found, Westmoreland's troops would destroy the village and take any munitions they found. This tactic would be repeated numerous times throughout the war. One of the largest search and destroy missions was Operation Cedar Falls. From 8 to 26 January 1967, U.S. forces swept the VC stronghold known as the Iron Triangle. While the metrics were impressive with an estimated 720 killed and 300 captured, 6,000 Vietnamese became refugees. An optimistic Westmoreland nonetheless remarked, "The ranks of the Viet Cong are thinning. The end begins to come into view."

Lansdale was a proponent of the "hearts and minds" approach. He believed in the use of subterfuge and force, but he rejected "search and destroy" tactics. Such tactics, as Lansdale saw it, only alienated the population. Advocating for what he called "civic action," what was crucially missing for a counterinsurgency program to work, as Lansdale pointed out, was a government to which the population could feel loyalty. At any rate, confusion and indecision characterized the U.S. position. In the end, Lansdale's method was the road not taken.

A major hurdle that could not be negotiated was the Army senior leadership's predisposition to offensive employment of its forces against military targets. Failing to see the people themselves as the contested "land," and as such, the center of gravity, John Nagl argues,

This mindset prevented U.S. efforts from achieving Mao's and Galula's principal COIN fundamental of winning the popular fight. The U.S. Army's overarching goal was the destruction of enemy combatants and resources--a contradiction to the historically regarded goal of securing the support of the populace. The Army's failure to understand the dynamics of the operating environment resulted in its inability to positively influence the Vietnamese populace. This also meant that it could not accomplish the objective of defeating the enemy insurgency because the enemy had in large part become the populace.[14]

Late November 1965 also witnessed the first major battle between the United States and the NVA regulars of PAVN. From November 14-18, 1965, in what would be the first large scale helicopter air assault, 1st Battalion, 7th Cavalry Regiment, engaged a large North Vietnamese force in the Ia Drang Valley that was poised to launch an offensive on U.S. Forces in Pleiku. Notably, for the first time, B-52 Stratofortress strategic bombers played a tactical support role. Marking the first major encounter between U.S. troops and North Vietnamese regulars, the battle became a significant turning point in the war, intensifying the conflict, along with becoming a tactical blueprint for U.S. forces, using air mobility, artillery fire, and close air support to accomplish battlefield objectives.

Once the VC got a taste of this new American blueprint, they adopted what would be called "cinch the belt" tactic. As John Nagl observes,

Once the Viet Cong experienced American firepower, they quickly changed their tactics, hugging American units to preclude the use of close air support and artillery strikes. Unlike their American enemy, the Viet Cong demonstrated tactical flexibility and willingness to admit and learn from their mistakes.[15]

On March 2, 1965, seeking to cull North Vietnamese support to the burgeoning insurgency, Johnson launched Operation Rolling Thunder. This bombing campaign aimed to destroy North Vietnam's infrastructure, transportation system, industrial base, and military bases. However, according to U.S. Air Force historian Earl Tilford, "the targeting was uncoordinated, and the targets were approved randomly— even illogically." In all, the U.S. would fly 36 million sorties throughout the war. Before the war's end, the United States would drop more bombs on North Vietnam and the Ho Chi Minh Trail than it did during the entire Second World War.

Additionally, bombs often fell into the empty jungle, missing their targets. In time, the North would develop an air defense network that

---

[14] John A. Nagl, *Learning to Eat Soup with a Knife: Counterinsurgency Lessons from Malaya to Vietnam* (Westport, CT: Praeger, 2002), 37.
[15] Ibid., 123.

would shoot down hundreds of American aircraft. In terms of the human toll, the bombing strategy killed 90,000 Vietnamese, including 75,000 civilians.

Studies and Observation Group (SOG)

As another line of effort in the war against Hanoi, the innocuously named Studies and Observation Group (SOG) had taken the fight across the fence. Operating in North and South Vietnam, Laos, and Cambodia, SOG activities covered a variety from gathering intelligence, conducting raids, psyops, personnel recovery, bomb damage assessment, maritime operations, and even sent agents into the North. Organized January 24, 1964, to take over the CIA's mission, SOG was composed of members from the Special Forces, Navy SEALs, Recon Marines, Air Commandos, and the Central Intelligence Agency. SOG teams conducted thousands of missions on the Ho Chi Minh Trail, providing an estimated 75% of intelligence, and executed numerous cross-border raids to disrupt enemy safe havens.

While building a legendary reputation in its capacity as a unit specializing in reconnaissance, sabotage, and guerilla operations in the enemy's back yard, SOG also practiced impressively malicious subversion and deception operations, often with the delayed but ultimate objective of creating fresh grievance amongst the North Vietnamese populace against their government. SOG member John L. Plaster elaborates on this skillset in his book *SOG: The Secret Wars of America's Commandos in Vietnam*:

> Simpler sabotage operations with subversive intent included forging North Vietnamese currency and distributing it to cause inflation and loss of confidence (or left on the body of a dead NVA to make him look like a dead traitor and instill paranoia in his commander,) booby-trapped US equipment, rifle or mortar rounds (a program called Eldest Son) designed to kill or maim users left behind in the enemy's caches or left in magazines on their bodies that were blamed on poor manufacturing in China, soccer balls or clothing made in the colors of South Vietnamese flag designed to be confiscated from children by North Vietnamese

officials, and so-called Poison Pen Letters designed to incriminate Communist officials as US spies or collaborators, to name a few.[16]

But the truly dirty tricks were the long cons. As Plaster relates, "the key to a successful black propaganda campaign was to develop a general theme upon which to hang all sorts of individual operations."

Taking as a single example from many, Project Humidor was a multi-faceted, highly detailed operation aimed at reinforcing the North Vietnamese government's paranoia of its own citizens' loyalty, leading to mistreatment. The scam went like this: SOG would send trained Vietnamese commandos on speedboats to kidnap North Vietnamese fishermen, blindfold them, and spirit them to Cu Lao Cham Island in South Vietnamese waters. Due to the blindfolds and the much faster speed of SOG's boats, the fishermen would believe they were still in North Vietnam. The commandos would then introduce themselves as members of the Sacred Sword of the Patriot League, a fake resistance element created by SOG supposedly operating in North Vietnam out of the fake guerilla base SOG had constructed. The fishermen would then spend two weeks being indoctrinated into the fake organization, being trained in its fake doctrine as well as simple sabotage and clandestine methods of communication, all while feeding them (false) information of the extent of the organization and its operations.

At the end of their stay, the detainees would be provided gifts including a transistor radio set to a frequency on which SOG broadcast its fake "Radio Hanoi" and dropped back on the North Vietnamese coast. In a twist, the intent of the operation was not to initiate an actual resistance aimed at toppling the North Vietnamese government; an action from which SOG was restricted by higher command. The goal was for each of the new "members" of the organization to be detained by the Communist authorities, searched and interrogated. A thorough search would reveal the marked silk maps, coded messages, and other incriminating evidence sewn into the seams of the fishermen's clothing surreptitiously sewn into their clothing during their stay. Regardless of loyalty, the fishermen would break under interrogation and reveal all their "knowledge" about

---

[16] John L. Plaster, *SOG: The Secret Wars of America's Commandos in Vietnam* (New York: Simon & Schuster, 1997), 297.

the organization, which the North Vietnamese would then be obliged to misallocate time, men and resources to hunt.

Note: interestingly and despite SOG's proscription on starting resistance elements, it's very common for deceptions involving fictional units or resistances to lead to those units forming in reality. For instance, the British SAS was originally the creation of Major Dudley Wrangel Clarke, who created a fictional "Special Air Service Brigade" poised to conduct deep reconnaissance and sabotage operations against Rommel's then-supreme forces in North Africa. Only later did David Sterling use the name for his actual Special Operations unit, in much the way Clarke had written.

Project Humidor was only one of many long-term and diabolical ruses and ploys used by SOG during the conflict, many of them dove-tailing into each other or riffing off the product of the last. Commenting on the effectiveness of such deception operations, Major General Jack Singlaub, Chief of SOG, once observed:

The impact on the war of having a guerrilla force in the North destroy a truck is not related to the loss of materiel in the truck so much as it is an indication that people cannot be trusted to be loyal to the Communist cause. And that forces them to expend a great deal of extra energy, manpower and resources for internal security.[17]

By 1965, the U.S. counterinsurgency strategy in Vietnam could be described as a primarily attrition-based approach. Relying heavily on firepower and large-scale military operations, the aim was to pacify the countryside and wear the enemy down through sustained combat rather than focusing on winning the hearts and minds of the local population. After a trip to South Vietnam in November 1965, Secretary of Defense Robert S. McNamara told Westmoreland, "I don't think we have done a thing we can point to that has been effective in five years. I ask you to show me one area in this country that we have pacified."

Adding further complexity, the war managers had to contend with growing media attention. For the first time ever, American households could see battles unfold on their television screens in real-time. This significantly impacted public opinion about the conflict. As body counts

---

[17] Ibid.

grew, so did anti-war sentiment. With growing media attention, college campuses became major hubs for anti-war protests. The draft system, which disproportionately affected young men, fueled much of the anti-war sentiment. The war at home against the war had begun.

Witnessing the growing American sentiment against the war, and realizing how it could generate significant domestic opposition and political pressure, Le Duan, who by this time had become Party Secretary, advised as follows:

We can fight a protracted war where the U.S. cannot. Although the U.S. can immediately send 300,000 to 400,000 troops at once. Why must the U.S. do it? Because it's afraid of protracted war and even more afraid of the American people's opposition.

In time, Le Duan's prescience would prove true. The war manager's strength in Vietnam was directly correlated to the level of public and political will it had back home.

Sometimes the best programs are not top down driven. That was certainly the case with the Marine Corps' Combined Action Platoon (CAP) program. Launched in 1965 by a Marine battalion commander, the impetus of CAPS was in response to the need to improve security in villages. A platoon drawn from 3rd Battalion, 4th Marines, under Lieutenant Colonel William W. Taylor was the first of 114 to operate between 1965-1971. Designed to gain the trust of villagers in rural areas, and to deprive the Viet Cong of their support and bases, Talyor's CAP program was a key element of the Marine Corps' counterinsurgency strategy in Vietnam. Each CAP was made up of a 13-member Marine rifle squad, a U.S. Navy Corpsman, and a Vietnamese militia platoon. Combined Action Platoons lived in the villages, patrolled both day and night, provided medical services, helped with construction, and trained the local militiamen.

CORDS (1967-1972)

By the end of 1966, it is evident that Johnson understood that Westmoreland's attrition-based approach alone could not defeat the North and its VC surrogate. He thereby authorized the creation of the Civil Operations and Revolutionary Development Support (CORDS) program. Its objective was to gain support for the government of South Vietnam from its rural population which by this time was largely under the sway of the Viet Cong. Estimates suggest that by late 1967, over a third of the rural populace supported the insurgents.

As an advisory organization, CORDS assisted the South Vietnamese government in getting their jobs done in preparation for when U.S. assistance was no longer available. The strength of CORDS was its unified structure combined military and pacification efforts. As one of Westmoreland's three deputy commanders, Robert W. Komer was appointed as the director of CORDS on May 9, 1967. There had been military advisory teams in all of South Vietnam's 44 provinces and most of its 243 districts. But there were two separate chains of command for military and civilian pacification efforts, making it particularly difficult for the civilian-run pacification program to function. With the military and civilian architecture brought together under a single manager, for the first time in the war, the pacification program in Vietnam made headway. Johnson increased U.S. spending on pacification and economic programs. Between 1966 and 1970, these programs increased from $582 million to $1.5 billion.

CORDS achieved considerable success in supporting and protecting the population and in undermining the communist insurgents' influence and appeal. Komer left South Vietnam in 1968, going on to serve as the U.S. ambassador to Turkey. Succeeding him, William Colby's style of leadership meshed much more positively than Komer's bombastic manner that often alienated the Vietnamese. As a strong proponent of the "hearts and minds" approach, Colby believed that the way to defeat the VC and the North Vietnamese was to help make the government of South Vietnam responsive to the needs of its people.

CORDS was responsible for directing all counterinsurgency operations in Vietnam. Following Colby's lead, CORDS took on a locally run approach to COIN. Maintaining a native face, CORDS included projects that spanned every aspect of the U.S. effort including refugee assistance, economic aid, healthcare support, land reforms, democratic elections, amnesty programs, and numerous other education, and agriculture programs. Complementing these positive developments, another turning point of the war in favor of South Vietnam came in 1967 with the election of President Nguyen Van Thieu and return of some measure of governmental stability.

As a counter to the South Vietnamese government, the Viet Cong Infrastructure (VCI) constituted a shadow government, essentially acting as a clandestine political and administrative network within South Vietnam to undermine the established regime. Within this covert network, VCI operatives embedded themselves in nearly every facet of South Vietnamese society. Using the familiar carrot and stick approach, while they offered medical treatment, and justice, they also collected taxes, managed the local economy, distributed propaganda, ran reeducation programs, and drafted young people into the National Liberation Army. Operating beneath the surface, they used espionage, sabotage and terrorism and resorted to kidnappings and assassinations to gain influence over the local population. Difficult to identify and dismantle, the VCI played a significant role in the Vietnam War by providing the Viet Cong with crucial intelligence, logistical support, and popular backing within South Vietnam. This covert apparatus afforded a serious challenge to the government of South Vietnam. To fight the war on this level, the government developed a special program called Phoenix.

Integrating intelligence collection, the Phoenix Program was designed to identify and attack the Viet Cong Infrastructure. Phoenix used paramilitary teams to target undercover communist operatives in villages throughout South Vietnam. Identifying these operatives for arrest, Phoenix teams employed controversial methods, including torture, and assassination. The program was heavily criticized for its coercive methods, and for the number of neutral civilians killed. Colby consistently insisted that such tactics were not authorized by or permitted in the program. Labeled as a "civilian assassination program," Phoenix was responsible for killing some 26,000 suspected Viet Cong. Adding credence to its feared nature, the program also captured or caused the surrender of another 55,000 it targeted.

Although highly successful, the CORDS counterinsurgency project proved to be in the estimation of its first leader, Robert W. Komer, "too little, too late." Unfortunately, Komer was right. Notwithstanding the success of CORDS under Colby, in the words of Robert Komer, "Everything ran in its own compartment. I think that far more than people realize, Vietnam was a tragedy of bureaucratic inability to adapt to unconventional situations."

Military actions only have value if they contribute to the political goal. As Thomas Hammes cogently opines, "The comparative body count and control of a specific geographic area is not significant. The only issue is whether the action affects the political strengths of the combatants."[18] Hammes' comment accurately describes the effects of Hanoi's Tet Offensive of January 1968. On January 30, 1968, during the lunar new year, or Tet holiday, Viet Cong and North Vietnamese People's Army of Vietnam (PAVN) launched a surprise attack against ARVN and U.S. military and civilian command and control centers throughout South Vietnam. The attacks came against almost every major city and town and were pushed back, nearly destroying the armed wing of National Liberation Front with an estimated 50,000 enemy soldiers and guerrillas killed. Tet proved to be a political victory for the Communists, who showed they could mount major attacks against the technologically advanced U.S. forces. While it crippled the Viet Cong insurgency, in contrast it strategically damaged the political will of the United States, jolting the American public, by which commitment for the war began to wane. Hanoi's pyrrhic victory worked. Following Tet, the entire pacification program took a back seat, as the allies scrabbled to keep the communists from taking entire cities and towns. Tet had far-reaching effects. Johnson announced he would not seek reelection.

Withdrawal (1968-1973)

In July 1968, General Creighton Abrams took over command of the war in Vietnam. He changed the strategy to "Vietnamization." This strategy, which was met with a great deal of resistance from the American command structure and organizational culture, involved training and equipping South Vietnamese forces to take on a larger combat role while gradually withdrawing American troops. However, as it would soon become evident, South Vietnam could only be preserved with an unsustainable U.S. military commitment.

---

[18] Hammes, *The Sling and the Stone,* 58.

The new American president, Richard Nixon, who pledged to bring America an honorable end to the war in Vietnam, after the Tet Offensive, realized that a quick military victory in Vietnam was not possible. He focused on a strategy of "Vietnamization" to gradually withdraw American troops while building up South Vietnamese forces to fight the war on their own. Essentially acknowledging that a complete military victory was unlikely, Nixon sought a way to exit the war with some semblance of "peace with honor." In June 1969, Nixon announced the withdrawal of U.S. troops from Vietnam that would transpire in a series of steps over the course of his presidency.

Notwithstanding these overtures to peace with honor, Nixon resumed bombing North Vietnam, which had been suspended by Johnson, and also ordered secret B-52 bombings of North Vietnamese base camps in Cambodia. Additionally, Nixon authorized the invasion of Cambodia in 1970 with the goal of attacking the North Vietnamese Central Office for South Vietnam (COSVN) and eradicating the Viet Cong's cross border sanctuaries. Nixon and his Secretary of State Kissinger went to great lengths to keep the missions secret. While the U.S. strikes in Cambodia did significantly damage Viet Cong sanctuaries, they did not completely destroy them. The resilient VC relocated and rebuilt their camps and continued operations from within Cambodia throughout the remainder of the war. Nixon's war in Cambodia also caused large scale civilian casualties. This in turn destabilized the Cambodian government ultimately paving the way for the Khmer Rouge takeover.

Nixon's Vietnamization policy called for the bolstering of South Vietnamese troops to take over the war without U.S. involvement. In accordance with his promise, Nixon gradually reduced the number of U.S. troops in Vietnam, from over 500,000 in 1969 to 50,000 when he was reelected in 1972. By the time the air strikes ended in May of 1970, 3,857 B-52 sorties had been flown, dropping 108,823 tons of bombs.

General Giap's prescience was beginning to be realized: "The enemy (the West) does not possess the psychological and political means to fight a long, drawn-out war."[19] As understood by Giap, political power provided the source for military power. As the American war was now

---

[19] Bernard Fall, *The Two Vietnams: A Political and Military Analysis* (New York: Routledge, 1984), 413.

approaching its twentieth year, the U.S. military force had eroded the basis of its political support. By the end of 1972, most U.S. forces in Vietnam were withdrawn. The Paris Peace Accords in January 1973 then established a ceasefire and ended the direct U.S. military involvement in the Vietnam War. The last American combat soldier left on March 29, 1973; 2.7 million service men and women having served in Vietnam.

With the U.S. withdrawn, the war became primarily a struggle between the conventional military forces of South and North Vietnam rather than against the Viet Cong insurgency. When the fighting resumed shortly after the ceasefire in 1973, South Vietnamese forces acquitted themselves reasonably well, only to succumb to the final North Vietnamese offensive in 1975. Facing impeachment, Nixon resigned in 1974. This left the new president, Gerald Ford, to organize final evacuations. The scene in Saigon became apocalyptically chaotic as thousands of Vietnamese civilians raced for the airport trying to avoid the NVA who were closing in. With the airport shut down, hundreds of civilians descend on the U.S. Embassy, frantic to get out on one of the helicopters that transported them off roofs. Then, on April 30, 1975, Saigon fell to little resistance which was renamed Ho Chi Minh City. In the end, Communist conventional forces, not the insurgents, defeated the South Vietnamese.

Summary and Implications

By way of the five laws of irregular warfare, the following is an analysis of the American-led US-South Vietnamese counterinsurgency effort:

Political objective (s) – On the heels of the Korean War, America's strategy in Southeast Asia was one of containment, to prevent the spread of communism and to defeat the NVA and their Viet Cong insurgency. The sustained political goal was to maintain an independent non-communist South Vietnam. For their part, Hanoi followed Mao's three phases of guerrilla war – the strategic defensive, stalemate, and strategic offensive over a thirty-year period. In the end, America learned their "Mao Zedong" the hard way.

As for the Viet Cong, they were not trying to persuade Vietnamese peasants that communism was good, rather that their lives would be safer and more secure if they came to their side. Most peasants did not understand the ideological issues behind the struggle, i.e., Western capitalism versus communism. Understanding this, the Viet Cong targeted the widespread discontent among South Vietnamese peasants, particularly regarding land ownership and mistreatment by the government. These were useful levers to gain popular support.

A key political issue for the United States was that it believed that North Vietnam was merely a Sino-Soviet puppet state. However, neither the Soviet Union nor China had much control over North Vietnam. The Soviets increased their aid to North Vietnam, but their influence did not increase proportionally. Not realizing the true political association, the United States was particularly wary of escalating military action against North Vietnam fearing direct intervention from either the Soviet Union or China, provoking a wider conflict, as the U.S. had experienced in Korea.

In terms of COIN dynamics, the U.S.-led GVN failed to diminish the motives. Likewise, regarding integration and intelligence, it was not until CORDS that the U.S. had a coherent and doable COIN approach. Arguably, early on, if the U.S. had such a strategy and an effective South Vietnamese leader, then the U.S. could have been successful. However, things being as they were, as specified by John Nagl,

One of the reasons for the failure to develop a comprehensive plan for the achievement of national objectives was the inability or unwillingness of the many organizations involved in the counterinsurgency effort to coordinate their programs at least until the creation of CORDS in 1967.[20]

North Vietnam realized the weakness of the critical U.S. strategic will. As such, General Giap remarked:

The war was fought on many fronts. At that time the most important one was American public opinion... Westmoreland did not believe in human beings, he believed in numbers... he believed in weapons and material. Military power is not the decisive factor in war. Human beings! Human beings are the decisive factor.

Regarding the political will of the North Vietnamese and their VC surrogate, in terms of blood and treasure, the war had a disastrously high price tag. But they were willing to pay it. An estimated 1,100,000 North

---

[20] Nagl, *Learning to Eat Soup with a Knife*, 11.

Vietnamese and Viet Cong fighters were killed, along with 2 million civilians on both sides.

Legitimacy – Legitimacy is the degree to which a population accepts that government actions are in its interest.

As demonstrated by numerous COIN studies, the two critical variables related to the host government's impact on the success or failure of counterinsurgency operations are the host nation's governing capacity and the capability of its forces. The Diem government, followed by the general's regime, failed in regard to both. As argued by David Ahern,

> Had Diem been able to articulate a constructive political program and been able to build the administrative machinery to implement it he might have succeeded in the long term but his reliance on family and personal loyalties to project his authority together with the arbitrary quality of his governing style reduced almost to nil the prospects of American sponsored institution building and political reform. [21]

Without a doubt, one of the greatest U.S. mistakes was the American encouragement of the South Vietnamese Generals' coup against President Ngo Dinh Diem, and the assassination of both Diem and his brother Nhu. The political chaos following their assassinations led directly to an increase in communist attacks and to the introduction of large numbers of U.S. ground forces to prevent the collapse of South Vietnam. It is fair to say that despite significant U.S. support, Diem and his successors never seriously worked to develop state legitimacy.

Another significant factor contributing to the American-South Vietnamese defeat in the war was their inability to effectively "win the hearts and minds" of the Vietnamese population. This was due primarily to issues like struggling to understand Vietnamese culture, employing tactics that caused civilian casualties, and supporting a government viewed as corrupt by many South Vietnamese people.

Adaptability – While it may be said that the CORDS program was succeeding, unfortunately, by the time this change in strategy was made, U.S. public opinion had begun to shift against the war. John A. Nagl argues that the U.S. Army failed in Vietnam because it didn't become a

---

[21] Thomas L. Ahern, *CIA and the Hose of Ngo: Covert Action in Vietnam: 1954-1963* (Langley, VA: CIA Staff History, 2009), 214.

learning institution that could adapt to the insurgency threat in Vietnam. He goes on to say that after the war, the U.S. Army focused on conventional wars instead of learning from its defeat.

The overall relative weight of the various tactics employed by the U.S.-led counterinsurgency effort could be surmised by the following graph:

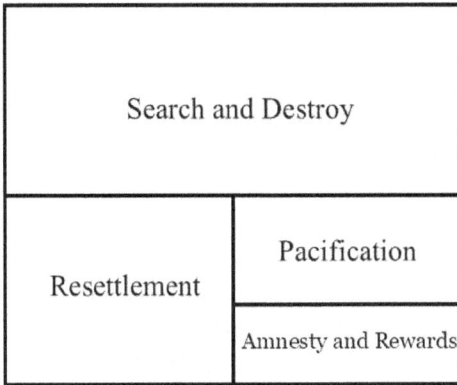

Search and Destroy	
Resettlement	Pacification
	Amnesty and Rewards

In contrast, North Vietnamese forces and the Viet Cong frequently shifted between conventional and unconventional tactics throughout the war, using guerrilla warfare, when necessary, while also engaging in larger, more traditional battles with the U.S. and South Vietnamese forces depending on the situation. Such flexibility made them a challenging adversary for the American military.

Influence – Perception is influenced by interpretation. Influence is the essence of what leadership is and does. It's fair to say that Diem failed to gain the consent of the governed. The U.S. military likewise struggled to understand and influence the Vietnamese culture and employed tactics that caused civilian casualties. Notwithstanding these setbacks, arguably by 1972, CORDS had won the "People's War" with its pacification efforts in the countryside. By that time, however, it was too late.

Native Face – Throughout the war, the Viet Cong maintained public support through regimentation and propaganda. By contrast, a common suggestion as to why the U.S. failed is that the U.S. media misrepresented the war to the public and gave small scale tactical events strategic significance through graphic reporting. While not necessarily "misrepresenting" the Vietnam War outright, the U.S. media, particularly television, played a significant role in shaping public opinion against the war by providing unfiltered coverage of the conflict's brutality and seeming lack of progress. Initially, media reporting on the war was largely positive, reflecting the government's optimistic stance on the conflict. Ultimately, media coverage did contribute to a shift in public sentiment against U.S. involvement in Vietnam.

Many argue that the Vietnam War was unwinnable and that America's involvement in Vietnam was, therefore, a tragically misguided policy. Unmistakably, the cause of the U.S. defeat is complex and multifaceted. In terms of blood and treasure, the U.S. lost more than 58,000 killed and 300,000 wounded and spent over a trillion.

After America's withdrawal from Vietnam, Harry G. Summers stated, "It is indicative of our strategic failure in Vietnam that almost a decade after our involvement the true nature of the Vietnam War is still in question."[22] They could see the problem, but they could not solve it. It's fair to say that in Vietnam we fought three wars. One in the village, one in the jungle, and the other back home.

For Further Reading:

1. *The Pentagon Papers* by Neil Sheehan.
2. *Vietnam Declassified* by Thomas L. Ahern.
3. *SOG: The Secret Wars of America's Commandos in Vietnam* by John L. Plaster.

---

[22] Harry G. Summers, *On Strategy: A Critical Analysis of the Vietnam War* (New York: Random House, 1995), 122.

9

The Cold War Next Door: America's Irregular Wars in the Western Hemisphere

"Among the things under which the earth trembles and cannot bear up are a slave who becomes a king." – The Wisdom of Agur, Proverbs 30:21

Special Forces advisor training Salvadoran troops.

On January 20, 1953, as the lifeforce of the Hukbalahap Rebellion was in its death throes, Dwight D. Eisenhower was sworn into office as the 34th President of the United States. Seven months later, on July 27, 1953, an armistice was signed, ending combat operations on the Korean Peninsula, dividing Korea into North and South at the 38th parallel. The former General and Supreme Allied Commander's hard stance on Communism was renowned. As such, Eisenhower spoke of the need to "wrest the initiative from the Kremlin, and, if possible, liberate' areas from Communist control."[1]

Central America. Courtesy AP.

Seeking to sap U.S. abilities, and relying on Communist guerrilla organizations, Stalin's goal was to foment insurrection in various war-torn countries. Echoing this sentiment, Soviet Premier Nikita Khrushchev would later pledge to support "national wars of liberation" throughout the world. The Soviet Union sponsored a significant number of national wars of liberation, including major conflicts like the Vietnam

---

[1] Richard H. Immerman, *The CIA in Guatemala: The Foreign Policy of Intervention* (Austin: University of Texas Press, 1982).

War, the Algerian War, the Angolan Civil War, various wars in African nations. The Cold War came next door when the Soviets began providing arms and training to leftist guerrilla groups in countries like Nicaragua (Sandinistas) and El Salvador (FMLN), aiming to overthrow pro-U.S. governments in the region. This chapter will discuss three Soviet-sponsored wars of liberation in Cuba, Nicaragua, and El Salvador, and highlight the PLAIN laws of political objective and adaptability.

The Cuban Revolution

The Cuban Revolution is intriguing since the country was not a war-torn candidate. Moreover, due to its proximity to the United States, along with her stabilizing role, the island nation experienced socio-economic stability; at least for many. Even so, the Cuban revolution turned, which began in 1953, turned into a communist revolution in 1960. In the end, due to the Cuban government's ineptitude, and failed U.S. foreign policies, the Communist flag was raised over the island that lies a mere ninety miles from Florida.

Like its distant cousin, the Philippines, the roots of Cuba's problems lie in the soil of Spain's colonial exploitation. The Island's earlier failed revolutions of 1865, and especially 1898, led directly to America's involvement in the Spanish-American War (see chapter 4). In the decades leading up to war with Spain, the United States had profited from its interests on the island. American investors bought up large tracts of land to harvest the sugar crop. Repressive measures against Cuban peasants leveled sentiment against Spanish colonial presence. Tipping the scales, Spanish General Valeriano Weyler began implementing a policy of reconcentration by which troublesome populations were resettled.

Meanwhile, by the time Grover Cleveland took the oval office, tensions between America and Spain were growing. Cleveland declared that the United States might intervene should Spain fail to end the crisis in Cuba. Then, on April 25, 1898, following the sinking of the *USS Maine* in Havana harbor, the United States declared war on Spain. Two days after America declared war, the Teller Amendment was added which asserted that the U.S. would not attempt to exercise hegemony over Cuba. Commodore Dewey then sailed from Hong Kong with Aguinaldo on board, fought the Battle of Manila Bay, and assisted Generals Merritt and Otis, who began their conquest of the Philippines.

"Remember the Maine" was the war cry that would marshal thousands of U.S. troops to the island. In June 1898, a force of some 17,000 landed on the shores of Cuba, and in a lightning campaign, routed Spanish troops in major battles like the Battle of San Juan Hill, resulting in a decisive American victory and the eventual surrender of Spanish forces on the island. After one of the quickest victories in history, representatives of Spain and the United States signed a peace treaty in Paris on December 10, 1898, which established the independence of Cuba, ceded Puerto Rico and Guam to the United States, and allowed the victorious power to purchase the Philippines Islands from Spain for $20 million. The war had cost the United States $250 million and 3,000 lives, of whom 90% had perished from infectious diseases.

After defeating Spain, the U.S. occupied Cuba until 1902 and forced Cuba to grant the Platt Amendment, which gave the U.S. the right to intervene in Cuban affairs and led to the creation of a U.S. naval base at Guantánamo Bay. During the 1920s and 30s, Americans flocked to Cuba. Tourism to the island boomed. U.S. companies and mob bosses profited from the growth, building casinos and lavish hotels. According to historian Robert Asprey, America "tried to reap the fruits of colonialism without accepting responsibilities for colonialism. It was the opening chapter in the USA's imperialistic fling, and no one quite knew how to interpret it."[2]

By the 1930s, the Cuban government of Ramón Grau San Martín, who gave the mob free rein, was corrupt and marked by a lack of strong centralized power and instability. Imperialistically, the U.S. supported a military coup to force the resignation of San Martín. With the sanction of U.S. ambassador Sumner Welles, and enlisted men who allied themselves with student activists in the University of Havana, Cuban Army Sergeant Fulgencio Batista led the called the Sergeants' Revolt of 1933. Batista then disposed of some 900 officers from their commands and became the head of the armed forces. Five days after the uprising,

---

[2] Robert B. Asprey, *War in the Shadows: The Guerrilla in History*. Vol. 1. 2 vols. Garden City, NY: Doubleday, 1975), 936.

Batista promoted coconspirator Ramón Grau to the role of President, while he remained the real boss. Thus began the Batista era.[3]

The next six Cuban "presidents" held office under Batista's Army dictatorship. When Batista was elected president in 1940, American commercial dominance entered a fresh new chapter. With huge U.S. loans, the Cuban economy surged forward.[4] In 1944, having filled his coffers with an estimated $20 million, Batista relinquished his dictatorship and retired to Florida. The next two presidents ruled in his style yet lacked his effectiveness. This prompted Batista to first return as a Cuban senator in 1948, then, on 10 March 1952, in another coup, he seized power from the serving President, Prio Socarrás. Batista was once again dictator over six million people. While he had been relatively progressive during his first term, in the 1950s, Batista proved far more dictatorial and indifferent to popular concerns.

The Cuba of the 1950s was every bit like that of its distant cousin the Philippines. Despite the country having one of the highest per capita incomes in Latin America, many Cubans, especially in rural areas, struggled to find full-time employment. Of a population of 6 million, Cuba had about 500,000 peasants. U.S. companies monopolized Cuba's natural resources, and the country became a haven for organized crime syndicates. These transgressions aroused the anger of many including a young lawyer named Fidel Castro.

Castro put together a group of 138 men and on July 26, 1953, in the hopes of securing more weapons, attacked an isolated army barracks in Moncada. The attack was an unmitigated disaster. The rebels who were not killed were captured, including Castro and his brother Raul. Although the attack failed, storming the Moncada Barracks catapulted him onto the Cuban national stage. At his public trial, Castro argued against Batista's dictatorship and said, "history will absolve me." Castro was sentenced to 15 years in prison. Albeit, under increasing international and domestic pressure for reforms, Batista released Castro and his fellow revolutionaries after just two years. Castro and Raul fled to Mexico. Joined there by Argentinean, future T-shirt character, Che Guevara, Castro made new plans for the revolution.

3 Ibid., 940.
4 Ibid., 941.

While in Mexico, the revolutionaries got word of Batista's increasingly oppressive measures. Batista declared martial law. The revolutionaries purchased a small yacht and with 86 men set sail for Cuba. Aware of their "invasion," Cuban troops attacked on the beach. Only a handful of them survived to escape up into the mountains. From his mountain base in eastern Cuba, Castro gathered new supporters and launched guerilla attacks on military targets. Castro allowed foreign journalists to visit. Among these was Robert Taber. In his book, The War of the Flea, Taber, channeling Castro opines:

The guerrilla fighter is primarily a propagandist, an agitator, a disseminator of the revolutionary idea, who uses the struggle itself-the actual physical conflict-as an instrument of agitation. His primary goal is to raise the level of revolutionary anticipation, and then of popular participation, to the crisis point at which the revolution becomes general throughout the country and the people in their masses carry out the final task-the destruction of the existing order and (often but not always) of the army that defends it.[5]

Castro's cause was nearly identical to the anti-colonial call to arms of Aguinaldo. His movement's main objectives were distribution of land to peasants, nationalization of public services, free elections, and large-scale education reform. Additionally, to increase his city dweller base, Castro made concessions to restore the 1940 constitution. Through such means as his pirate radio station, Castro's M26, Movement of July 26) was given international attention and increased popularity. From his mountain guerrilla base in the Sierra Maestra, Castro launched fresh attacks. Through these the rebels gained weapons and other supplies while the government lost clout. In a 1957 manifesto, Castro and his followers vowed "to put an end to a regime based on force and the violation of individual rights." Castro called for "free elections, a democratic regime, a constitutional government, and a free, democratic and just Cuba." Other protests, unrelated to Castro, broke out all over the island leading Batista to suppress freedom of assembly and expression on August 1, 1957.

In 1958, an angry and desperate Batista sent a large army of some 12,000 to the mountains to try and flush the rebels out once and for all.

---

[5] Robert Taber, *The War of the Flea* (New York: Potomac, 2002), 12.

Against Castro's 3,000 guerrillas, this seemed like overkill. However, several weighty factors foredoomed the government assault. To begin with, 7,000 of General Eulogio Cantillo's army of 12,000 including new recruits. Second, Castro has the full support of the local populace. And third, by this time, Castro's guerrillas were hardened fighters. In preparation for the government advance, Castro and Che had the mountain trails mined.

Between July 11 and August 8, Castro's guerrillas retrograded and ambushed their way up into the Sierra Maestra Mountains. Che's battle-hardened guerrillas faced off with Cantillo's green troops at Estrada Palma Sugar Mill. Albeit, at one point, it looked as though Castro's goose was cooked. Incredibly, General Cantillo agreed to a truce with Castro, and the revolutionaries would eventually escape the government noose. The rebels won a propaganda victory by escaping.

The failed offensive showcased the weakness of the Batista government. As the Cuban Army withdrew, the rebels followed, and marched westward toward Havana. As they passed through towns and villages, the revolutionaries were largely welcomed, demonstrating to the world that Batista's regime was on the brink of falling. As Castro's guerrilla army of 9,000 neared the capital, the international community persuaded Batista to flee. Batista's last official act was to hand over the country to the hapless General Cantillo. He fled the island with an amassed fortune of more than $300 million. Then, on January 1, 1959, Castro moved into Havana relatively unopposed.

Acts of revolutionary justice followed. Political parties and the free press were banned. Elections were suspended. Hundreds deemed loyal to Batista were arrested and killed. Large numbers of Cubans fled to the US. In fulfillment of his vows, Castro seized all land owned by U.S. companies, shut down the mob's casinos, and set about reforming. As few could have expected, with a rag-tag guerrilla army of less than 3,000, Castro had overturned a US-sponsored dictatorship and would eventually create a socialist republic at America's front door. A crucial factor in Batista's defeat was the decision by an exasperated U.S. government to withdraw political and military support just a few months before Cantillo's operation. This left the already demoralized Cuban army further weakened and ripe for collapse.

Castro's first order of business was a series of land reform laws. The purpose of these agrarian reforms was to abolish all large estates, limit the size of landholdings, and parcel out expropriated land to farm workers. The United States, worrisome of the new leader's left-leaning policies, cancelled the importation of Cuban sugar, a huge blow to the Cuban economy. To survive economically, Castro began a courtship with the Soviet Union. Apt to gain a toehold in the Western Hemisphere, the Soviets agreed to buy Cuban sugar in exchange for Soviet oil, and began a thirty-year program, pumping more than $29 billion into the new dictatorship. US-Cuban relations began spiraling down, until their final break in January 1961. Castro's expropriation of U.S. economic assets and his developing close links with the Soviet Union necessitated a U.S. response.

Bay of Pigs

On March 17, 1960, President Dwight D. Eisenhower directed the Central Intelligence Agency to develop a plan for the invasion of Cuba and overthrow of the Castro regime. The stated objective was "to bring about the replacement of the Castro regime with one more devoted to the true interests of the Cuban people and more acceptable to the U.S. in such a manner as to avoid any appearance of U.S. intervention."[6] Code named, *Bumpy Road*, the operation's success rested upon the premise that a bold attack by Cuban exiles would trigger a popular uprising against what Americans felt was a hated regime.

Beginning with a recruitment drive in Miami, the CIA was given $13 million to organize and train what became known as Brigade in secret camps in the Florida Everglades. By the fall of 1960, the brigade's strength rose to some 500. To facilitate a better training environment, the Cuban exile brigade was moved to a new base in the Sierra Madre in Guatemala. By November 1960, more than 1000 Cuban exiles were being trained in Guatemala and Nicaragua by Green Berets from 77th Special Forces Group (now 7th SFG). Following his election in November 1960, President John F. Kennedy reluctantly gave tacit approval for the CIA-planned clandestine invasion of Cuba to proceed.

---

[6] Michael McClintock, *Instruments of Statecraft,* 398.

The date of the invasion was set for April 15, 1961. The close proximity of the Cuban exile community in Miami to Cuba made it difficult to maintain secrecy around the invasion plans. Cuban exiles living in Florida widely discussed the invasion plans. Intercepted by Castro's spies, loose lips significantly contributed to the operation's disastrous failure. As the day of the invasion approached, Kennedy became anxious that the operation would reveal American involvement. He therefore ordered the operation to be scaled down significantly.

On the morning of April 15, 1961, six B-24 bombers with Cuban markings took off from Nicaragua. Their mission was to destroy the Cuban Air Force. In the lone sortie, only three of Castro's planes were destroyed and seven civilians were killed. Kennedy, now facing international scrutiny for the bombings, canceled the crucial air support. It has been suggested that CIA planners went along with this decision because they assumed the president would change his mind if the operation ran into trouble. Launched from Guatemala, the operation went wrong almost from the start. Alerted to the coming attack, as components of Brigade 2506 landed at the Bay of Pigs, Cuban forces easily contained the repatriation force on the beach. Remaining Cuban fighter planes easily destroyed the Brigade's ammunition-laden supply ships. As the situation deteriorated, JFK refused to authorize the U.S. military to intervene in any way, and within two days, Castro's forces marched nearly 1,000 Cuban exiles to prison. Following the failed Bay of Pigs invasion, Castro openly declared himself Marxist-Leninist, and his ties with the Soviet Union grew ever closer.

The failed invasion strengthened the position of Castro's administration. It also led to a reassessment of Cuba policy by the Kennedy administration. Kennedy established a committee under former Army Chief of Staff General Maxwell Taylor and his brother Attorney General Robert Kennedy to examine why the operation had been so ineptly bungled in every phase of execution. The assessment led Kennedy to implement a new covert program in Cuba, codenamed Operation Mongoose. Designed to do what the Bay of Pigs had failed to do, Operation Mongoose aimed to remove the Communist Castro regime through a coordinated program of political, psychological, military, sabotage, and intelligence operations. However, throughout the spring and summer of 1962, tensions came to their highest point when the

Soviets introduced ballistic missiles with nuclear warheads into Cuba. In October 1962, in the face of the mounting threat, Kennedy suspended Operation Mongoose. After being interrogated, the surviving members of Brigade 2506 were finally released after 20 months in captivity. Castro agreed to exchange the 1,113 prisoners for $53 million.

The Nicaraguan Civil War (1978-1990)

In the wake of dictator Castro taking power in Cuba, U.S. policy makers were determined to avoid having another "Cuba" in the hemisphere. U.S. policymakers implemented a counterinsurgency strategy throughout Central America to thwart the spread of communism. Washington used all means and its disposal, including CIA-sponsored regime change, as in Guatemala (1954), Ecuador (1963), Brazil (1964), Chile (1964), and Bolivia (1964). From 1936-1979, the U.S. sponsored the Somoza political family, enriching the dictatorship to the tune of $900 million. In 1972, an earthquake hit the capital and principal city of Managua. The quake killed 10,000 of the city's 400,000 residents and left another 50,000 homeless. When most of the $400 million U.S. aid package was diverted into the Somoza pocketbook, the revolutionary spirit of the Sandinista movement ignited. Founded in 1961, by the mid-70s the Sandinista National Liberation Front (FSLN) or Sandinistas could boast of a guerrilla army of some 3,000. Then, in 1975, Somoza censored the press and threatened all opponents with internment and torture. Somoza's National Guard also increased its violence against people and communities suspected of collaborating with the Sandinistas. Many of the FSLN guerrillas were killed, including its leader and founder Carlos Fonseca in 1976, who had established ties with Castro.

The revolutionary powder keg ignited on January 10, 1978, when the National Guard murdered Pedro Joaquín Chamorro, the editor of the opposition newspaper *La Prensa*. President Jimmy Carter did relatively little to stop the rebel movement, partly because his government did not foresee a Sandinista victory. Yet in 1979, after receiving aid and support from both Cuba and the Soviet Union, Sandinista guerrillas surrounded Managua, and President Anastasio Somoza Debayle resigned and subsequently fled the country. After their seizure of power, the

Sandinista government followed a similar trajectory to that of Castro in Cuba. Their policy advocated a social democracy and a mixed economy. Left-leaning as they were, the U.S. policy makers viewed the Sandinista government as undemocratic and totalitarian, falling within the same Soviet-Cuban model.

The Nicaraguan revolution threatened to worsen an already unstable and violent situation in neighboring El Salvador where the Marxist–Leninist Fuerzas Populares de Liberación Farabundo Martí (FPL) began conducting small-scale guerrilla operations.[7] Consolidating the power, the Sandinistas nationalized businesses, made good on their land reform vows, but also conducted mass executions, and oppressed the indigenous population. They quickly proved to be intolerant of dissent and brutal to those who dared speak out against them. Before long, the mismanaged economy suffered from skyrocketing inflation. Sandinistas further strengthened their ties with other Marxist revolutionary movements in other Latin American countries such as the Farabundo Martí National Liberation Front or FMLN in El Salvador.

Upon ousting the Somoza regime, Daniel Ortega became leader of the ruling multi-partisan Junta of National Reconstruction. Courting Soviet support, Ortega received $10 million in Soviet Bloc assistance in 1980. Sandinista port facilities began offloading Warsaw pact tanks, armored carriers, machine guns, rifles, grenade launchers, enabling them to field an army of some 120,000. The bulk of supplies came through the port of Corinto. In 1982, with Cuban assistance, military airfields were constructed, for potential Soviet military use.

Acting almost immediately against the new government, Enrique Bermúdez, known as Comandante 380, raised and commanded the rebels, known as contrarrevolucionarios or Contras for short. A former National Guard officer, Bermúdez created the 15th of September Legion, the primary Contra movement. This body of Contras operated out of Honduras. To the north, in Guatemala City, ex Guardsman formed the Nicaraguan Democratic Force (Fuerza Democrática Nicaragüense) or FDN. The new FDN merged with the 15th of September Legion, elevating Bermudez to overall command.

---

[7] The FPL was the oldest of the five groups that merged in 1980 to form the Frente Farabundo Martí para la Liberación Nacional (FMLN).

Upon taking office in 1981, President Ronald Reagan condemned the FSLN and authorized the CIA to begin financing, arming and training the Contras. Consisting largely of the remnants of Somoza's National Guard, the contras received about $80 million in support from 1981-1987. This aid took the form of aircraft, ships, Green Beret trainers, and CIA contract agents. Green Berets trained contras in cross border camps in Honduras and Costa Rica. The CIA sought support from Honduras and Argentina for training and resources expanding the rebels' tasks to include espionage and paramilitary operations within Nicaragua.

In April 1982, Eden Pastora split from the Sandinista regime and organized the Democratic Revolutionary Alliance (ARDE), which declared war on the Sandinista regime. Pastora based his Contra group out of Costa Rica and operated along the southern border of Nicaragua. It therefore became known as the Southern Front. Pastora alienated himself by refusing to work with Bermúdez, citing Bermúdez's former status in the Somoza regime made him politically tainted. The CIA decided to support Bermúdez and not Pastora. While the contras were quite active throughout the country, they were unable to take and hold territory within Nicaragua and declare a provisional government. Instead of achieving this aim, as John Nutter observes,

Their counter revolution descended to looting, rape, and terrorism against coffee pickers, teachers, local officials, and anyone riding in a vehicle. Dozens of Nicaraguans disappeared.[8]

The unsavory nature of contra brutality could not be disguised. An interview with a Green Beret sergeant who trained Contras was most enlightening. "Looking back, of all the training we gave them, anything on the law of armed conflict was sadly missing."

Despite the negative press, Adolpho Calero, installed by the CIA as the political leader of the Nicaraguan Democratic Force, was compared to George Washington. Other contra leaders were compared to the founding fathers. In reality, they behaved more like the terrorists that the Reagan administration was so quick to condemn. The contras became so unpopular that by December 1982 the House of Representatives passed

[8] John J. Nutter, *The CIA's Black Ops: Covert Action, Foreign Policy, and Democracy* (New York: Prometheus, 2000), 65.

the first Boland Amendment, prohibiting the CIA or the Defense Department from using any funds for the purpose of overthrowing the government of Nicaragua. That year, Congress had authorized $22 million. However, the CIA could continue to support the contras if the purpose was something other than to overthrow the government.

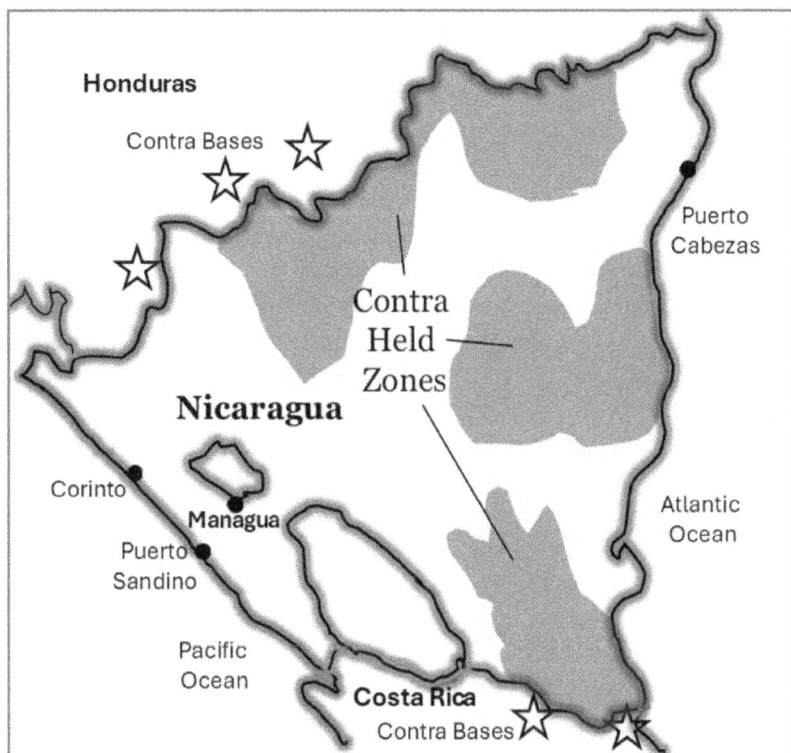

U.S. Backed Contras 1981-1987.

An additional $24 million for aid to the Contras was authorized in 1984, an amount significantly lower than what the Reagan administration wanted. Reagan directed National Security Advisor Robert McFarlane to somehow get the needed funds. McFarlane convinced Saudi Arabia to contribute $1 million per month to the Contras through a secret bank account set up by Lt. Col. Oliver North. Then, in October 1984, a second Boland Amendment took effect which prohibited any military or paramilitary support for the contras. As a result, the CIA and Department of Defense (DOD) began withdrawing personnel from Central America. However, McFarlane and North continued to provide support to the Contras.

By 1984 the Contras were hardly making headway. Frustrated by the lack of progress, Director of Central Intelligence William Casey proposed the idea of mining Nicaraguan harbors. This would forestall Soviet Bloc war shipments and destabilize the government. Limpet mines were introduced to the ports of Corinto and Puerto Sandino. The anticipation was for shipping to come to a standstill. Undeterred, shipping kept coming in, albeit it ran the gauntlet. The net result was the Nicaraguan fishing industry suffered a setback, damaging eight ships, and sinking a Japanese freighter. All told, the episode gave more unwelcome attention to the mounting American role in the conflict.

The US-sponsored insurgency continued to be funded, albeit in rather unconventional ways. Shell organizations, such as the Institute for Democracy, Education, and Assistance (IDEA) secured a State Department grant for $50,000. An additional $230,000 set aside for humanitarian aid to Nicaragua was repurposed. While North kept up his search for a more substantial funding solution, opportunity knocked. In 1985, seven American citizens in Beirut were nabbed by Hezbollah, a radical Islamic group sponsored by Iran. While the details are disputed, North developed the scheme to sell Iran weapons it needed for its ongoing war with Iraq, despite the fact that other elements in U.S. intelligence were simultaneously shipping money, equipment, and intelligence to Iraq. It was hoped that this arrangement might curry favor with the Iranians, thereby influencing them to release the hostages.

The first shipments of arms to the Iranians came from Israeli inventories with the proviso that they would be replaced. TOW missiles and Hawk anti-aircraft missiles comprised the bulk of the shipments. A sizable portion of the $48 million that Iran paid for the missiles went to the Contras. Initially, three American hostages were released. All was going well until the Iranian middleman, Manocher Ghorbanifar, was left out of the loop as another shipment was attempted. Ghorbanifar took his revenge by making the affair public.

When news of the Iran-Contra Affair broke several dozen Reagan officials, including Secretary of Defense Caspar Weinberger, National

Security Adviser Robert McFarlane, and National Security Council staff member Oliver North, were indicted on charges ranging from obstruction of justice to perjury and withholding evidence, and eleven officials were convicted. The remaining American hostages were later released.

By 1990, the Nicaraguan economy was destroyed, and Violetta Chamorro, the widow of slain Pedro Joaquín Chamorro, defeated the Sandinistas in the elections of 1990.

The Salvadoran Civil War (1980-1992)

El Salvador in the 1970s had become one of the most socioeconomically disparate nations in the world. It was basically a feudal system with a tiny ruling class. Percentage wise, 77% of the arable land belonged to .01 % of the population. At the top of this feudal pyramid were the so-called Fourteen Families, or las catorce familias in Spanish. These wealthy families controlled El Salvador's politics and economy during the country's "Coffee Republic" period from 1871 to 1927. The families, such as De Sola, Llach, Hill, and Salaverria, maintained control over El Salvador's land and coffee resource since the Spanish arrived in the early 16th century. This politically powerful oligarchy controlled the economy and military for much of the country's history. Whenever the peasants revolted, the Fourteen Families sent in the military to crush the revolt. In 1932, the Central American Socialist Party was formed leading to the uprising of peasants and indigenous people against the government. One of the leaders was Farabundo Martí. The revolt was heel-stomped when the barracks-bound military massacred 30,000 peasants. The Farabundo Martí National Liberation or FMLN, named in honor of Farabundo Martí, offered significant change to the kleptocratic and corrupt Salvadoran government.

On October 15, 1979, a military coup ousted dictator Carlos Humberto Romero. Since his 1977 inauguration, Romero had basically enacted martial law to quell the political unrest brought on by what was perceived to be a fraudulent election. The conspirators then formed the Revolutionary Junta Government. The day after the coup, the Civilian-Military Junta began an agrarian reform program. President Jimmy

Carter immediately recognized the junta's legitimacy as the government of El Salvador and seeking to foster stability, pledged support. For their part, the coffee elite, who saw their land threatened, turned to the junta to reverse the legislation. The Junta lackeys readily complied, formed death squads and paramilitaries units to attack peasants and leftist militants. In light of the atrocities, Archbishop Oscar Romero wrote a letter to Carter asking him to withhold military support from the junta. He warned that increased U.S. aid would worsen the political repression and injustice in El Salvador. During mass, on March 24, 1980, Archbishop Romero and six Jesuits were murdered. The archbishop's murder plunged the nation into a civil war. However, because of its anticommunist stance, U.S. aid to the junta continued unabated.

The situation in San Salvador became like that of Saigon in 1963 following the coup that ousted Diem. The left-leaning chairman of the junta, Majano Ramos, resigned and was later sent into exile. His resignation allowed Gutiérrez Avendaño to become commander-in-chief and chairman of the junta in May 1980. When the dust settled, former mayor of San Salvador José Napoleón Duarte became commander-in-chief and chairman of the junta in December 1980.

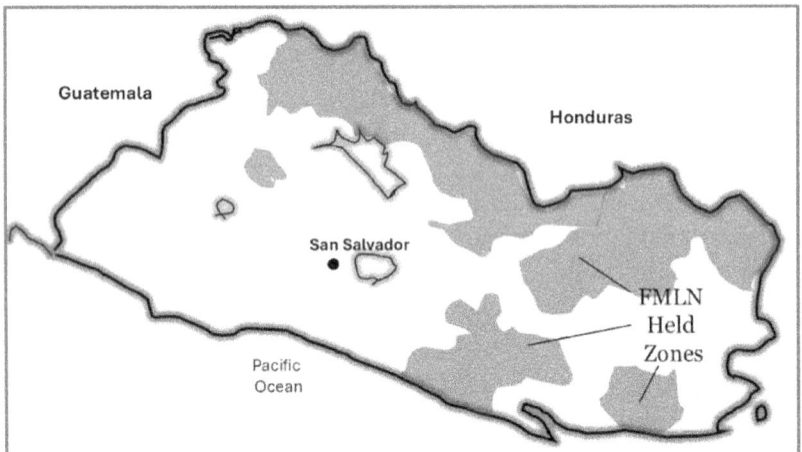

Farabundo Martí National Liberation Held Areas 1981.

212

The FMLN responded by launching an all-out attack on the government on January 10, 1981. By this time, numbering around 5,000 fighters and armed with weapons supplied by Cuba, the Sandinistas in Nicaragua, and the Soviets, the FMLN launched a three-week campaign. Armed with modern rifles, and under Castro's patronage, rebels captured radio stations in San Salvador, burned buses in the cities, and in the countryside, encouraged passengers to join the revolution. Shortly before leaving office, Carter authorized additional military aid, including rifles, grenades, and helicopters.

Seeing the FMLN as a threat to U.S. national security, Reagan noted,

San Salvador is closer to Dallas and Dallas is to Washington DC. It is at our doorstep, and it's become the stage for a bold attempt by the Soviet Union, Cuba, and Nicaragua to instill communism by force throughout the hemisphere.[9]

For Reagan, failure to confront communism in El Salvador, and Nicaragua, would lead to a domino effect in Central America. The rationale for defending the fledgling and corrupt Salvadoran junta was therefore like the rationale to defend South Vietnam in the 1950s and 60s. Like Nicaragua, U.S. goals in El Salvador were to stop the spread of communism by supporting anticommunist governments. For the next eight years, the nation would receive roughly a million a day in economic aid. The Reagan administration hoped that a counterinsurgency strategy of military and social reform could quell the rebellion and lead to a better state of the peace. As such, one of the primary U.S. goals was to improve the performance of the Salvadoran military. The counterinsurgency tactics implemented by the Salvadoran government often targeted civilians. Most notably in the El Mozote massacre of 1981, in which the Atlacatl Battalion wiped out an entire village of 978 people, including 533 children. The Salvadoran military had to be reformed, or it would lose.

To stave off collapse of the Salvadoran government, and to reverse Soviet gains across the globe, newly elected president Ronald Reagan sent an aid package of $25 million, including advisors. Green Berets of 7th Special Forces Group were dispatched to train the Salvadoran military to fight the burgeoning FMLN insurgency. Congress imposed a 55-man

---

[9] U.S. Department of State, *American Foreign Policy Current Documents: 1984* (Washington, DC: Department of State, 1984).

"force cap" on U.S. forces in El Salvador, limiting the number of trainers in the country. Under the provisions of the rules of engagement, the U.S. trainers were not allowed to engage in active combat with the Salvadoran Army. Notwithstanding this force cap, policy loopholes existed allowing for additional Special Forces soldiers to be assigned in-country. At any given time, there were between 80 and 100 "snake eaters," complimented with an additional 50 or more CIA paramilitary advisers training Salvadoran security forces. Many of these were Vietnam veterans of MACV-SOG.

Learning from its Vietnam experience, U.S. policy makers decided on a counterinsurgency strategy that balanced providing military hardware, economic aid, security guarantees, and diplomatic backing, while keeping a small footprint on the ground. Under the Green Beret's tutelage, the Salvadoran military grew from 12,000 to 55,000 men becoming a highly lethal counter-insurgency force. Despite the best U.S. efforts, Salvadoran security forces adhered to their brutal tactics. Salvadoran tactics amounted to a "drain the sea" approach. In one of its characteristic sweeps, to eliminate the insurgency's support base in the countryside, the Salvadoran Army cleared the rebel-held Chalatenango province. In a nineteen-day "cleanup" operation, women were raped and murdered, and suspected FMLN were dragged from their homes into the street and then executed.

In many respects, by being more professional or better motivated, the insurgents were individually superior to the COIN force. Receiving Cuban, North Vietnamese, and East German advisers, the FMLN were proving to be far more organized, and better funded and trained than originally anticipated. By 1982, FMLN attacks began to escalate. The FMLN attacked the Ilopango Air Force Base in San Salvador, destroying six of the Air Force's 14 helicopters, and 8 other aircraft. Sabotage increased exponentially. There were over 1,000 attacks involving explosives or arson. At its peak of strength in the early 1980's, the FMLN controlled somewhere between 25 and 33% of the country. However, despite their popular support, the insurgency never grew into a mass uprising. Moreover, while government actions contributed to substantial new grievances, in fact, in some areas they were perceived as worse than insurgents, they nonetheless were not of sufficient strength to root out the insurgents.

The year 1984 was a turning point in the war. Sánchez Cerén became the commanding General of the FMLN and in the elections, José Napoleón Duarte won the presidency. For the FMLN, the year witnessed an increase in Soviet Bloc aid. For the counterinsurgents, the relatively free and fair election signaled Duarte's stand against corruption. U.S. policy makers were beginning to see significant progress. By now, through the massive amounts of USA aid, the Salvadoran government increased its legitimacy, while the military increased its competence and improved its respect for human rights. As the U.S. support continued, the Salvadoran armed forces grew in competence and the FMLN were forced to withdraw from the cities and adjust their strategy to guerrilla type prolonged war.

From 1981-1989, the U.S. spent more than $6 billion on El Salvador. As the government increased its strength and competence, the insurgents withdrew to the countryside, and lost access to mass media. Their transition to guerrilla operations also forced the FMLN to resort to forced recruitment, further alienating the population. By 1988, the conflict was stuck in stalemate. However, with the end of the Cold War, proxy support for the FMLN dried up, helping to push the conflict toward a negotiated settlement. Then, in 1989, in what would be their last-ditch effort, the FMLN launched a major offensive on the capital city. While managing to seize control of many of San Salvador's poor neighborhoods, the insurgents were eventually forced to retreat from the city following weeks of fighting and aerial bombardment. Then, in 1992, with the signing of the Chapultepec Peace Accords in Mexico City, the FMLN disarmed and demobilized as part of a settlement and amnesty agreement.

The 12-year civil war resulted in the deaths of an estimated 75,000 people and the displacement of thousands more. The government made pledges to redress grievances, revise the constitution, and create a new civilian police force. The FMLN even took political party status through which it gained some representation in the government. In a post analysis, Brian D'Haeseleer states, "Even though the FMLN did not achieve its goal of overthrowing the government, and failed militarily, scoring a political victory."[10] Overall, the U.S. backed counterinsurgency in El Salvador remains a case study in legitimacy. It may be said that the Salvadoran security forces did poorly in combat. In fact, its only significant successes were in intimidating and massacring the civilian population.

As to using the Salvadoran counterinsurgency as a model, Brian D'Haeseleer cautions tacticians, arguing:

> A meticulous study of the conflict should give cause for concern. The U.S. experience in El Salvador confirms that outside intervention in civil wars exacerbates an already volatile situation and extends the bloodshed. U.S. aid prolonged the conflict by encouraging a military solution to defeat the FMLN, not a political or diplomatic resolution.[11]

D'Haeseleer's analysis stands up to scrutiny. Similar to the US-backed Batista regime, external support to the Salvadoran security forces, primarily in the form of military equipment and weapons, served only to prolong the conflict.

According to scholarly consensus, El Salvador "presents a fascinating and unique model for a negotiated settlement to a brutal civil war."[12]

Summary and Implications

By way of the five laws of irregular warfare, the following is an analysis of the American COIN effort:

Political objective (s) – The consistent U.S. policy objectives throughout the conflicts in Cuba, Nicaragua, and El Salvador were to oppose communism and avoid another Vietnam in Central America. As for the insurgents, the primary goal was an all too familiar need for social equality and the end of an oppressive government. The Nicaraguan and Salvadoran Civil Wars were caused by decades of military-dominated rule and social inequality. The wars ended with a peace agreement that mandated the reduction of the armed forces, the dissolution of guerrilla

---

[10] Brian D'Haeseleer, *The Salvadoran Crucible: The Failure of U.S. Counterinsurgency in El Salvador, 1979-1992* (Lawrence, KS: University of Kansas Press, 2017), 152.
[11] Ibid., 164.
[12] Russell Crandall, *America's Dirty Wars*, 334.

units, and the creation of a new civilian police force. As for Cuba, Castro's revolutionary movement began with the goal of deposing Batista's capitalist dictatorship and evolved into communist dictatorship of his own making. As for the war in Nicaragua, the partisans who ran the effort suggest that Congress interfered just as success in the conflict was coming into view.

Legitimacy – Legitimacy is the degree to which a population accepts that government actions are in its interest. It is notable that Batista's regime collapsed as a result of internal factors rather than being directly defeated by Castro's insurgency.[13] This demonstrates how the U.S. COIN efforts on Cuba served only to prop up a failed state.

Adaptability – For Castro's insurgency, terrain played a key role. The Sandinistas' unique contribution to insurgent warfare was to eliminate the need for the final conventional military offensive.[14] As Hammes observes, "[Che] did not understand that their success was based on the unique conditions of Cuba, which included a pending collapse of the Batista regime. Che paid for his mistaken theory when he tried to apply it (elsewhere)."[15]

Influence – As stated before, success in COIN requires an operational design that stabilizes the affected area by identifying with the people, while working with them to create a better state of the peace, winning their allegiance, and if possible, redress grievances that led to the insurgency.

Native Face – Limiting the number of U.S. trainers in El Salvador likewise limited the capabilities of the COIN effort. U.S. administrators' insistence of the limited footprint stymied the Green Berets efforts.

---

[13] C. Paul, et al, *Paths to Victory: Detailed Insurgency Case Studies* (Santa Monica, CA: RAND, 2013), 110.
[14] Thomas X. Hammes, *The Sling and the Stone: On War in the 21st Century* (St. Paul, MN: Zenith Press, 2004), 88
[15] Ibid, 77.

For Further Reading:

1. *The Salvadoran Crucible* by Brian D'Haeseleer.
2. *The CIA's Black Ops* by John Nutter.
3. *War in the Shadows* by Robert B. Asprey.

Part Four

Global War on Terrorism (2001-2021)

10

The Afghanistan War

"The insurgencies in Iraq and Afghanistan were not, in truth, the wars for which we were best prepared in 2001. However, they are the wars we are fighting, and they clearly are the kind of wars we must master."
– General David Petraeus

SecDef Donald Rumsfeld speaks at Kabul Air Base, Afghanistan, May 2003. Courtesy AP.

All knowledge is either discovery or recovery. The war in Afghanistan tested the U.S. military in ways not seen since Vietnam. There, the U.S. military had to relearn hard-won lessons. Amongst other things, it had to relearn how to fight against insurgents who blended in with civilian populations. At its peak, the Afghanistan War witnessed the commitment of over 50 allied nations that contributed to the war, making it one of the most impressive alliance efforts in history. In respect to its duration and complexity, the Afghanistan War remains the greatest challenge to America's military strength and warfighting prowess. This chapter highlights the PLAIN law of political objective and adaptability.

On September 11th, 2001, the Islamic terrorist organization, known as Al Qaeda, carried out coordinated attacks on the United States by hijacking four commercial airplanes and crashing them into the World Trade Center, the Pentagon, and a field in Pennsylvania. These attacks left the world stunned and angry and brought America together as a nation. Standing at ground zero, in the ruins of the World Trade Center, President George Bush vowed revenge on the attackers. During his "bullhorn" speech, someone in the crowd yelled, "I can't hear you." Bush replied, "I can hear you... the rest of the world hears you, and the people who knocked these buildings down will hear all of us soon." The assembled crowd erupted with a resounding, "USA! USA! USA!" In the wake of the September 11th Attacks which killed nearly 3,000 Americans, the United States invaded Afghanistan with the goals of killing Osama bin Laden, dismantling Al Qaeda, and denying Islamist terrorists a safe base of operations by toppling the Taliban government.

Established by Osama bin Laden in 1988, Al Qaeda's initial aim was to resist the Soviet occupation of Afghanistan. The United States could have eliminated bin Laden several times during President Bill Clinton's tenure as the U.S. had actionable intelligence and opportunity to do so. Thoroughly invested and entrenched in what became the Soviet's Vietnam, under bin Laden's leadership, and spurred on by the success of the Soviet departure, Al Qaeda widened its aperture in its fight against the occupying infidel. In the wake of the Soviet withdrawal in 1989, bin Laden's organization evolved into a global fundamentalist insurgency that sought to unite the Muslim community under a united political and religious authority. The U.S. strategic objective was to ensure that Afghanistan would never again be used to launch a terror attack.

Operation Enduring Freedom (OEF)

In a televised address to the nation, Bush announced, "On my orders the United States military has begun strikes against Al Qaeda terrorist training camps and military installations of the Taliban regime in Afghanistan." Launched as Operation Enduring Freedom (OEF), the war officially commenced the evening of October 7, 2001. The first U.S. forces on the ground were members of the CIA's Special Activities Division (SAD) which brought real-time accuracy to the developing mission. SAD teams facilitated the insertion and link-up of 5th Special Forces Group's teams with the Northern Alliance. Known as the United Islamic Front for the Salvation of Afghanistan, the Northern Alliance had been fighting to oust the Taliban since they came to power in 1996. The Northern Alliance was led by Ahmad Shah Massoud, a veteran mujahideen commander and master of guerilla warfare.[1] To fracture the Northern Alliance's diverse makeup of Tajiks, Uzbeks, and Hazaras, bin Laden targeted Massoud for assassination. This was carried out by two Al Qaeda operatives posing as journalists on September 9th, 2001. However, bin Laden's aim to sever the Northern Alliance failed. Massoud's death instead served to cement the Northern Alliance resolve under Abdul Rashid Dostum and Mohammed Fahim Khan.

Led by Colonel John Mulholland, 5th Special Forces Group formed a Joint Special Operations Task Force (JSOTF) code named Task Force Dagger. Dagger would be the first special operations force on the ground in Afghanistan to link up with and support the Northern Alliance. Dagger's mission was to support the Northern Alliance in their fight against the Taliban regime by providing training, advice, and coordination. Based out of a former Soviet airbase in Karshi-Khandabad (K2), Uzbekistan, Dagger launched its first teams into Afghanistan on the night of October 19, 2001. Mulholland decided that Dostum at Mazar-e Sharif would get Operational Detachment Alpha (ODA) 595 and ODA 555 would go to Khan at Bagram.

Mounted on horseback, the men of ODA 595 took part in their first combat engagement when Dostum's force took the village of Bishqab from the Taliban on October 21, 2001. This was followed by three more successful engagements in which Dostum's forces, along with ODA 595's coordinated airstrikes, pushed the Taliban out of the valleys. Joined by ODA 523 and General Atta's forces, the burgeoning Allied force fought their way up the Balkh Valley, pushing toward Mazar. The key battle

---

[1] Mujahideen is an Arabic term for those who engage in jihad.

came when the retreating Taliban brought all available forces forward and prepared defensive lines at Soviet-built defensive positions near the village of Bai Beche. What followed was one of the most unconventional and one-sided battles in history.

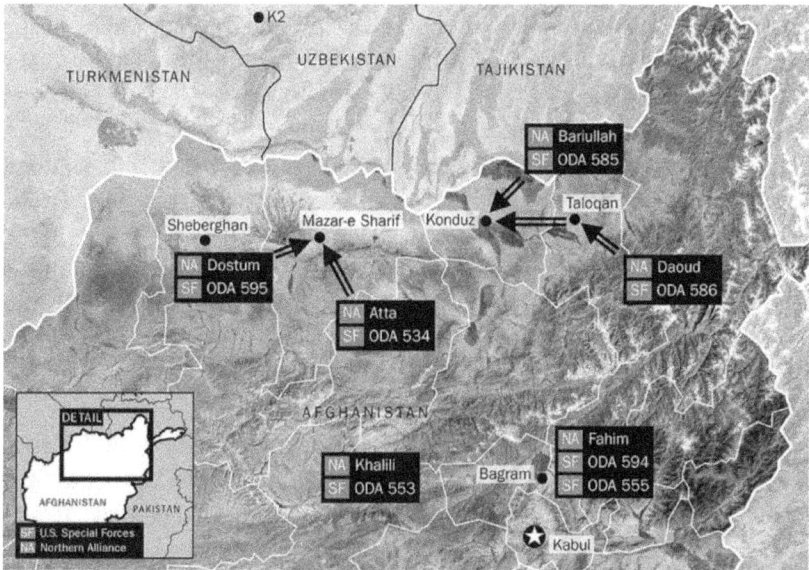

U.S. Invasion of Afghanistan 2001. Courtesy AP.

On November 5, 2001, ODA 595 coordinated a B-52 airstrike with a cavalry charge by Dostum's forces to overrun the Taliban defenses. With their center of gravity fractured, the routed Taliban retreated in disorder, unable to make any effective stand. The victory at Bai Beche opened the way for Dostum and Atta's Northern Alliance forces to capture Mazar-e Sharif which fell to them on November 10. The fall of Mazar-e Sharif effectively ended any Taliban resistance in the north.

To the east, ODA 555 linked up with the SAD team in the Panjshir Valley and were taken to a safe house to link up with Fahim Khan's Northern Alliance fighters. With Kahn's men as guides, ODA 555 arrived at a point overlooking Bagram airbase. Sprawling out in the Panjshir Valley floor were literally hundreds of Taliban and Al Qaeda targets, all under camouflage for protection from aerial observation. The Taliban defensive works, armored vehicles, guns and tanks were all densely clustered and could hardly be numbered. What followed was a systematic destruction of the Taliban. Strike aircraft and heavy bombers came in wave after wave, dropping precision guided munitions incinerating the Taliban positions and vehicles. Bolstered by the newly arrived ODA 586, the combined force annihilated Taliban forces in the Panjshir Valley, then took Kabul on November 13, 2001. There they cleared the U.S. embassy complex and set up the first American mission in Kabul since the Soviet invasion of 1979.

In the west, Taliban near Bamian engaged ODA 553 and General Khalili's forces. The Taliban surrendered the city on November 11. The ten-man ODA 553 was inserted into the Bamian Valley on November 2. As the cities fell in rapid succession to the Allied forces, Mulholland focused his forces on the last Taliban stronghold of Konduz. ODA 586 deployed there to assist Khan's forces. Following an 11-day airstrike, the demoralized Taliban in Konduz surrendered to Khan's men on November 23.

In the south, ODA 574 linked up with Pashtun leader Hamid Karzai and his fighters with the mission to oust the Taliban from Kandahar. Key to that aim was taking the town of Tarin Kowt. In response to the approach of the Allied force, the inhabitants of the town of Tarin Kowt revolted and expelled their Taliban administrators. Not giving up without a fight, the Taliban marshaled a force of some 500 men to retake the town. ODA 574 and Karzai's fighters deployed in front of the town to block their advance. Coordinated airstrikes attacked a Taliban column of about 80 vehicles, destroying nearly 30, with an approximate 150-200 enemy dead. In the seven-hour battle, using superior technology and guts, ODA 574 and their mujahadeen allies managed to halt the Taliban advance and drive them away from the town. This would prove to be an important victory for Karzai, who used it to recruit more men to his fledgling guerrilla band. Eventually, his force would grow to a peak of around 800 men. From there, ODA 574 and Karzai's force advanced on Kandahar.

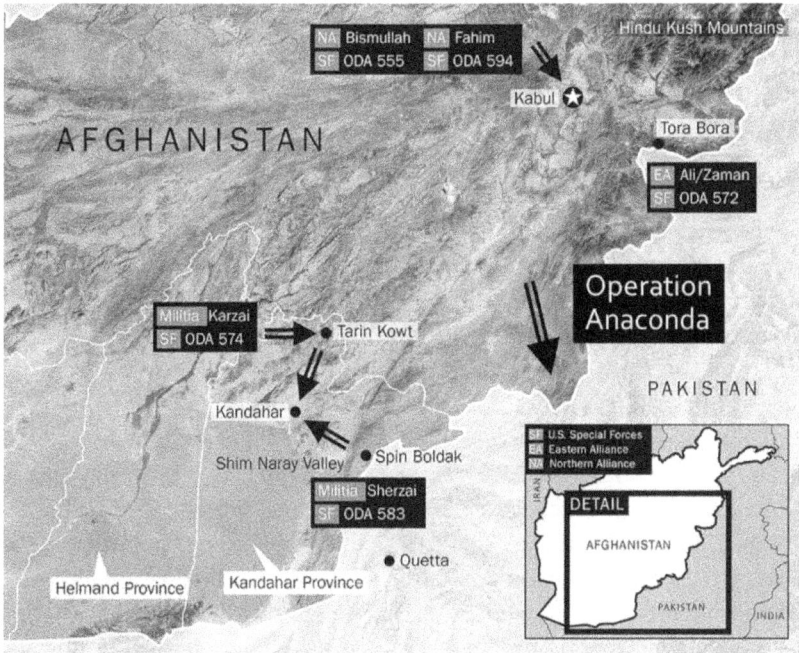

U.S. Invasion of Afghanistan 2001. Courtesy AP.

Not everything went so well in the U.S.-led invasion. Many of the Taliban and Al Qaeda fighters who surrendered in Mazar-e Sharif were detained as prisoners in a 19th century mud fortress known as Qala-e-Gangi just west of the city. A revolt broke out on November 25, 2001. The fortress had previously been used by Dostum's forces as a weapons and ammunition depot. The 300-500 prisoners at Qala-e-Gangi expected to be freed after surrendering, were improperly searched by the NA. Some carried weapons and grenades into the prison. The prisoners managed to break into the fort's armory and arm themselves with an array of small arms and RPGs. However, Dostum's men managed to contain the revolt until a quick reaction force arrived. The fighting continued for four days while 5th Group Green Berets called in airstrike after airstrike. The siege ended on November 27 after the last remnants of resistance were crushed. Some 86 prisoners survived the ordeal. General Dostum later claimed responsibility for the security failure.

Following their victory at Tarin Kowt, Karzai and ODA 574 moved on Kandahar. Karzai gathered local Pashtun fighters as they drew nearer. For two days, ODA 574 called in precision airstrikes on dug-in Taliban positions on the approaches to the city. Tragedy struck on December 5, 2001, when a 2000-pound joint direct attack munition (JDAM) fell short of its intended target, killing three members of 574, about 20 members of Karzai's force, and wounded five ODA members, including Karzai. A casualty evacuation force composed of ODB 570, ODA 524, along with a Marine CH-53, were deployed to evacuate the wounded. Five minutes after the blast, Hamid Karzai received a satellite phone call informing him that he had been selected to lead Afghanistan's new interim government.

Following Dagger's overwhelming success, Al Qaeda fighters and the remnants of Taliban fled east to a region known as Tora Bora in the White Mountains which border Pakistan. Coalition human intelligence (HUMINT) suggested that significant numbers of enemy targets were congregating there. Tora Bora, south of Jalalabad, meaning "black cave," offered Al Qaeda a system of caves and defenses. The cave network had been developed by the mujahadeen during the war against the Soviets. Beginning on December 3, 2001, about 20 SAD operatives, along with ODAs 563 and 572, were inserted in Jalalabad to begin an operation against Al Qaeda forces in the Tora Bora caves. This task force would advise mujahadeen under the control of two warlords, Hazrat Ali and Mohammed Zaman. Some 2,500 to 3,000 militiamen, paid for by the CIA, were recruited for the operation to isolate and destroy Al Qaeda forces in the Tora Bora caves.

As SOF units led Afghan militias with varied success, U.S. Air Force combat controllers or CCTs called in hundreds of airstrikes on Al Qaeda defensive positions. In fact, from the end of 2001, CCTs would call in 688,000 pounds of bombs on Afghanistan. With TF Dagger stretched thin, operators from 1st SFOD-Delta were brought in to bolster the attack. Small teams attached themselves to the militia, taking over tactical command from the CIA. Continuing at a steady advance through the difficult terrain, and backed by air strikes, the combined force suffocated the entrapped Al Qaeda forces. However, on December 12, Mohammed Zaman (incredibly) negotiated a truce with Al Qaeda, giving them time to surrender their weapons and escape over the mountains into Pakistan.[2] Among those who escaped was bin Laden. Little did America know at that point they faced a long war, one in which over 800,000 would serve. Allowing Al Qaeda to flee into Pakistan would prove to be a strategic blunder. It allowed the Taliban and Al Qaeda to lick its wounds and resurge from this cross border safe haven. By 2006, the Taliban would gain sufficient force to control parts of southern Afghanistan.

Notwithstanding this failure, TF Dagger managed to wrench Afghanistan from the control of the Taliban in only 49 days. The key to success in the early combat operations in Afghanistan was due to the by-with-through operational philosophy of the Special Forces ODAs. Working with their Afghan allies, using precision airstrikes, the Special Forces soldiers drove the Taliban into the mountain valleys of eastern Afghanistan. This initial campaign ended with the establishment of Combined Joint Special Operations Task Force-Afghanistan (CJSOTF-A) and the formation of a duly elected Afghan government.

Operation Anaconda, which took place in early March 2002, utilized the combined resources of the U.S. military and CIA. Working with allied Afghan military forces, and North Atlantic Treaty Organization (NATO) forces, the operation attempted to destroy the remnants of Al Qaeda and Taliban forces. SF intelligence sources believed surviving Al Qaeda forces, upwards of 4,000 in size, to be gathering in the Shahi-Kot Valley, some 60 miles south of Gardez. Anaconda was to be the first large-scale battle in the U.S. invasion of Afghanistan. Additionally, it was the first operation in the Afghanistan Theater to involve a large number of U.S. conventional forces participating in direct combat activities.

The Rise of the Taliban

Due to its strategic location, a gateway between Asia and Europe, Afghanistan has a long history of domination by foreign conquerors. King Darius extended the Persian Empire into what is now Afghanistan around 522 BC. As part of his war against Persia, Alexander the Great likewise extended his Macedonian empire into Afghanistan. After the

---

[2] Gary Berntsen, *Jawbreaker: The Attack on Bin Laden and Al Qaeda* (New York: Penguin, 2006), 123.

rise of Islam, the 11th century conqueror Mahmud of Ghazni sought to create an empire from Iran to India and incorporated Afghanistan into his realm. Genghis Khan then came in the 13th century. Various other conquerors came and went. The modern Afghan state took shape in the 1740s with the Duranni Empire under Pashtun ruler Ahmad Shah Durrani. Ahmad Shah united all the Pashtun tribes.

As part of a geopolitical chess match called the "great game," looking to protect its Indian empire from Russia, British troops invaded in 1838. For the next hundred years Britain attempted to annex Afghanistan, resulting in a series of British-Afghan Wars (1838-42, 1878-80, 1919-21). Beleaguered by World War I, Great Britain ended the Third British-Afghan War (1919-21). Afghanistan then became an independent nation under Amir Amanullah Khan who proclaimed himself king in 1926. Despite Khan's three-year attempts at modernization, his policies fail and under immense pressure he abdicates and flees Afghanistan in 1929. Pashtun Mohammad Zahir Shah then became king. In an effort to bolster the economy and bring modernization, Shah's cousin, General Mohammed Daoud Khan, becomes prime minister and looks to the Soviet Union for assistance. Shah also introduces a number of social reforms including allowing women a more public presence. Then in 1965, the communist party came to Kabul. Led by Babrak Karmal and Nur Mohammad Taraki, the party gained seats in the government. Following a coup under General Khani, King Shah is overthrown, and the Republic of Afghanistan is established with firm ties to the USSR.

Khan proposed a new constitution that granted women rights and worked to modernize the largely communist state. He also cracked down on opponents, forcing many suspected of not supporting him out of the government. The unpopular Khan was then deposed in 1978 coup orchestrated by Taraki, who becomes president. The new republic is plagued by internal strife. Taraki opposes another prominent communist leader Hafizullah Amin and is killed. Under the auspices of buttressing the fledgling communist state, Moscow invaded Afghanistan by a rapid seizure of major cities, airstrips, and existing bases. In the lightning coup, they executed Hafizullah Amin, foisted Babrak Karmal in his place, then fanned out from the cities and bases in motorized columns to garrison the Afghan landscape. Droves of Afghans mujahideen swelled the ranks of several insurgencies sworn to engage in jihad against the communist government and purge the infidels from the land. The mountainous terrain and the enemy were entirely different from what the Red Army had prepared for. Soon the Soviets' glorious invasion devolved into a long counter guerrilla morass that lasted nearly ten years. Before it would all end, the Soviet occupation would exact a heavy toll of 1.5 million Afghan dead, 5 million wounded, and a further 5 million refugees, out of a 1989 population of 15.5 million. On February 15, 1989, the last Russian soldier left Afghanistan.

Following the Soviet withdrawal, mujahadeen continued their resistance against the Soviet-backed regime of Mohammad Najibullah. Added by the Soviets and the hardware they left behind, some 1500 tanks, 800 armored personnel carriers, 4500 artillery pieces, 120 modern fighter-bombers, and 14 attack helicopters, Najibullah's regime managed to stay in power. This period would come to be known as the First Afghan Civil War (1989-1992). After the Soviet Union was dissolved in late 1991, Russian fuel shipments and support to Afghanistan ceased. By 1992 the Afghan regime of President Mohammad Najibullah began to collapse. Then, in April 1992, the forces of Ahmad Shah Masood took Kabul. Left in yet another vacuous situation, mujahadeen groups under various warlords fought over the future of Afghanistan. The strife among these internally warring factions threatened another civil war. The precarious peace of the new Islamic state was maintained with the placing of Persian-speaking Tajik Burhanuddin Rabbani as president. All recognized the new government except Gulbuddin Hekmatyar who received operational, financial, and military support from Pakistan.

To the south of the battered capital, in Helmand Province, outside of governmental control developed a political-religious force that became known as the Taliban. Led by Mullah Mohammed Omar, the Taliban, Pashtun for "students," began with 50-armed madrassah students. Omar fought against the Soviet occupation and was known for destroying Russian tanks with his RPG-7 weapon. Omar reportedly received a vision of a woman who told him that he should rise up and end the chaos in Afghanistan and that Allah would guide him to victory.

Backed by Pakistan, the Sunni Muslim Taliban were of the Pashtun people and constituted around 13% of the total population of Pakistan. Omar decided to stick with Pakistan and in doing so eventually made an enemy of Rabbani. In September 1994, some 200 Taliban fighters attacked the border town of Spin Buldak. By the end of 1994, the Taliban had occupied almost the entire Kandahar province. By mid-1995, they had captured most of southern Afghanistan. After taking Kabul in 1996, to solidify his position, he declared himself the leader of the new Islamic emirate of Afghanistan. In the northern districts, warlords Ahmad Shah Massoud and Abdul Rashid Dostum opposed his leadership and vision. Becoming head of the Northern Alliance, Massoud, would be killed September 9, 2001, by Al Qaeda assassins posing as journalists.

Afghanistan's Tribes. Courtesy AP.

Under the Taliban, foreign ideas were dangerous. In addition to Afghan's historical artifacts that they destroyed, music, television, art, photography, and the use of the internet were banned. The Taliban imposed the strictest interpretation of Islam and Sharia law. There was no tolerance for any form of dissent. Anyone caught violating the will of the Taliban were accused of violating the will of Allah and severely punished for even the most minor of offenses. Everyone under the

Taliban spent their lives having to prove themselves every minute of every day for fear of reprisal. It was especially difficult for women. The Taliban instigated what could be described as gender apartheid. Girls' schools were shut down. Professions of any kind were barred for women. The female population found itself effectively under house arrest unable to leave their homes without the accompaniments of a male family member. Women were forced to conceal themselves from the public at all times; their entire bodies being shrouded in the burqa with only a small metal mesh over the eyes for them to see.

Life under the Taliban was thrust into the world spotlight on February 6, 2000, when nine Afghan men hijacked a Boeing 727 aircraft from Kabul to Mazar-i-Sharif Airport and forced it to land at Stansted Airport in Essex, England. The hijacking ended on February 10. After a lengthy negotiation process, 180 passengers and crew were freed when the Afghan men surrendered. After their arrest, during their questioning by the British, the men explained that they had taken this dramatic and dangerous step in order to escape the Taliban regime. It was later agreed by a British court that extraditing them back to Afghanistan was a violation of their human rights as it would certainly be akin to torture and death.

In its attempt to consolidate control over northern and western Afghanistan, the Taliban committed ethno-religious atrocities. For example, upon taking Mazar-i-Sharif in 1998, some 4,000 civilians were executed. The majority of those killed were Hazaras, a Persian-speaking Shi'a ethnic group, targeted for their religious identity. As Omar was consolidating his grip on Afghanistan, he summoned Osama bin Laden to him. The two developed a close relationship while fighting against the Soviet occupation in the 1980s. Omar is reported to have asked bin Laden to tone down his Anti-U.S. jihadist rhetoric. Bin Laden at any rate chose to ignore him. That Omar permitted this slight is understandable. Bin Laden is thought to have at least partially funded the Taliban's takeover of Afghanistan.

After the 1998 United States embassy bombings, President Clinton ordered cruise missile attacks against bin Laden's training camps within Afghanistan and demanded that bin Laden be extradited to stand trial. Omar would not hand him over, citing Pashtunwali tribal customs that require a host to protect guests. As many have suggested, Omar may have felt that if he bowed to the United States by handing over bin Laden, then the U.S. would try to further influence Afghanistan and attempt to meddle in its religious matters. The Taliban would continue to provide a safe haven for bin Laden and his Al Qaeda network in the years leading up to the 9-11 attacks which Osama bin Laden orchestrated from Afghanistan. Omar's refusal to hand over bin Laden precipitated the U.S. invasion of 2001. On the eve of the U.S. invasion on October 7, 2001, the Taliban had an estimated core strength of 45,000 fighters and controlled about 90% of the country.

The Karzai Regime

Following Tora Bora, with what was left of the Taliban in full flight, Mullah Omar, who had survived a drone attack, likewise disappeared into the south. U.S. planners were astounded by the rapid collapse of the Taliban. They had anticipated a long-drawn-out counterinsurgency. The scramble for post-invasion options was reminiscent of the U.S. capture of Manilla a hundred years before. The United States was now responsible to install a new government and provide Security force assistance to keep the peace while the new government found its footing. With the risk of the war-ravaged nation falling back into the hands of rival warlords, President Bush directed Secretary of State Colin Powell to organize an international conference. The conference which would be held in Bonn, Germany invited nearly everyone accept the Taliban.

Following the Bonn Agreement of December 5, 2001, Hamid Karzai was chosen to lead the new sixth-month interim government of Afghanistan. As a framework for building the new state, the Bonn Agreement established a roadmap for a new constitution by 2004, judiciary system, free elections by 2004, and a centralized security sector. It also included provisions for the protection of women and minority rights. While the agreement was based on previous successful post-conflict models, it was not well-suited to Afghanistan's cultural and political history.

Foremost, the agreement created internal conflict and political instability because it did not adequately distribute power between the Northern Alliance and the Pashtun faction. Namely, it largely favored the Northern Alliance, which had been instrumental in the fight against the

Taliban. This left the Pashtun community feeling marginalized, fueling resentment. It seems the biggest mistake of all was to not invite the Taliban to the Bonn Conference. It's fair to say the U.S. underestimated the ability of the Taliban to resurge. With failure thus baked in, in time, the sum of these issues would plague the future of Afghanistan. Be that as it may, Karzai received a resounding endorsement of his presidency in June 2002, after 1,550 delegates from Afghanistan's 364 districts elected him as interim leader. Karzai then went on to choose the members of his government who would serve until 2004, when the government was scheduled to organize elections.

Despite this initial support of the Karzai government, the terms of the Bonn agreement created a highly centralized state. Karzai was supposed to run the whole country from Kabul. According to Erik Schnotala:

By investing so much power in the executive, Afghanistan's political system raised the stakes for political competition and reignited long-running tensions between an urban elite eager to modernize and conservative rural populations distrustful of central governance.[3]

This executive arrangement, which in time proved to be unattainable, gave the president too much formal power and ignored Afghan traditions of local autonomy.

Turning next to security and defense, the U.S. and NATO established the International Security Assistance force or ISAF. The primary mission of ISAF was to train and equip the Afghan National Security Forces (ANSF) which would consist of an Afghan National Army (ANA) and the Afghan National Police (ANP). By design this force would combat the remaining Taliban insurgents in the country and help secure the new government to ensure Afghanistan would not be a haven for terrorists. The Afghan National Security Forces (ANSF) had a total growth target of 352,000 troops. It was hoped that the ANA would have 70,000 troops by the end of 2002. The government would have to settle for 10,000. To facilitate the creation of a new Afghan National Army, and undermine his political rivals, Karzai disarmed the tribal militias.

As first steps in Bush's redevelopment plan for Afghanistan, the U.S. military created Provision Reconstruction Teams or PRTs. Congress had

---

[3] Eric Schnotala, *Why the Afghan Government Collapsed*, 2022.

appropriated $4.5 billion in humanitarian and reconstruction assistance to Afghanistan. Coordinating with NATO and non-governmental organizations, these civil affairs teams worked to stabilize the nation and expand the authority of the Kabul government. With money pouring in, projects abounded. Included in those were those meant for Afghanistan's marginalized women and girls. Once again women found room in the workplace. Girls received education. In August, the first civilian passenger aircraft to fly non-stop from Europe to Afghanistan since the Soviet invasion landed in Kabul. In a further move toward security, various militia were disarmed and recruited in the ANA. Likewise, various warlords surrendered tanks and artillery to the Afghan Army. Notwithstanding these positive steps, by the close of 2003, many Afghans viewed the International Security Assistance Force as an occupation force. Moreover, Karzai asked Washington for increased aid, estimating that $15–$20 billion would be required to rebuild the Afghani economy.

Some ominous signs occurred in the summer of 2002 showing that the road ahead would be deadly and difficult. On July 1st, 2002, U.S. planes bombed a wedding north of Kandahar city killing nearly 30 people, including children. U.S. forces blamed the attack on the guests firing AK47 rifles into the air in celebration, mistaking this for an attack. This only made ISAF's job all the more difficult, particularly when it came to winning over the hearts and minds of the local population. Five days later, terrorism returned to Kabul. The Taliban struck back at Karzai's government on July 6th, 2002, when Karzai's vice president Abdul Kadir was gunned down. This was followed up with a car bomb on September 5, 2002, killing 30 when it exploded in front of the Ministry of Information. In addition to these troublesome acts, 2002 was permeated by hundreds of isolated attacks on ISAF bases, including Taliban ambushes on government convoys. Coalition checkpoints became targets for car bombs. Complicating security matters was the fact that the enormous border with Pakistan seemed to be completely open, enabling the Taliban to cross over at will.

The United States however viewed these as the acts of the remnants of the Taliban, and that given time, the situation would stabilize. Washington at this time was already planning its next major campaign. In early 2002, Pentagon planners began shifting military and intelligence resources away from Afghanistan in the direction of Iraq, which was named as part of the axis of evil and a chief threat in the "war on terror." In large measure, the Iraq War would make the Afghan War a side show. But those who believed the Taliban were finished were about to get a rude awakening.

Being in no position to fight back, the Taliban went underground and began to build up their support infrastructure. Helped in large measure by Afghanistan's lucrative opium trade, the Taliban gained arms and new fighters. Their recruitment experienced a gratuitous boost when forces of the United States and Great Britain began their controversial invasion of Iraq on March 20, 2003. The Bush administration's decision to invade Iraq diverted military and intelligence resources from Afghanistan and allowed the Taliban to regroup and resume the war. October 2004 witnessed Afghanistan's first democratic elections since the fall of the Taliban. Hamid Karzai was elected president with 55% of the vote. NATO, having taken over the ISAF mission, expanded its mission in Afghanistan, establishing four new Provincial Reconstruction Teams in the north of the country, and increasing its troop presence from 5,000 to around 10,000. This uptick followed an increase in Taliban attacks which were becoming nuanced with new tactics such as suicide bombings and improvised explosive devices (IEDs). From their sanctuary in Pakistan, the Taliban brought in Iraqi insurgents who introduced IEDs which were previously uncommon in Afghanistan. Between January 2005 and August 2006, there were 64 suicide attacks. Then, on June 28, 2005, three Navy SEALs were killed in combat operations in support of Operation Red Wings. While attempting an emergency extraction of the Red Wing team that same day, eight Navy SEALs and eight Army Night Stalkers were killed when their helicopter was shot down by enemy fire in the Tangi Valley. The Taliban were surging back.

With the United States thoroughly engrossed in Iraq, and as it plodded along in Afghanistan into 2004 and 2005, it became apparent that the conflicts didn't seem to have any clearly defined objectives which the U.S. and their coalition partners could aspire to, and the war was beginning to look like a war without end. Additional NATO ground forces started arriving. The prevailing mindset was that it was going to be a peacekeeping mission. Everyone believed that the war was over. That mindset quickly backfired when several of their patrols were savagely attacked resulting in numerous casualties. By July 2006, NATO's International Security Force (ISAF) expanded its operations to the south and east, raising troop levels from about 10,000 to around 65,000. ISAF troops were now represented by 42 countries. From April to July, a combined Dutch Australian force cleared the Chora Valley of Taliban. The Taliban struck back, launching a series of attacks against international troops, including suicide bombings and raids. On September 30, 2006, a suicide bomber detonated himself near the Ministry of Interior office, killing 12 civilians and injuring another 42.

The Taliban Offensive of 2006

By 2006, the Taliban's core strength rose to an estimated 60,000. This was a significant rebound after its near annihilation in 2001-2002. The Taliban commander, Mullah Dadullah Lang, was intent on capturing the provincial capitals of Lashkhar Gar and Kandahar City and to use these cities as a launching point to conquer the remainder of Afghanistan. In March 2006, the 82nd Airborne Division pulled out of country creating a vacuum. Mustering a force of about 4,000 fighters, the Taliban moved into Zharey in April and attacked police forces, resulting in several clashes. Then, in Helmand, where the Afghan government's hold was tenuous, the Taliban moved to recover an area known for growing poppy, a major source of income. Likewise, control over Sangin was strategically important for the Taliban because it facilitated the transportation of poppy from north to south. The Taliban offensive likewise witnessed spikes in the number of suicide attacks. Vehicle borne IED or VBIED attacks blossomed from 27 in 2005 to 139 in 2006, while remotely detonated bombings more than doubled, to 1,677. Notwithstanding these gains, when the Taliban moved into Panjwayi in the spring to gain support from the Noorzai and Ghilzai tribes, the tribes told them to leave.

To quell the Taliban resurgence, NATO launched Operation Mountain Thrust, a joint U.S., Canadian, British, and Dutch operation. Establishing a presence in Helmand, Coalition forces created 12 new combat outposts and expanded several others. Seeking improved governance, Coalition forces moved to construct new roads, district centers, and schools, committing over $500 million to projects. Despite Coalition best efforts, the operation did not manage to wrest the Taliban completely from Helmand. In what would be an action repeated several times, Coalition forces managed to clear several remote districts and then pulled back.

The Battle of Kaika

To the northeast, in the Panjwai Valley, a mixed force of 65 Americans and Afghans under the command of Captain Sheffield Ford fought their way out of encirclement and destruction in a savage battle that was protracted over a three-day period from June 23-25, 2006. Seventh Special Forces Group's ODA 765 was on a patrol to get to know the local population and build rapport with their Afghan counterparts. As it turned out, ODA 765 was soon to be the only U.S. ground force left in Kandahar. As part of Operation Mountain Thrust, ODA 765 received authorization to conduct what was named Operation Kaika, a joint operation between U.S. Special Forces (SF) and Afghan National Army soldiers (17 U.S. and 48 Afghans). Ford's operational design of Kaika was first to kill or capture Haji Dost Muhammad, a Taliban improvised explosive device (IED) facilitator believed to be in the area, to be followed by a systematic sweep of the Panjwai Valley, clearing it of Taliban and restoring order. Haji Dost Muhammad was responsible for a growing number of NATO casualties. The plan called for 765 and their Afghan counterparts to establish a footprint in the village of Pashmul to be followed days later by Assullah Kalid's paramilitary force of Afghan National Police (ANP) who would then link up with 765 and begin the sweep. Ford's intent was to establish a patrol base in the abandoned village of Pashmul, approximately 12 miles southwest of Kandahar City.

On the evening of June 22, Ford's element drove into the village of Pashmul having navigated its way 12 miles southwest along the alternate route from Firebase Gecko. The Team searched the village but found it abandoned. The Taliban had already moved into the area. Ford and his Team Sergeant, Master Sergeant Thomas D. Maholic, decided to establish a patrol base and begin a rest plan. At the time, unbeknownst to them, a large force of Taliban was loitering just outside the village poised to attack. As the last light of the setting sun shined upon the nearby hills, the quietness of dusk was shattered by an eruption of small arms and RPG fire from three different directions. The heavy volume of fire was sustained for over 20 minutes. Mortar rounds soon followed, walking in toward the center of the patrol base indicating a level of expertise the team had not yet encountered. By the heavy volume of fire being received, Ford knew his team was "outnumbered, outgunned, and surrounded."

Ford immediately reported troops in contact to Battalion HQ (King 65) and requested emergency close air support (CAS). The American Afghan force took cover and fought back with everything in their arsenal: M2 .50 caliber machine guns, M240 machineguns, AK-47s, M4s, AT-4s, etc. However, it soon became apparent to 765 that the Taliban force was much more disciplined and organized than the enemy they had encountered on previous engagements. Following the initial maelstrom of fire, the Taliban commander directed three separate elements to begin alternating their fires to support his assault elements who were bounding forward. The Taliban managed to scale the walls of the team's perimeter. With violence of action, Ford and the ANP managed to repel the attack, pouring fire into the advancing Taliban. About 50 minutes into the battle, an A-10 Thunderbolt aircraft arrived on station and disgorged its 500-pound bombs, eliminating some 40 Taliban fighters.

The fighting continued into the next day. The team engaged the enemy with a 60-mm mortar. Despite 765's devastatingly accurate defensive fires and CAS, the Taliban continued to maneuver onto the American Afghan position. Two additional A-10s dropped more 500-pound bombs, destroying enemy reinforcements. At one point, the Team's medic, Sergeant First Class Brendan O'Connor, crawled hundreds of meters under withering fire to save a wounded teammate. Close air support hammered the Taliban enough to allow O'Connor and his wounded to get inside the perimeter. If it wasn't for the tenacity of that element's command team, coupled with self-sacrificing courage, it would've surely been annihilated. Still the Taliban pressed the attack. Ford took stock of his situation: They would get no quick reaction force or QRF. He was running low on ammunition, food and water. The ANA resolve was teetering on the brink of collapse, and his CAS was running low on fuel. Ford realized at this point he had to get his men out of there. He had a small window of opportunity as the Taliban guns had fallen silent. They appeared to be reconsolidating.

Ford notified higher command of his intent to return to Firebase Gecko. With the next period of darkness coming, Ford would use the aircraft to paint an infrared corridor out for them. Time was of the essence. Under enemy fire, Ford had his force mount up and navigated their way through the narrow wall-brimmed dirt roads. Having suffered two American KIA, including the Team Sergeant, ODA 765 drove out of Pashmul while an AH-64 covered their movement home. Through sheer determination and guts, the Team extricated themselves from utter annihilation, removing 120 Taliban fighters from the battlefield. Members of the Team were awarded four silver stars, five bronze stars with "V" device, and Sergeant First Class Brendan O'Connor was awarded the distinguished service cross.

The Taliban Insurgency

Contrary to what many have believed, the Afghanistan insurgency was not monolithic but rather consisted of several groups that were subsumed under the Taliban umbrella. Two of these main groups included the Haqqani Network and the Hezb-e-Islami Gulbuddin or HIG. These formations of 5,000 and 1,500 respectively, maintained their own distinct command and control, operating, in large measure, in concert with Mullah Omar and his Quetta Shura, the principal Taliban formation (see diagram below).

In terms of support networks, the predominantly Pashtun and strictly Sunni Muslim Taliban had a steady stream of volunteers to recruit. These came primarily from the many Saudi-financed madrasas in Pakistan's Federally Administered Tribal Areas or FATA. These madrasas, which served the refugee communities in the FATA, gave military training to males between the ages of 13 and 17. The Taliban received recruits from as far away as the United States. For example, the American John Walker Lindh, who joined the Taliban, met Osama bin Laden and fought alongside his al-Qaida "brothers" as bombs fell on his training camp in Afghanistan.

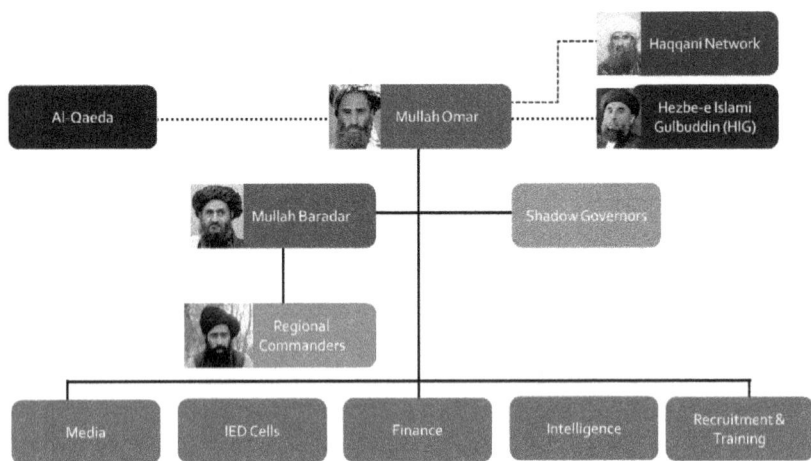

The Afghanistan Insurgency.

Perhaps the greatest unsolvable quandary for the U.S. was Pakistan. Though repeatedly denied by the government of Pervez Musharraf, it was a known fact that the Pakistani Inter-Services Intelligence or ISI was actively supporting the Taliban. Chief among this external support was the cross-border sanctuary in the FATA. Aware of the usefulness of having some degree of control over the Taliban, ISI kept tabs on the more prominent Taliban leaders and even housed them without the knowledge of the U.S. As it would be discovered later, ISI agents were well aware that Osama bin Laden himself was hiding in Pakistan and could have given him up to the Americans long before 2011. Pakistan also covertly supported the Taliban. As a member of the international coalition against terrorism, Pakistan secured over $32 billion in U.S. aid while empowering the very forces the U.S. was fighting.

Another conundrum was the opium trade. Afghanistan's soil is particularly suitable for poppy as is evident from the nation's long history of opium cultivation. While this lucrative crop was suppressed under the Taliban's rule from 1994-2000, producing about 11% of the world's opium, during the Taliban's insurgency to recover its power, the illicit harvest rose to a point where it consisted of 90% of the world's market. The proceeds from opium exports went to district officials, insurgents, warlords, and drug traffickers, with only a quarter going to opium farmers. Other sources of income come from wealthy Saudis, corrupt Afghan officials, Zakat donations, and road tolls. The Taliban were well financed. All told, from all sources they brought in about $200 million a month.

Trained by Pakistani ISI, the Taliban became increasingly skilled in using IEDs, a roadside menace that had claimed thousands of Coalition lives in Iraq. Iran's Quds Force, which at one time funded and supported Ahmad Shah Massoud's Northern Alliance against the Taliban, began financing and training the Taliban insurgents against the NATO-backed Karzai government. While Iran actively supported multiple insurgent and terrorist groups in Iraq, they likewise gave direct support in the form of weapons and advisers to the Taliban, increasing their lethality. Iran's Quds Force supplied the Taliban with thousands of tons of explosive flying projectiles or EFPs. As they did in Iraq, these exacted a high casualty toll. Between 2006 and 2021, IEDs were responsible for about half of all U.S. deaths.

Taliban Shadow Government

Early on, U.S. military partnerships with abusive and corrupt Afghan warlords was a logical consequence of the light U.S. military footprint and need for proxy fighting forces. Nonetheless, in many cases, U.S. actions served to legitimize and empower a class of strongmen who had conflicted allegiances between their own power networks and the nascent Afghan state. In this way, the U.S. helped to lay a foundation for corruption and the weak rule of law. This corrupt bureaucracy at the local level was perhaps most apparent in the town of Sangin.

Sanguin consisted mainly of two major tribes: the Alikozai and the Ishakzai. Since the invasion of 2001, the Alikozai tribe held influential positions, including that of the district governor and key administrative roles. Conversely, the Ishakzai tribe was marginalized and not afforded such positions within the local government. Exploiting their government positions, the Alikozai levied taxes and harassed and stole from the Ishakzai. Consequently, when Taliban commander Mullah Dadullah Lange initiated his 2006 offensive, the Ishakzai sided with the Taliban in hopes of removing what it perceived as the Alikozai yoke.

In a House Foreign Affairs Committee meeting, John Nagl remarked that it's a commonly accepted principle of counterinsurgency that, "if you are losing a counterinsurgency campaign, you're not being outfought, you're being out-governed." In areas under their control, the Taliban established their own governance structure, appointing provincial and district governors and even implementing a justice system based on Islamic law. What was happening in towns like Sangin was, instead of the Karzai government, it was the Taliban who were providing the essential services to the Afghan people, along with some degree of security and justice. These are fundamental ways of leveraging influence and legitimacy. Mohammed Issa, the Taliban governor of Kandahar, explained the Taliban's justice system in a 2010 interview:

In Kandahar City, the government courts stand empty and its judges that have been appointed in Kabul do not have the boldness to work here. The people understand that to solve their problems the only recourse is the Islamic Emirate (Taliban) Court.

Nonetheless, it would be a mistake to paint the Taliban in philanthropic ways, they would just as likely take over villages, kicking people out of their homes and turning the place into an armed camp. Unambiguously though, by the end of 2006, the Taliban were achieving what David Galula would call a position of strength as their power was issuing from the popular support they were achieving among the villages of Helmand and Kandahar Provinces.

As the Taliban contended with ISAF over these villages, and following the localized success of various ISAF operations such as Operation Mountain Thrust, the Taliban adapted their tactics. When outmatched, they adopted Fabian tactics of giving ground and melting into the nearby villages in favor of wearing down Coalition troops through ambush and IED attacks. In a manner much like that of Mosby's Rangers, the Taliban began operating in small groups, blending in with locals and hiding in family compounds and mosques. Weapons caches were emplaced around the countryside. This allowed them to move about without being identified as combatants. Taliban fighters could then arm themselves from these caches and launch attacks on ISAF forces.

The Taliban also adapted their communication methods. During periods in the conflict when their guerilla tactics were most effective, they would shift from technical to nontechnical and clandestine methods of communication, frustrating Coalition targeting methods. To the extent they utilized handheld radios for tactical facilitation, they spoke in simple code and riddled their communications with disinformation to be picked up by Coalition interpreters.

According to David Galula, the police are the eye and arm of the government. As such, they rightly constitute the first counterinsurgent. This being realized by the Taliban, an aspect of the Afghan government's security apparatus that was especially vulnerable to Taliban attack were the Afghan National Police. Reconstituted in early 2002 with a goal of 62,000, most Afghan National Police or ANPs lacked motivation, having joined primarily for the salary rather than out of a sense of duty or patriotism. Raising to about 149,000 officers in 2017, the ANP were poorly trained and poorly equipped, with only 15% possessing functional Kalashnikov assault rifles. Whenever they were attacked, they normally abandoned their posts. It's fair to say that functionally the ANP was made innocuous due to widespread corruption.

One thing the Taliban offensive of 2006 demonstrated was that the people of the affected provinces had little to no connectivity with their government in Kabul. Most in Afghanistan did not even know who was in charge of the government. Their territorial gains in Helmand and Kandahar provinces also allowed the Taliban to recruit directly from the villages they controlled.

Obama, McChrystal, and the Surge

In 2008, the newly elected President Barak Obama announced he would end the war in Iraq so we could get back to the war in Afghanistan. By the time Obama took office, the war was costing the American taxpayers an amount several times larger than Afghanistan's entire GDP, not to mention the U.S. was still searching for a viable strategy. In a televised address, Obama said, "Afghanistan is not lost, but for several years it has moved backwards. There's no imminent threat of the government being overthrown, but the Taliban has gained momentum." As Obama looked toward the exit ramp, he replaced General David McKiernan with General Stanley McChrystal and directed the new ISAF commander to produce a winnable strategy.

In McChrystal's analysis, the operational focus needed to be more on people-centric counterinsurgency and not on seizing terrain or merely destroying insurgent forces. The objective must be the population. Subsequently, as McChrystal saw it, there was also an obvious disconnect with the population and Karzai's government. According to his initial assessment:

The Afghan government has not integrated or supported traditional community governance structures, historically an important component of Afghan civil society, leaving communities vulnerable to being undermined by insurgent groups and powerbrokers. The breakdown of social cohesion at the community level has increased instability, made Afghans feel unsafe, and fueled the insurgency.

A grassroots effort would be needed to be implemented soon or the U.S. would lose the war. As part of his people-centric strategy, ISAF would shift to what may be termed counterinsurgency-plus, incorporating new but old ways of counterinsurgency alongside the uninterrupted counterterrorism campaign, an activity for which McCrystal had specialized in Iraq. In line with his new population-centric approach, the new commander also scaled down U.S. air strikes. "We must avoid the trap of winning tactical victories, but suffering strategic defeats, by causing civilian casualties or excessive damage and thus alienating the people," said the general.

Realizing that the United States was losing and that something drastic had to be done, in a September memo to Obama, McChrystal adamantly wrote, "This new strategy must be properly resourced and executed through an integrated civilian-military counter-insurgency campaign that earns the support of the Afghan people and provides them with a secure environment. Without extra troops," said McChrystal, "the mission will likely result in failure." To resource his new strategy, McChrystal asked for a 60,000-troop increase. Obama gave push back. At that time, there were around 150,000 U.S. troops in country. McChrystal also called for more Afghan security forces, including 16 new Afghan National Army companies to deploy to Helmand province. By the fall of 2010, he wanted 134,000 Afghan soldiers and just over 100,000

Afghan police. These numbers would roughly equate to the needed 1:50 ratio of coalition forces to population, which was then about 30 million.

However, sensing the quagmire that Afghanistan was becoming, and hopeful that the new resources could turn the tide of the war, Obama succumbed to military pressure. In December 2009, the president announced a troop surge of 30,000 and, incredibly, in the same speech, a demobilization date 18 months later. This blunder made it clear from the start that the surge would be a temporary measure and gave the Taliban a pretty good idea when the United States would begin to pull out. As critics noted at the time, telling your adversary exactly when you were going to quit was hardly the best way to persuade them to give up the fight. Obama told the Taliban exactly how long they needed to hang on to wait us out. To succeed, the surge would have had to be far larger and much longer in duration. It seemed to Washington that Afghanistan simply wasn't worth that level of effort.

Village Stability Operations

As part of his vision to realign the U.S.-led COIN efforts, McChrystal mandated two major changes. The pursuit of a population-centric COIN approach with an improved unity of command toward that effort. Regarding the first mandate, a new but old tactic, from a bottom-up approach, was to have great success. Known as Village Stability Operations or VSO, the venture would be command and controlled by Combined Forces Special Operations Component Command–Afghanistan (CFSOCC-A). It was led by newly promoted Brigadier General Edward Reeder, former Commander of 7th Special Forces Group (A) and previous USSOCOM executive officer for Admiral Olson. Reeder and his staff drafted what would eventually be called VSO/ALP. Major Sheffield Ford was the driving force behind turning "tribal engagement" into VSO. This approach was grounded in the idea that counterinsurgency must address local needs. By embedding Special Forces teams in Afghan villages, it attempted to create bonds of trust and legitimacy between the people and the "central government." The intellectual foundation from Major Jim Gant's *One Tribe at a Time* highlighted the potential of tribal dynamics and grassroots governance in countering insurgency. Drawing inspiration from the historic blueprint of programs like the Civilian Irregular Defense Group or CIDG, which was successfully employed by Green Berets in Vietnam to shield the Montagnard tribes from the Viet Cong, VSO offered a solution to securing the population from the Taliban and a viable means of connecting the 30,000 or so Afghan villages with Kabul. Seventh Group's ODA 7224 launched the pilot program in the central Afghan village of Nili. Starting in July 2009, the VSO Program grew to 46 sites by March 2011, and 103 by the end of 2012. Following the tenet that counterinsurgency is local, VSO promoted good governance. In reality, VSO was the Coalition's first real attempt at clear-hold-build. For example, Coalition forces had cleared Helmand provinces numerous times without maintaining a footprint. The people-centric concept embedded Special Forces ODAs at the village level to increase security and provide stability. The 'bottom-up' execution worked well with the 'top-down' planning and management to support the ISAF COIN strategy.

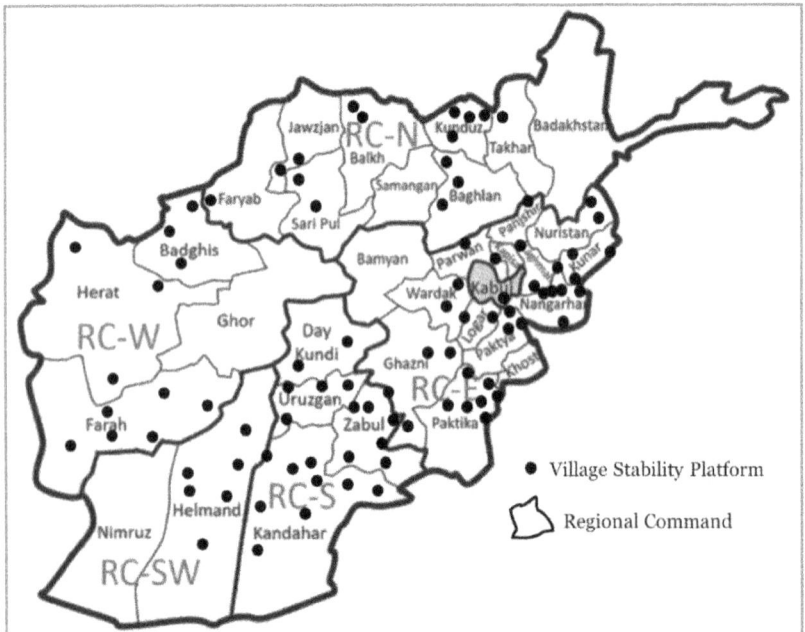

Afghanistan Provinces within ISAF's Regional Commands (RCs).

Drawn from the surplus population of young village men, Afghan Local Police (ALP) were vetted and trained by Coalition forces. These men had a stake in the welfare of the villages from which they were recruited. Once vetted and trained, ALPs were armed and consisted of the first layer of security in the village. Additionally, Green Berets were given enablers, such as Civil Affairs, PSYOP, medical, and logistics support personnel to assist as needed. In 2011, select female soldiers, trained as Cultural Support Teams or CSTs, were attached to some ODAs to work with women and children. Additionally, in some remote sites, conventional infantry squads beefed up security. VSO teams, living in the small towns and villages, worked closely with local leaders to improve health, education, and economic conditions.

The beauty of the VSO concept was that it encompassed and incorporated a wide array of SOF practitioners, Special Forces, Civil Affairs, and MISO (Psyops) in the capacity for which they were designed. By its design, operations were conducted at the grassroots level in villages in towns where the contest between the people and the insurgents is an everyday reality. VSO teams focused first on bringing security, then worked to develop the connection of local traditional governance to the formal governance at the district level and above, all the way to Kabul. Arguably, VSO offered the first real answer to the Taliban's shadow government.

As village stability platforms (VSPs) took shape, it metastasized in varying ways depending on the region. As it was said, "If you've seen one VSO site... you've seen one VSO site." One vignette demonstrates this characteristic hallmark. Under the watchful eye of Green Berets, Marine Raiders, and Navy SEALs, the VSO program flourished. Many provinces shifted from non-permissive to permissive environments. According to a 2012 RAND study, within less than a year of VSO/ALP site establishment, violence levels decreased by significant amounts. Such was the case for the VSP at Aryob Zazi.

Astride a heavily rutted mountain road, a mere mile and a half from the Pakistan border, the VSP was constructed near combat outpost Herrera to interdict a major Taliban supply route. Prior to the VSP, Taliban had gained recruits, intelligence, and finances from the village. In the fall of 2012, one 3rd Special Forces Group ODA ripped out with another ODA, taking ownership of the VSP in the village. Focusing on security, the Team conducted patrols with their ALP counterparts, intent on stretching their control to the more outlying areas of the province. Check points were set up to gain both security and social atmospherics, particularly near the border crossing point from Pakistan. One area the Team worked hard at getting into was just north of Zazi. Called Rokuon, it was a personnel and logistics hub the Taliban were using. The Taliban would defend its canalized approach with mortars.

One Special Forces Medical Sergeant summing up the complexities and dangers of VSO related the following:

Of my five combat trips to Afghanistan, the one where the risk was most constant and where I almost died the most was VSO. It was a risky mission because we lived and worked with the populace. On one occasion, an IED nearly killed half the ODA, and on another a complex attack composed of a suicide bomber, 2000-pound VBIED and follow-on assault nearly succeeded in wiping out the entire team and district government. That second attack only failed due to the VBIED bomber's lack of familiarity with his own initiation device.

It's important to understand that a lot of the risk associated with VSO was due to constant exposure. If you think back to the concept of zonal security, in VSO you're plopping directly into the insurgency's B Zone and must assume that all of your movements are being reported by their early warning. You also have to assume that you're interacting on a daily basis with their C Zone, the insider threat. That includes everybody from the armed guards you're hiring for base security and local workers managing the camp to your partner force.

That same trip our sister team was rendered combat ineffective by an ANP officer that turned his mounted machine gun onto the team during their pre-op commo circle. A single 18D was the only teammate who wasn't killed or wounded before the shooter was killed by other members of the partner force, and his precise link to the enemy and motivations for the attack are unknown. Aside from the danger of an insider attack, there's a high likelihood that the insurgency knows you are coming before you ever leave the base due to their reporting. There are challenges intrinsic to VSO and COIN in general involving maintaining the appropriate level of awareness, life-preserving suspicion, and readiness to eliminate a sudden threat while also investing in and interacting with the local populace in a way that effects your goals. That second part's important, because the only measure of real security you have is by making friends with enough of the right locals that your own B zone can keep you alive. That same tension can extend within the A zone, the VSO camp itself.

The Special Forces Medic related another vignette from a second VSO under construction:

We had an instance where the workers were building the camp's defenses and one of our teammates left his portable speaker and phone out. This being Afghanistan, somebody stole them. When I came back a couple of my teammates had all the workers lined up, with their stated plan being to pick one and beat him up as an example to the others. Despite my stating the idiocy of the idea (the workers lived with us and could see to our death by saying the right things in the market 50 meters from where we were standing) I finally waved my teammates off by threatening them with imminent bodily harm. I explained the importance of getting the phone back to the workers and asked them to ensure its return as a personal favor and left them alone. I'd built a good relationship with the workers by working with them on camp construction and figuring out ways to make their job easier, so the phone and speaker reappeared within 30 minutes. There's an undercurrent in military culture that the only thing people from certain parts of the world understand is violence. I think that's a big part of why we lost in Afghanistan and Iraq. And just saying, but if you enjoy categorizing other men's masculinity with the two or three of Greek letters you picked up through osmosis, you may be misusing pseudoscience and you're probably a chode.

Notwithstanding the inherent dangers, VSO changed the operational culture as SOF teams connected with the Afghan people and brought stability. One of the chief aims of the VSO project was for it to provide the connecting link with Kabul's central government. By design, village stability was to work backwards. Stability was to be first established in the villages, then village governance was to be connected to the districts, then the provinces, and then Kabul. Toward that end, the Team worked in close concert with the village elders and the police chief. While connectivity between villages like Zazi with the Afghan central government in Kabul was not always measurable, for the most part, the villagers involved in the village stability program were aimable and even loyal to their Green Beret guests.

On one occasion, the Team had a detainee brought to them. Bloodied, the villagers claimed he had been with the Taliban and was returning home. The villagers had stoned him. The Team medic treated him and gave him money to go get a higher level of care. During his tour in the village, the Special Forces Medic saw a lot of injuries, including Afghan children, and a wedding party gone bad. The groom had been shot through the back of the head at his own wedding by his brother. He showed up in the back of a station wagon gurgling blood. After searching him, the medic kept him alive, giving him a cricothyroidotomy, and put him on a medevac flight to forward operating base Shank in Logar Province. Notwithstanding all the progress, the Team medic later learned that after the VSP was turned over to the Afghan Security Forces, as part of the planned transition, the Taliban recaptured the village and executed the police chief and the ALPs.

As well as the VSO program was trending, several factors laid the seeds to its eventual failure. The first problem was the condensed timeline on which the program was made to operate. Obama's plan was for the U.S. to hand over the war by 2014. Like its CORDS counterpart, the VSO program was a good nine years into the conflict. And again, like it's Vietnam counterpart, by the time VSO was implemented, a U.S. president was looking for an exit ramp, and time was not on its side. Arguably, the second problem, related to the first, was that a tremendous sense of haste involved the selecting of platforms and the vetting of ALPs. Driven by the timeline, many VSO sites bypassed safety measures such as the shura vetting of ALPs. Additionally, lack of SOF oversight for some sites left newly created local police operating independently. In many sites in the country's north, warlords and local commanders reflagged their existing militias under the ALP program. Such actions counteracted efforts to empower district governors and Afghan National Army commanders.

Additional problems were related to non-SOF troops augmented to serve as VSO security. On March 11, 2012, Staff Sergeant Robert Bales, part of infantry "uplift" to village stability platform Balambai, in Panjwai District, left the base, and on his own lone mission, murdered 16 civilians.[4] Following the attack, President Karzai demanded U.S. troops be pulled from villages, but later acquiesced.

The Counterterrorism Campaign

Joint Special Operations Command or JSOC was created to unify the special mission units tasked with America's toughest missions. In Afghanistan, JSOC would conduct one of the longest counterterrorism campaigns in history. JSOC special mission units or SMUs include the

---

[4] In August 2013, Bales was convicted and sentenced to life imprisonment without parole.

U.S. Navy's SEAL Team 6 or DEVGRU, the U.S. Army's 1st Special Forces Operational Detachment – Delta or Delta Force, the U.S. Army's 75th Ranger Regiment, the U.S. Air Force's 24th Special Tactics Squadron, and the U.S. Army's 160th Special Operations Aviation Regiment that provides helicopter support. JSOC pitted its SMUs against Al Qaeda and Taliban fighters in some of the world's most inhospitable and austere environments.

During the early stages of the Afghanistan invasion, a joint operation involving U.S. Army Rangers and Delta Force operators conducted a raid on a compound near Kandahar believed to be the residence of Mullah Omar. Under intense enemy fire, 300 Rangers parachuted onto a 6,400-foot airstrip while Delta operators seized the compound, removing a dozen or so Taliban from the battlefield. Following the Battle of Tora Bora in December 2001, the next major JSOC raid in Afghanistan was Operation Anaconda, which took place in the Shah-i-Kot Valley in March 2002. The operation targeted a large concentration of Al Qaeda and Taliban fighters. In the first two weeks of March 2002, a U.S. force of 1,700, augmented by 1,000 pro-government Afghan militia, battled 1,000 Al Qaeda and Taliban fighters for control of the valley.

Like Tora Bora, the Shah-i-Kot Valley was an insurgent enclave, which the mujahedeen had used to their success in their fight against the Soviets in the 1980s. The defensive works included caves, bunkers and entrenchments. Task Force Dagger developed a plan that involved Zia Lodin's Afghan fighters as the main offensive punch driving into the valley supported by precision airstrikes and close air support orchestrated by 5th Special Forces Group's ODAs 594 and 372. The attack would force the enemy forces to squirt into the valley while an additional Afghan force led by Kamal Khan and Zakin Khan, supported by the 101st Airborne Division and the 10th Mountain Division, would hold key blocking positions.

During the 17-day battle from 2-18 March 2002, the U.S. dropped 3,450 bombs on Taliban and Al Qaeda positions. U.S. aircraft, which included F-15E strike fighters, F-16 fighter bombers, A-10 ground attack jets, and B-52 strategic bombers, flew some 950 sorties. While the operation managed to clear the valley of the enemy, removing some 800 Al Qaeda and Taliban fighters from the battlefield, eight U.S. troops were killed, 82 were wounded, and two MH-47 helicopters were lost. Worst of all, a large majority of Al Qaeda, including Osama bin Laden, managed to escape. Following the battle, the trail of bin Laden and his associates grew cold.

Following Operation Anaconda and the early 2002 battles in Afghanistan, the next major operation conducted by JSOC was the U.S.-led invasion of Iraq in March 2003. There, JSOC formed Task Force 20 to conduct covert operations in Western Iraq before the main ground offensive began. As Delta became absorbed in the Iraq mission, Afghanistan became the domain of SEAL Team 6 (ST-6).[5] Aided by a Ranger platoon and supported by three of Task Force Brown's MH-47s, ST-6's mission was to hunt Al Qaeda along with providing personal security for President Karzai. In one mounted patrol, conducted on April 22, 2004, former NFL football player Corporal Pat Tillman was killed. The Rangers and JSOC reported that Tillman was killed by enemy fire. The controversy later ended when Tillman's family was notified that he died at the hands of his fellow Rangers.

In October 2003, Lieutenant General Stanley McChrystal became JSOC commander. He would lead the unit until 2008, when he replaced General David McKiernan. McChrystal implemented several changes in JSOC battle rhythm. He directed his force to widen its aperture to include Taliban high-value targets or HVTs. Up to that point JSOC's SMUs had focused primarily on Al Qaeda. The environment was indeed target rich, albeit, as Sean Naylor observes,

Even with authorization to target the Taliban, the task force never reached the operational tempo of its Iraq counterpart, for several reasons. One was the rural nature of the insurgency in Afghanistan, and the size of the area in which the Taliban operated, which prevented quick turnarounds of the sort possible when striking several targets in the same Baghdad or Fallujah neighborhood.[6]

In 2005, JSOC began collaborating with the CIA's ground branch operatives to form Omega teams. Taken from the CIA's playbook, Omega teams consisted of Afghan militia who were trained and led by ST-6 or Ranger noncommissioned officers. Most of the militia that joined this

[5] Sean Naylor, *Relentless Strike: The Secret History of Joint Special Operations Command* (New York: St. Martin's Griffin, 2015), 351.
[6] Ibid., 355.

formation had been previously recruited by the Agency in 2001 to work with the Special Forces and chase the Taliban out of Afghanistan. The Agency retained several teams at its Afghan bases. At one point, this covert paramilitary force, also known as counterterrorism pursuit teams or CTPTs, numbered over 3,000. Under the control of the CIA, the aim of these hunter-killer teams included everything from gathering intelligence, going after "minor insurgent and criminal kingpins," to cross-border raids into Pakistan in concert with predator drone strikes.[7] The CTPT cross-border raids in Pakistan remain classified.

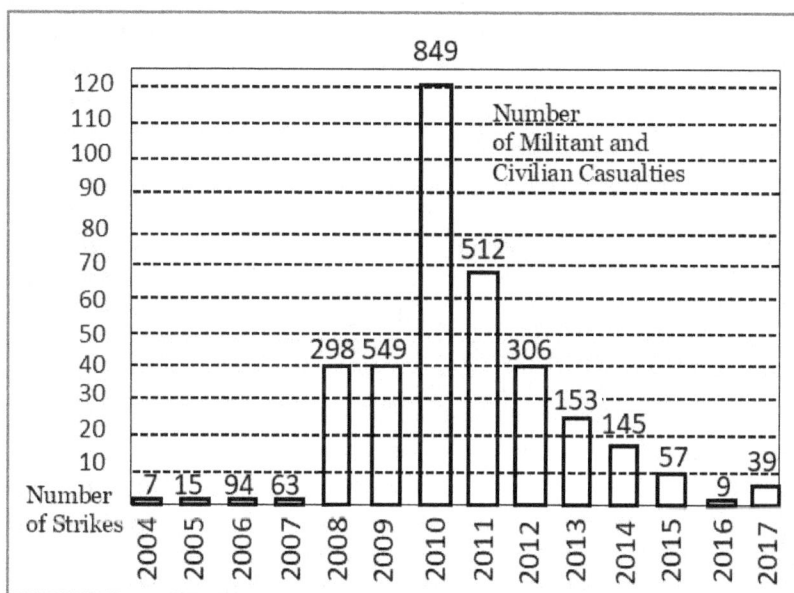

U.S. Drone Strikes in the FATA by year. Bureau of Investigative Journalism.

Beginning on June 19, 2004, the U.S. launched a covert "drone war" on thousands of targets in Afghanistan and northwest Pakistan's FATA. The program utilized unmanned aerial vehicles (MQ-9B Sky Guardian drones) operated by the U.S. Air Force under the operational control of the CIA's Special Activities Division. The covert program extended into Yemen and Somalia. Drones became the weapon of choice in Pakistan's

---

[7] Ibid., 353.

FATA as it gave the U.S. the means of decapitating Al Qaeda and Taliban while not putting any American soldiers in harm's way. Remotely piloted MQ-1 Predators armed with AGM-114 Hellfire missiles could conduct armed reconnaissance and interdict critical targets. It's no overstatement to say that predators liquidated the Al Qaeda and Taliban hierarchy. There was some collateral damage. Research completed by *The Long War Journal* on drone strikes found that from 2004-2011, approximately 108 civilians were killed in drone strikes while 1,816 Taliban and Al Qaeda extremists removed from the battlefield.

As the war in Iraq was winding down, more JSOC assets were freed up to prosecute the war in Afghanistan. JSOC SMUs took the fight to the enemy, removing something like 10-20 Taliban from the battlefield per operation, which under McChrystal, was nearly every night. Accordingly, in 2008, JSOC conducted 550 raids, removing some 1,000 terrorists from the battlefield.[8]

The intensity continued unabated after Vice Admiral William McRaven replaced McChrystal as JSOC commander in 2008. Some operations, however, created more enemies than they removed. Such was the case of one high-profile raid on a suspected insurgent bomb-maker compound in Gardez, Paktia province. The intelligence turned out to be faulty. Early on February 12, 2010, local police detective Mohammed Daod Shahabuddin was awakened by the raid force in his yard. As he went outside to investigate, he and his teenage son were shot dead. By the time the Rangers finished clearing the compound, seven Afghans were killed, including two pregnant women and a teenage girl. Adding insult to injury, ISAF insisted that the men killed were insurgents and the women were victims of "honor killings" by the "insurgents."[9] McRaven himself visited the family's house two months after the raid. Offering a sheep as a traditional Afghan condolence, he apologized and accepted responsibility for the deaths.

Due to this and several other incidents which claimed civilian lives, JSOC became required to conduct a "call-out" prior to a raid. While this gave anyone inside a chance to surrender before the assault, it also gave insurgents ample opportunity to not only destroy sensitive materials, but

---

[8] Ibid., 356.
[9] Ibid., 369.

it also degraded the element of surprise the raid could normally have achieved. Another requirement was for JSOC to work with the Afghan Partner Unit or APU. Trained by the Special Forces, and led by a Ranger, APU troops were largely taken along for show.

As an additional Afghan CT asset, CJSOTF-A stood up the Afghan National Army Commando Corps. Organized to mirror the structure and function of the Ranger Regiment, the first two battalions were formed in 2007. The initial Afghan cadre received training from the Special Forces in 2006 in Jordan. Comprising seven percent of the Afghan National Security Forces, the Afghan Commandos nonetheless conducted about 80% of the fighting. In 2010, further building out the ANSF, the Afghan National Army Special Operations Command was created. The first Afghan Special Forces team was trained and graduated in May 2010. The soldiers for the first team were selected from the ANA Commandos.

The momentum in the GWOT turned again on 2 May 2011, when ST-6 killed Al Qaeda leader Osama bin Laden when they raided his compound in Abbottabad, Pakistan. The joint CIA-JSOC raid ended a decades-old manhunt. Following its success, Obama announced that all U.S. forces would be out of Iraq by December 31, 2011, and out of Afghanistan by December 31, 2014, at which time the Afghan government would assume responsibility for its own security.

Leadership of the war changed again following a controversial Rolling Stone article in which McCrystal and his aides were quoted criticizing Obama's administration. Hours after the article appeared, McCrystal offered his resignation to the president. Obama promptly replaced McCrystal with General David Petraeus. In a gesture of diplomacy, Obama stated, "This is a change in personnel, not a change in policy." As head of the military's Central Command and architect of the 2007 Iraq surge, Petraeus certainly seemed the best candidate to command a victory in Afghanistan.

In the next two years, the U.S.-led counterinsurgent forces had their fair share of faux pas. On February 20, 2012, U.S. soldiers on Bagram airbase burned Quran pages that had been used by Taliban prisoners to communicate with each other. Afghan workers at the base noticed the burning pages and doused the flames with water and jackets. Not surprisingly, the following day thousands of Afghans protested the outrage, including many who worked at the base. Protesters threw rocks,

Molotov cocktails, and burned cars. The protests spread to other cities, including Kabul, where large-scale riots erupted. In the aftermath, 30 Afghans and five U.S. soldiers were killed and over 200 people were wounded. The incident led to international condemnation and three separate investigations.

The years 2012 to 2013 saw sweeping changes. In 2012, as the 2014 turnover milestone got closer, the U.S. and Afghan governments signed an agreement that required the Afghan government to approve all future night raids. Before the agreement, Afghan forces were involved in 97% of night raids, but only led about 40% of them. Then, in 2013, upon the death of Mullah Omar, Mullah Akhtar Mansour succeeded as leader of the Taliban. Moreover, the elections of 2014 elected Ashraf Ghani to the Afghan presidency. Then, in 2014, Obama changed policies, ending counterinsurgency and pivoting to Security Force Assistance. The coalition would now support the Afghan military while the Afghans did the fighting.

Operation Freedom's Sentinel

Beginning January 1, 2015, the U.S. strategy in Afghanistan was to hand everything over to the new Ghani government. U.S. troop strength significantly decreased from 194,000 in 2011, to 10,000 in 2015. Likewise, NATO reduced its footprint to about 13,000 troops. The U.S. would continue to train, advise, and assist Afghan security forces while its counterterrorism campaign and drone war went on unabated. As late as 2022, drone strikes still accounted for most insurgent deaths. On July 31, 2022, Al Qaeda leader Ayman al-Zawahiri, one of the perpetrators of the 9-11 attacks, was killed in one such strike.

As the year 2015 matured, hopes for stability were dashed as the government lost control of significant territory to the Taliban. On the morning of September 28, 2015, a rapid advance by some 500 Taliban insurgents was able to displace a force of about 7,000 government troops from Kunduz city, which fell in a matter of hours. The next day, supported by Special Forces, the ANSF regained the provincial capital. The attack, which caused an estimated 900 civilian deaths, woefully demonstrated the ANSF's inability to protect the populace. The Kunduz attack served as a precursor to coming events. By 2015, a resurgent Taliban controlled nearly one-fifth of the country's 412 districts.

Kunduz was also the site of one of war's more horrific friendly fire mishaps. On October 3, 2015, a U.S. Air Force AC-130 gunship attacked the Doctors Without Borders hospital in Kunduz. Following the Taliban's attack on the city in September 2015, the hospital found itself on the front lines. As Taliban who were pressing their way into Kunduz, the ANA requested AC-130 close air support. Human error misidentified the hospital as a legitimate military target. Coupled with the garbled communications between the aircraft and the ANA, 211 rounds were fired into the hospital, killing 42 and injuring over 30.

As the Afghan war entered its 16 year, newly elected President Donald Trump vowed to end the so-called forever war. The Afghanistan War Trump inherited carried the same dilemma: Defeat the Taliban to build a state and build a state to defeat the Taliban. Though his original instinct was to pull out, he later reversed his calls for a speedy exit. Like his predecessors, it was evident that Trump did not want to be the president who lost the Afghanistan War. When Trump took office, there were more than 10,000 troops in Afghanistan. In the fall of 2017, Trump directed Secretary of Defense, retired Marine Corps General James Mattis to undertake a comprehensive new strategy in the war.

In a speech regarding this new strategy, Trump outlined three core tenets, which at the most basic psychological level amounted to the desire to justify America's expenditure of blood and treasure. And in a manner reminiscent of America's desire to depart Vietnam with honor, Trump said that a hasty withdrawal would create a vacuum that terrorists, such as ISIS and Al Qaeda, would instantly fill. Trump said a core pillar of America's new strategy would be a shift from a time-based approach to one based on conditions. "Conditions on the ground," said Trump, "not arbitrary timetables will guide our strategy from now on." Arguably, however, like his predecessors, Trump instead pressed ahead with an open-ended military commitment to prevent what he called the emergence of "a vacuum for terrorists."

Straining India-Pakistan relations, and wanting others to take more of the bill, Trump also invited India to play a greater role in rebuilding Afghanistan. Demonstrating the complexity of the war, as part of its great game with Pakistan, India supported the government of

Afghanistan with military equipment. In 2014, India spent billions buying weapons, including Mi-24 Hind attack helicopters, from Russia to boost the strength of the Afghan National Army. For their part, Pakistan favored a Taliban victory, as bringing back the Islamic Emirate would create advantageous strategic depth against India. In 2018, India provided over $3 billion in assistance to the Ghani government, while Pakistan, who continued to receive aid from the US, continued its support to the Taliban. In full realization this conundrum, Trump sought to bring the Taliban to the negotiating table, but a viable plan remained out of sight.

In the collective memory of the Afghan people, lies the legendary account of the destruction of General Elphinstone's British Army of the Indus. In January 1842, after negotiating a withdrawal, under terms that included surrendering most of their weaponry and leaving Kabul with little to no protection, the British began their ill-fated retreat. The force of some 4,000 British soldiers, encumbered by 12,000 civilians, attempted to reach the safety of Jalalabad – a 92-mile trek through the Hindu Kush Mountains. As the hapless force traversed the rugged mountainous landscape, they were mercilessly cut down by Afghan fighters led by Akbar Khan. When all was said and done, 16,000 corpses littered the route Elphinstone took out of Kabul. Only one man, the surgeon William Brydon, made it to Jalalabad and safety. The Soviets likewise experienced a costly withdrawal, losing some 10,000 men in their Afghan adventure. By 2019, the U.S. had suffered the loss of some 2,400 troops. The question was, how would America's withdrawal fare?

By the end of 2019, thousands of U.S. troops had already left following an agreement with the Taliban. Then, on February 20, 2020, seeking a comprehensive and sustainable peace agreement, U.S. and Taliban representatives met in Doha, Qatar. Brokered by U.S. envoy Zalmay Khalilzad, the Doha Agreement called for an initial reduction of U.S. forces from 13,000 to 8,600 troops, followed by a full withdrawal which would occur May 1, 2021. Under the conditions of the agreement, the Taliban pledged not to attack U.S. troops and likewise promised to prevent Al Qaeda from operating in areas under Taliban control. As a bargaining chip, by this time, the ANSF had 352,000 on paper. As events would unfold, however, this sizeable force was ill-prepared to sustain security following a U.S. withdrawal.

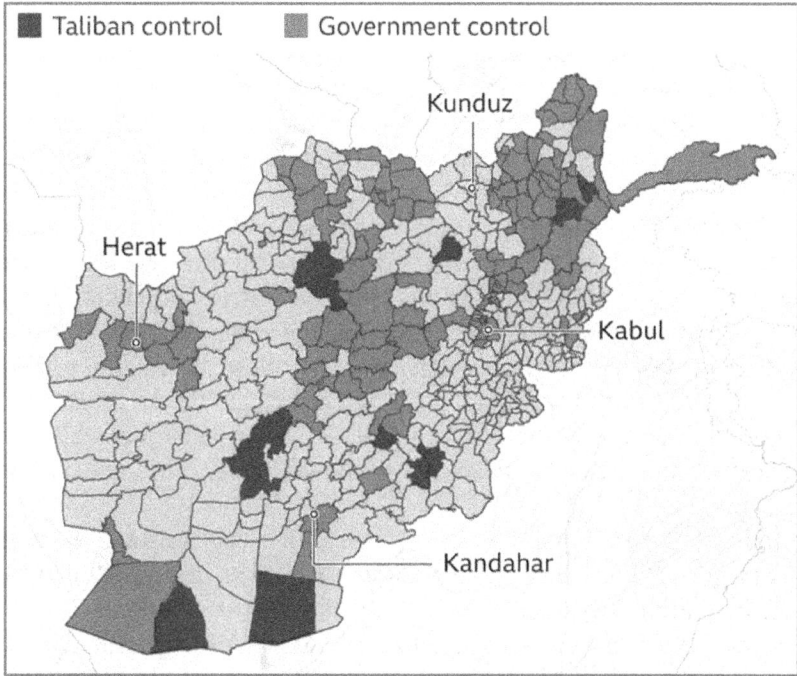

Areas of Taliban Control 2018. Courtesy AP.

With its exit strategy forecasted, from 2019-2021 the U.S. began its steady drawdown. As the U.S. strength in Afghanistan dwindled, incoming President Joe Biden was warned by NATO Secretary-General Jens Stoltenberg that withdrawing troops too soon would make Afghanistan a haven for terrorists, and the Islamic State could rebuild its caliphate. In an April 14, 2021, speech, Biden confidently remarked, "We will not conduct a hasty rush to the exit. We'll do it responsibly, deliberately, and safely. And we will do it in full coordination with our allies and partners, who now have more forces in Afghanistan than we do."

The 2021 Taliban Offensive

Following his election, Biden changed the 2021 U.S. withdrawal date from May 1 to August 31, then changed it again to September 11, presumably to coincide with the twentieth anniversary of the September 11th attacks. Then, beginning May 1, 2021, a final Taliban offensive got underway. Despite 20 years of training and billions of dollars of funding, wherever the Taliban attacked, they routed the Afghan security forces.

In many areas, Afghan units, charged to secure the populace, abandoned their posts to fend for themselves. Ignoring his military advisers, Biden abandoned a conditions-based withdrawal and abruptly abandoned the country, bringing about the worst strategic blunder in modern American military history. America watched in shock as the Afghan national Security Force collapsed. In some districts, to avoid further bloodshed, the authorities agreed to allow the Taliban to take over. In August, as the Afghan government toppled before the Taliban juggernaut, the U.S. military and State Department scrambled to evacuate some 124,000 embassy personnel, Americans, and at-risk Afghans.

In just ten days, from August 6-15, nearly all of Afghanistan's districts fell in the Taliban's swift march toward Kabul. As the Taliban noose tightened around the capital, Ghani fled the country. While the Afghan army collapsed without much of a fight, thousands of desperate Afghans descended on Kabul airport. Following the U.S. withdrawal from Bagram on July 2, Kabul airport was the last stronghold of U.S. security. What began has been accurately described as "one of the most egregiously incompetent self-inflicted debacles in modern military history." With the collapse of what was left of the Afghan government, hundreds of U.S. citizens were left hostage, and the Afghan people were delivered into the hands of the Taliban. In the words of Carter Malkasian, America's shambolic exit "seemed to mark the passing of the United States as the world's superpower."[10] In a hasty rush to the exit, Biden ordered U.S. troops to abandon $7 billion worth of military equipment which ended up in the hands of the Taliban. Additionally, according to the Special Inspector General for Afghanistan Reconstruction (SIGAR), the Taliban likely also gained access to approximately $57.6 million in funds that the United States had provided to the former Afghan government.

---

[10] Carter Malkasian, *The American War in Afghanistan: A History* (New York: Oxford University Press, 2021), 471.

With thousands of Afghan civilians attempting to escape the Taliban, a suicide bomber slipped into a huge crowd gathered outside the airport. The detonation killed 13 U.S. service members and 170 Afghan civilians. Biden's administration responded by targeting a car, killing 10 innocent civilians. General Milley called it a "righteous kill." Thus, ended America's longest war. In terms of blood and treasure, the Afghan War cost the United States 3,590 killed in action, over 35,000 injured and $2.3 trillion dollars. Other casualties of the war include, roughly 50,000 Afghan civilians, of which at least 7,792 were children and more than 3,000 were women, 3,846 U.S. contractors, 1,144 other allied service members, 70,000 Afghan military and police, about 67,000 people in Pakistan. The war removed an estimated 84,000 Al Qaeda and Taliban fighters from the battlefield. Afghanistan has since regressed to its pre-9-11 state.

Summary and Implications

The following is an analysis of the American IW effort in Afghanistan by way of the five PLAIN laws:

Political objective (s) –America's principal fault in Afghanistan was a strategic blunder. The U.S. entered a war in Afghanistan without a clearly defined operational design. Vengeance is not an effective policy. The U.S. entered Afghanistan intent on capturing Osama bin Laden and toppling the Taliban. However, the mission morphed into nation building, amongst other things. Arguably, America fought without a functional strategy. Bottom line: the American people will support a war when vital interests are at stake and there is a plausible theory of victory, but they grow weary of long wars in faraway places without a coherent plan and exit strategy.

It may be said, the American failure in Afghanistan was in large measure due to a prevalent lack of cultural intelligence. In his article, entitled *Learning Counterinsurgency: Observations from Soldiering in Iraq*, General Petraeus presented fourteen observations of the challenges of conducting counterinsurgency in a culture vastly different than our own. In his ninth observation, he identified cultural awareness as a force multiplier and stated that "people are, in many respects, the decisive terrain, and that we must study that terrain in the same way that we have always studied the geographical terrain."[11] In time, Army leaders caught on to this, but unfortunately it was too late in the game. America had to learn that Afghanistan is a mountainous country, with strong traditions of local self-government and autonomy, and significant ethnic differences.

After a stunning victory, an equally stunning lack of cultural intelligence (CQ) was exhibited by imposing a Western style government on a medieval society. A single man was positioned as head of state in a society that had been ruled by warlords for the past three millennia. We more or less tried to create a carbon copy of American-style democracy by imposing a strong central government in a nation that has never had one. A mistake the U.S. has repeated throughout history. The central government was likewise weak, and warlords and tribal elders held authority. For all intents and purposes, these traditional power brokers were more meaningful for most citizens than any centralized government in Kabul. As it turned out, introducing a Western model of democracy into such a society proved to be foolhardy.

It is rightly said that success can test one's mettle as surely as the strongest adversary. It must be said that the decision to go into Iraq while still engaged in Afghanistan was the height of hubris. After the U.S. invasion of Iraq, with resources being prioritized there, the Afghan War became a backwater, permitting the Taliban to regroup. Despite this setback, the U.S. overestimated its ability to "fix" Afghanistan.

Legitimacy – It is no understatement that the corrupt bureaucratic nature of the Afghan government foredoomed U.S. efforts to stabilize the country. Likewise, in many districts, the government failed to ensure that adequate services were provided, thus driving many to get their needs met by the Taliban shadow government. Simply put, no amount of money and strength by a superpower will change the outcome on the ground without a legitimate government in place.

---

[11] Lieutenant General David Petraeus. *Learning Counterinsurgency: Observations from Soldiering in Iraq. Military Review* (Jan-Feb 2006), 2

<u>Adaptability</u> – Arguably, the most innovative and stabilizing approach to the war in Afghanistan was the VSO concept. Recognizing the importance of governance at the local level and its connectivity to Kabul was a key linkage and a major step toward building legitimacy. The aim of the program was to secure the population and close the gap between Kabul and the rest of the country. It most areas, VSO was working well.

It stands to reason that VSO is a case of tactical brilliance redeeming failed policy. Not only that, but VSO was the greatest threat to the Taliban's shadow government. It must be said however that while VSO brought tactical victories, and in large measure achieved strategic depth, it nonetheless risked perpetuating a dependency on U.S. enablers (e.g., Special Forces, logistics, and funding), leaving the Afghan government unable to sustain these efforts independently. The question which will perhaps never be answered is: Like the Civilian Irregular Defense Groups (CIDGs) in Vietnam, might VSO have worked if it had been given the time to do so?

Arguably the two most employed IW activities in the Afghanistan War were counterinsurgency and counterterrorism. As history shows, the tension between these two approaches is often worked against each other. The question is: Was there ever a cohesive strategy that aligned these goals?

The following graph outlines the U.S. IW approach in Afghanistan:

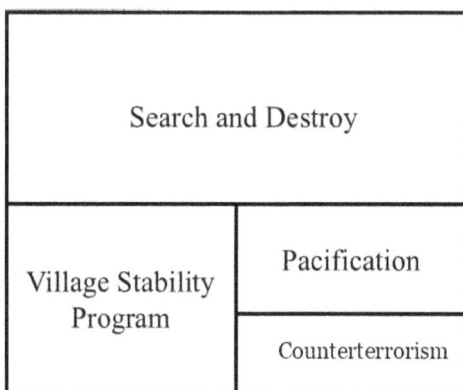

Another major problem throughout the war was opium. Despite the various campaigns to eradicate it or programs to replace it, the U.S. never found a way to deprive the Taliban from this major cash crop. Additionally, as the war in Afghanistan went on for four presidencies, the U.S. never solved the problem of the Taliban using Pakistan as a cross-border sanctuary. As it turned out, Pakistan was the key to the conflict. Albeit America was unable to solve the riddle.

As withdrawal of U.S. forces approached, and the time came for the U.S. to pivot to Security Force Assistance, placing the burden on Afghan forces, the question of whether VSO's grassroots momentum could survive without sustained foreign support was soon answered. It did not last long. Despite its success, it may be argued, it is highly doubtful that a program like VSO could have succeeded in a state that fundamentally lacked cohesion or a shared national identity. The very presence of U.S. forces, no matter how well-intentioned, was seen by many Afghans as an occupation. This it seems was the underlying reality that foredoomed the VSO project.

In the years before the end of the war, I diagnosed the problem, at least at the tactical level, as follows. The history of the U.S. involvement in Afghanistan may be compared to a tale of two commanders – commander X and commander Y. In preparation, commander X deploys his force to the Central Asian nation. Having done little to shape the perceptions of his troops, his force hits the ground running, gives little attention to the culture and people, if they bother to work with them at all, and spend their time jacking steel and drinking Rip-its, looking for a fight to drop tons of ordinance and kill "bad guys." By the end of his tour, the metrics his force has racked up are extensive, validating unit awards and the now impressive officer evaluation report of commander X. When the time comes for commander X to leave Afghanistan, the "foxhole" is essentially the same.

Over against the exceptionally kinetic nature of commander X's deployment is that of commander Y. Before "RIPing out" (relief in place) with commander X, commander Y's force has done a fair amount of familiarizing themselves with the culture and human geographic essentials of the particular region of assignment. Though the deployment includes some offensive strikes, they are surgical in nature. By the time comes for commander Y to RIP out, there is decidedly less dust to settle, the majority of the deployment having been spent with actions and goals that require time and development to bear fruit. Overall, commander Y's deployment looks less impressive metrically, but has made more of a lasting impact on the stability and development of Afghanistan, as well as the host Nation's legitimacy.

The truth is, in Afghanistan, we had more commander X than commander Y. The natural question is: If those on the ground in Afghanistan got more out of being commander X, who really wants to be commander Y? As Sun Tzu observed, "When you subdue your enemy without fighting, who will pronounce you valorous?"[12] But then again, perhaps that was why we were in Afghanistan for twenty years. We relearned, in counterinsurgency, you can't raid your way to victory. In the long run, it pays to be commander Y.

Influence – There are many similarities between America's experiences in Vietnam and Afghanistan. Like Diem, it's fair to say that Karzai likewise failed to gain the consent of the governed. The most basic problem was Karzai's penchant for courting the warlords. U.S. relations with Karzai started off as warm, then deteriorated into suspicion, and ended with hostility. By insisting on a central, Western-style government headquartered in Kabul, the U.S. essentially undermined the traditional power of the tribes and in essence isolated Karzai. Then, as relations deteriorated, Karzai turned to traditional Afghan power brokers who were nonetheless ruthless and corrupt.

The Afghan government further isolated itself through the use of various elements within the ANSF. For example, the Afghan Commandos' predilection to pillage and destroy private property during raids led to them to become a punitive threat to uncooperative villages. The government would threaten, "Help us institute security, or we'll have to send in the Commandos." While this was sometimes a useful negotiation tool with village elders, one of the second order effects of Commando raids was continuously revitalized Taliban recruitment.

Among the greatest U.S. missteps was the failure to include the Taliban in the 2001 Bonn Agreement. As argued by U.S. envoy Zalmay Khalilzad, there was no time to talk. "At the time (2001) we were still fighting the

---

[12] Sun Tzu, *The Art of War*, (New York: Oxford, 1971), 87.

Taliban. We were likewise fighting the Taliban when we sat down to negotiate terms of withdrawal later in 2018." Throughout the twenty-year war, following a penchant to lob the Taliban together with Al Qaeda, the U.S. likewise failed to accommodate those in the Taliban who wished to surrender. There was really no rehabilitation program in place for surrendering Taliban. Their choices were to fight on or die.

Native Face – Perhaps one of the greatest lessons of the Afghanistan War is, you cannot defeat an insurgency as an outside power. Indeed, the large U.S. troop presence often served to incite opposition and mobilize Afghans against the United States, as in the case of the Korengalis and the Safi tribe.[13] Simply put, counterinsurgents cannot want it more than the native force. As the U.S. failed to build a functioning competent state, it likewise failed to field a competent national army. Despite the billions spent and the countless hours of training, the Afghan Security Forces were not up to the task.

In the end, it seems that the United States tried to do the impossible. It attempted to build an effective, legitimate state, provide it with security and some degree of justice and economic potential, weed out corruption, and have its security forces perform with a modicum of success. While America assumed this enormous challenge, the Afghan government was in danger of becoming almost as unpopular as the Taliban. And while America won all the battles, the defeated Taliban could always slip back over the Durand Line into Pakistan to regroup. It seems America failed to heed the advice of General Dostum. After the stunning U.S. victory in December 2001, Dostum is reported to have said, "You need to leave now, or you will be here forever and lose."

As a footnote to history, the United States could have eliminated bin Laden several times during President Bill Clinton's tenure, as the U.S. had actionable intelligence and opportunity to do so. Had we done so then, there would likely not have been an Afghanistan War.

As Sun Tzu once opined, "Long campaigns can exhaust a country's strength, and there is no example of a country benefiting from prolonged warfare." In that vein, another great lesson from the Afghanistan War is

---

[13] Carter Malkasian, *The American War in Afghanistan: A History* (New York: Oxford University Press, 2021), 186.

longevity favors the insurgent. As Napoleon once put it, "time is the dominant factor." The insurgent doesn't have to win the battles. He simply has to stay in the fight until the counterinsurgent gives up and goes home. Hence, the Taliban expression, "you have the watches, we have the time." By simply not losing, insurgencies compel their opponent to choose either continue to fight, perhaps indefinitely, or quit and go home. America did the latter.

For Further Reading:

1. *The American War in Afghanistan* by Carter Malkasian.
2. *American Spartan* by Ann Scott Tyson.
3. *The Afghanistan Papers* by Craig Whitlock.
4. *The Only Thing Worth Dying For* by Eric Blehm.

11

Operation Enduring Freedom – Philippines

"To a very high degree the measure of success in battle leadership is the ability to profit by the lessons of battle experience." – General Truscott

Special Forces Soldier training Philippine Armed Forces on the Zamboanga Peninsula, March 2003.

On the morning of May 27, 2001, the Al Qaeda-affiliated Abu Sayyaf Group or ASG, kidnapped eighteen people, included two American missionaries from the Dos Palmas Resort on the Philippine Island of Palawan. The ASG terrorists hauled off their kidnapped victims hundreds of miles across the Sulu Sea to Basilan Island. Without a doubt, the ASG posed a great threat to the sovereignty of the Philippines. Their ability to readily field a battalion plus size units for battle and send a company plus size element anywhere in the archipelago at any time to terrorize people had to be eliminated. Upon the request of the Philippine government, Operation Enduring Freedom – Philippines or OEF-P was launched as a new front in the Global War on Terror. Beginning January 15, 2002, the operation targeted multiple terrorist groups in the Philippines and lasted until October 23, 2017. Though it received limited media attention, OEF-P's light U.S. footprint demonstrated an indirect approach to counterinsurgency that called for the Filipino security forces to take the lead. This chapter will highlight the PLAIN laws of political objective and native face.

I (Bob Ball) couldn't believe my eyes as I watched the plane fly into the second trade center tower on the television. I was with friends on leave in New Zealand and like the people around me in a state of shock. Finally getting through to Battalion on the phone I was told to carry on with leave. I had no idea what the future would bring. As events unfolded a half a world away, the situation began to change a lot closer to home. Situated only three and a half hours away by plane, 1st Special Forces Group's 1st Battalion had been tracking the kidnapping of Gracia and Martin Burnam from the Palawan resort for several months now. The two Christian missionaries had served with New Tribes Mission in the Philippines where Martin was a jungle pilot delivering mail, supplies, and encouragement to other missionaries, along with transporting sick and injured patients to medical facilities.

The Philippines has faced many security challenges since gaining its independence at the end of World War II. The Huk rebellion had been defeated in the 1950s but various other communist insurgencies lingered on. Additionally, a very serious Islamic separatist movement had grown in the south. The Islamic separatists were fractured into several groups with different leadership and political goals. This certainly wasn't a new situation as the Moro's had fought the Spanish, the Americans, and the Japanese. Their ferocity of their warriors in battle had gained them a worldwide reputation for bravery. In fact, the .45 caliber auto cartridge was said to have been developed to try and break up their sword attacks at close range.

What the new threats represented caused great consternation with political leaders concerned with keeping the integrity of the Philippines intact. The Moro National Liberation Front (MNLF) and Moro Islamic

Liberation Front (MILF) where the two principal separatist groups. While these two groups would go on to play significant roles in their continuing conflict with the government in Manila, this was largely seen as a Filipino internal conflict. In the aftermath of 9/11, Al Qaeda came to the forefront of global security concerns. Their representative in the Philippines was the Abu Sayyaf Group. Their bold and audacious kidnapping of Americans and Filipinos in Palawan and other terror exploits took on a new meaning and importance. While the Abu Sayaf was the primary antagonist, they were not our primary focus.

Once the Visiting Forces Agreement had been signed in 1998, the Philippines and the U.S. had returned to episodic SOF engagement as well as increased participation in Balikatan, an annual military exercise. During this exercise, we would reconnect with classmates and counterparts and reestablish liaison in several prime training locations. All these factors enabled us as Americans a greater awareness of the dynamic security environment in the Philippines and the political sensitives surrounding them. While we may have voiced our frustration in the past for lack of action, we were about to get very busy indeed.

Understanding the linkages between Al Qaeda and the Abu Sayyaf, the U.S. and Philippine Government began discussions of strategic options to address security concerns associated the ASG's activities in the southern Philippines. Armed with funding and authorities, there remained concerns between both parties about how it would be done. The primary concern was outlined in the Filipino constitution which forbid foreign troops from conducting operations against Filipinos in the Philippines. This concern was mitigated to an acceptable level by choosing U.S. Special Operations Forces to lead the U.S. effort. While modifications and challenges to the Visiting Forces Agreements would continue for years, the largest hurdles had been overcome. With the broad strategic vision identified and the legalities of foreign troops operating in the Philippines addressed, permission was given to begin more detailed operational planning. Special Operations Command Pacific (SOCPAC) would lead the effort, and 1st Special Forces Group would make up the main effort for the operation. SOF renewed engagement over the past two years in the Philippines would be vital in taking plans into action. In October 2001, while 5th Special Forces Group

ODAs were deploying to Afghanistan, operational plans were being developed to deal with the ASG.

One thing that was lacking at this time was a detailed assessment of the security environment of the southern Philippines. Most if not all our engagements with the Philippine Armed Forces (PAF) had been in the north and primarily if not exclusively on the island of Luzon. Again, due to the political sensitivities associated with the conflict in Mindanao and further south, a true picture of the situation was not available. Therefore, SOCPAC and 1st Special Forces Group conducted site surveys with AFP counterparts to confirm or deny factors affecting the overall situation and to gain greater insight into the current capabilities and capacity of the AFP to conduct operations against the ASG.

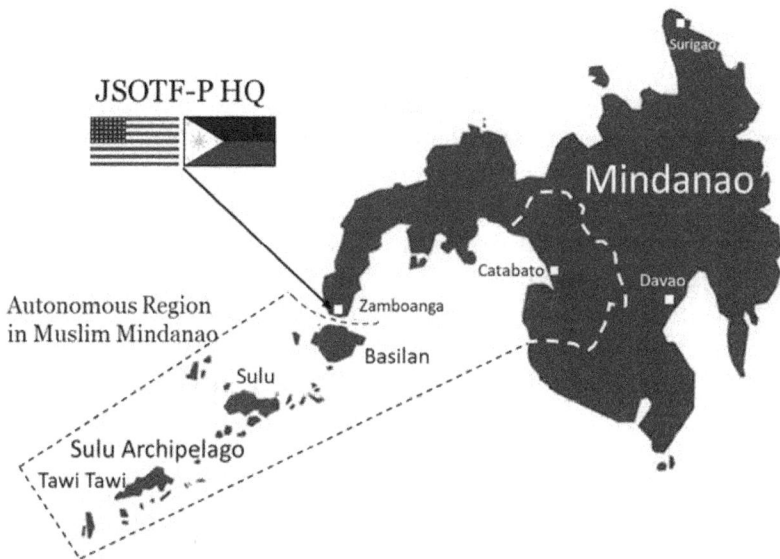

Mindanao. JSOTF-P.

These assessment site surveys answered a lot of questions and helped to form the framework of the plan. The main effort would be a 1st Special Forces Group led operation on Basilan Island. The island was believed to be infested with at least 1,200 ASG fighters. Access to that island at this time was by ferry or barge to the primitive ports at Isabela City, Lamitan, and Maluso, and several small dirt airstrips scatted throughout the island for very light aircraft. Electricity was provided by generators and only a very few miles of paved roads. A very austere yet breathtakingly beautiful operational environment to say the least. The island itself was made up of wooded mountains in the center surrounded by coastal plains of established coconut and rubber plantations. The inhabitants were a mixture of Christians and Muslims largely making a living by coffee, farming, fishing or work in the harvest of coconut and rubber.

The AFP had divided the island into three sections. The Headquarters and Scout Rangers in the Northwestern area, the Philippine Marines in the Southwest and the Army in the east. They resupplied themselves through some WWII landing craft into ports located in the vicinity of the battalion headquarters. Taking advantage of the suitable terrain, Abu Sayyaf situated itself in the hills to the east. Even with the brigade plus sized AFP unit on the island, they maintained freedom of movement across the Sulu archipelago and beyond.

After a short stay in Okinawa for training and planning, 1st Special Forces Group deployed to the Philippines. As part of JTF 510, the group established an operational headquarters at a Philippine airbase just outside of Zamboanga. First Battalion established a headquarters just outside of Isabela, and the remaining ODAs and B-Teams linked up with their Philippine counterparts. All told, 160 Special Forces trainers were assembled for the task of ousting the ASG. PACOM and SOCPAC provided supporting efforts to link the island with the administrative, operational, and logistical assets to support the operation. The AFP provided basing, intelligence, transportation, other supporting functions, and all the combat forces.

Once the ODAs arrived at their locations they got to work immediately improving the security situation and began training. Our primary effort here was not the enemy but building the capability and capacity of our counterparts. One of the principal challenges was breaking the barracks-laden mentality of the Filipino security forces. The shoulder-to-shoulder nature of the training helped fix the problem. We also provided improved communication and air assets for operational support. With emphasis on building individual as well as unit proficiency, the teams began to increase the proficiency of training. One additional controlling measure was that the ODA's could not participate in company and below operations. We could accompany the battalion commander and battalion headquarters units on operations, and we could participate in training exercises.

In response to the increase in readiness, the AFP Brigade Headquarters increased operations throughout the island. Many of these operations were medical, civil outreach and security operations. The largest civil operation was by U.S. Marine engineers, which improved all major roads on the island. Later this was replicated on Jolo and enabled AFP elements to increase mobility and combat effectiveness. Many Filipino villages supported the government and provided militia as local security. Medical outreach enabled hundreds of people to be treated and assisted by the local hospitals and Non-Governmental Organizations to care for people in need. The people of Basilan were extremely tired of the terror the ASG imposed on the local population, and they helped tremendously in riding the ASG off the island.

Meanwhile up in Luzon, other 1st Special Forces Group units were assisting the AFP with the creation of the Light Reaction Company (LRC). Within months the units were conducting operations in the southern Philippines. As Russell Crandall observes, "After a decade of relatively unchallenged impunity in the southern Philippines, the ASG were forced to retreat into hiding or flee to other islands like Jolo."[1] In one June 2002 operation, one of the ASG's key leaders, Abu Sayyaf, was killed. Having been relentlessly targeted, the ASG departed Basilan with their hostages. It would be several months before the ASG with their hostages was cornered in Mindanao. While the outcome was tragic, several hostages were saved and the ASG severely disrupted. It would take several years to finally track down the last of the ASG Leadership. Joint Special Operations Task Force the Philippines (JSOTF-P) was established and SOF continued to develop AFP counterparts as the search went on throughout the Sulu Archipelago. This included several AFP Division size battles in Jolo, an ASG initiated attack at Tipo Tipo, Basilan, culminating in Mindanao and the battle of Marawi and the capture and killing of foreign fighters in October 2016.

After fifteen years of campaigning, the AFP had finally neutralized the Abu Sayyaf. Fighting continues in the southern Philippines, and much of

---

[1] Russell Crandall, *America's Dirty Wars: Irregular Warfare from 1776 to the War on Terror* (New York: Cambridge, 2014), 400.

the supporting ASG infrastructure remains, however, this is now mainly a police problem.

Unfortunately, there were U.S. casualties in the operation and many more Filipino victims and citizens and soldiers who died defending their friends and families; lest we forget. What is also important to remember is "Balikatan." While the exercise with the same name gave us the ability to remain in the Philippines for an extended period, it still goes on today and is a great testament to depth of the Philippine and U.S. relationship. There is also the spirit of the term "shoulder to shoulder," that goes back to the Philippine resistance in World War II and one of the birth places of the Special Forces regiment; lest we forget.

Summing up the overall effectiveness of OEF-P, Russell Crandall suggests that it can be qualified as a COIN success story, and states:

While the U.S. campaign in the Philippines in 2002 did not generate headlines, they demonstrated that a hands-off approach to counterinsurgency, followed by continued military cooperation and development programs, could gradually make progress against insurgent groups and eliminate the gray areas that allows such organizations to form in the first place.[2]

Summary and Implications

The following is an analysis of the American IW effort in OEF-P by way of the five PLAIN laws:

Political objective (s) – The U.S. had a great Country Team, JUSMAG, and PACOM support. After several iterations and challenges, the Visiting Forces Agreement remains intact. The biggest concern at the tactical level was adhering to the Philippine Constitution forbidding foreign troops in combat operations against their own people. On Basilan, the people were very much ready to see the ASG go, and in Jolo and parts of Mindanao, large support areas remained.

Legitimacy – Balikatan was a positive platform which created an instant understanding with many Filipinos. However, in contentious areas on Basilan, Jolo, and Mindanao, the population is more concerned with

---

[2] Russell Crandall, *America's Dirty Wars*, 403.

family, clan, and tribe issues. Lastly, it is very difficult to bring and maintain government services in these areas.

Adaptability – Being uniquely suited to operate in difficult human and environmental terrain, SOF operations and Intelligence fusion brought together the AFP and U.S. planners.[3] In the words of OEF-P veterans, "You're constantly working on food, water, security/training, and medical issues. You are on an island where some of the population like you, and some are openly hostile." Working through the complex local issues, SOF improved operational units through better training and eventually integrated police and local government into operations.

Influence – As stated before, success in COIN requires an operational design that stabilizes the affected area by identifying with the people, while working with them to create a better state of the peace, winning their allegiance, and if possible, redress grievances that led to the insurgency. For the large part in the beginning, it was word of mouth. Cell phones are plentiful and radio if you can get the batteries. There were no smart phones. In some places there was absolutely nothing but people living there. There is a language problem and religious differences to overcome. In places there has been no government representation for years. It really came down to the relationships in the local areas. Despite the austerity of the local information environment there was a reasonable amount of national and international press coverage, reintegration efforts and CMO outreach in contested areas.

Native Face – The efficiency of a unit conducting COIN depends largely on its knowledge of the people and the terrain. Again, it was the name Balikatan that became a country wide symbol of cooperation between the Philippines and the United States. However, down in the southern part of the Philippines it was that young AFP Platoon leader, the nurses, and

---

3 David S. Maxwell, *Operation Enduring Freedom-Philippines, Routledge Handbook of U.S. Counterterrorism and Irregular Warfare* (Oxford, Routledge, 2023).

hopefully some NGO support, treating people and trying to solve basic problems having to do with food, water, security, and medical issues.[4]

The following analysis the American IW effort in OEF-P by way of the Ten IW Lines of Effort: **DISSECTION**

Diminish the Motives: On Basilan almost the entire population wanted the ASG gone. Increased security, civil, and medical efforts enabled the ferry service to resume, commerce began, and medical services improved. Daily contact with Zamboanga brought much needed supplies and resulted in an improved economy and growth. Electricity became more regular, and people were calmer and happy with the peace.

In Jolo and Mindanao it's difficult just to get there. When you do get there, it all comes down to basic human necessities. Over time, the AFP eventually made things better. Nonetheless, the environment remains austere, and the grievances remain.

Intelligence: Operations and intelligence fusion at all levels helped to flatten communications and improve operational efficiency. As awareness grew, so did the ability to share information; first locally and then nationally.

Security Forces: Over time, the AFP was able to provide enough security to stimulate economic growth in Basilan and parts of coastal Mindanao. Additionally, operational security OPSEC improved to the extent that several large operations and battles were fought in both Jolo and Mindanao with positive government outcomes.

Essential Services: The further away from Zamboanga, the more austere and the less essential service the government could provide. Over time the AFP improved these areas but everywhere the AFP was not the ASG were.

---

[4] Conversations with the following OEF-P veterans: Jeff Prough LTC, USA (R), Jeremy Lumbaca, LTC, USA (R), Steve Toth, CW3, USA (R), John LeCombe, MSG, USA (R), Steven Miller, MSG, USA (R), and Steve Soeder MSG, USA.

<u>Civil Control</u>: Civil control goes along with the people. If the people supported the government and were happy with the current efforts, then the people are controlling themselves in a functional society. The ASG took up these responsibilities in their stronghold.

<u>Tangible Support Reduction</u>: As events showed, however I think as time went on the ability to sustain this effort diminished. This forced the ASG to take more operational risks and bid for partnership if they were to survive.

<u>Integration</u>: One of the key highlights of the campaign was the growth of the Philippine National Police (PNP) force and their effectiveness in conducting operations with AFP units. During the 2013 Zamboanga Crisis, the police and military combined operations were very successful. This enabled the growth of PNP capacity and capability to conduct police operations against criminal characteristics typical of ASG operations. This was again demonstrated during the battle of Marawi in October 2016 to great success. The LRC is now the Light Reaction Regiment and part of a Joint Philippine SOF force.

AFP joint SOF capabilities improved greatly, and the formation and improvement of AFP SOF units gave the government of the Philippines additional options to counter threats (10). Integration of police forces and police force responsibility for the types of crimes ASG was committing is also a big part in improving AFP capacity to respond to national crisis. AFP gained a lot of success at all levels and across all services.

<u>Overall Quality of Force</u>: The overall quality of forces improved over time and enabled the AFP to successfully interdict all the ASG key leaders and weakened their capability to sustain themselves in the field.

<u>Native Government</u>: Balikatan informed the people in a way they understood at a national level. In the field it was that Filipino NGO nurse, soldier, sailor, marine or police that they would encounter and often we would be with them. That gave credibility to both countries.

From September 11, 2001, to October 23, 2017, the day the battle of Marawi officially ended, U.S. SOF partnered with the Security Forces of the Philippines. In my opinion we did a good job in their conflict. By, though, with, train, advise and assist, operations and intelligence fusion, and building partner capacity were the focus. JSOTF-P got outstanding support from the Country Team and PACOM. The greatest appreciation goes out to the Filipino people and all our counterparts who really did all the challenging work for a positive outcome. Operations in the southern Philippines continues. We may be gone, but new threats have emerged, and old threats remain. Lest we forget.

For Further Reading:

1. *Operation Enduring Freedom-Philippines, Routledge Handbook of U.S. Counterterrorism and Irregular Warfare* by David S. Maxwell.
2. *America's Dirty Wars* by Russell Crandall.
3. *Success in the Shadows: Operation Enduring Freedom–Philippines and the Global War on Terror, 2002–2015* by Barry M. Stentiford.

12

The Iraq War

"We know we're killing a lot, capturing a lot, collecting arms. We just don't know yet whether that's the same as winning." – Donald Rumsfeld

U.S. Marines passing the body of an Iraqi insurgent in Fallujah.

It has been suggested that a combination of fear and hubris drove the 2003 U.S.-led invasion of Iraq. As Russell Crandall suggests, the Bush Administration's decision to invade was "in part a result of the psychological impact of 9/11."[1] Twenty years later, there is no shortage of explanations as to why the war took place. The U.S. justification for

---

[1] Russell Crandall, *America's Dirty Wars: Irregular Warfare from 1776 to the War on Terror* (New York: Cambridge, 2014), 368.

war included alleged weapons of mass destruction or WMDs, a purported link between Saddam Hussein's government and Al Qaeda, and a desire for regime change to stabilize the region. This chapter will highlight the PLAIN laws of political objective and adaptability.

In what would be one of the most consequential strategic misdirections of history, Operation Iraqi Freedom, or OIF, was launched March 19, 2003, the second major campaign in the Global War on Terror. That evening, President George W. Bush announced, "On my orders, coalition forces have begun striking selected targets of military importance to undermine Saddam Hussein's ability to wage war. These are the opening stages of what will be a broad and concerted campaign." That initial effort consisted of massive airstrikes to decapitate Iraq's leadership, clearing the way for a ground invasion of about 160,000 Coalition troops, including the British 1st Armored Division, 1st Marine Division, U.S. Army V Corps, consisting of 3rd Infantry Division, 101st Airborne and 82nd Airborne Divisions, and 4th Infantry Division. The invasion force was about a third of the size of the 1991 U.S.-led invasion of Iraq, which was a response to Iraq's invasion of Kuwait in August 1990.

This leaner U.S.-led coalition included elements from three Special Forces Groups: 3rd, 5th, and 10th Special Forces Groups. In the west, 5th Special Forces Group (JSOTF-West), was tasked to seize airbases and conduct "Scud-hunting" to prevent the launch of Iraqi missiles against coalition forces and Israel. On the night of 19 March 2003, ODA 574 infiltrated via MH-53 helicopters into Iraq to capture Wadi Al Khirr airbase. The airbase would be used to bring in Coalition forces. To the west, on the Iraq-Jordanian border, fifteen 5th SF Group ODAs breached the 12-foot-high berm that lay on the border. The ODAs of TF Dagger ranged far and wide in the trackless desert, seizing numerous airfields, preventing the deployment of scud missiles and halting the reinforcement of Saddam's forces.

Also in the west, JSOC's Task Force 20 raided its way through the western expanses of Iraq. Commanded by Major General Dell Dailey, Task Force 20 consisted of units from the 1st Special Forces Operational Detachment-Delta (Delta Force), the 75th Ranger Regiment, Naval Special Warfare Development Group (DEVGRU) or SEAL Team 6 (ST-6), and the U.S. Air Force's 24th Special Tactics Squadron (24 STS). The

ST-6 component in TF 20 supported by 2nd Battalion 75th Ranger Regiment conducted a raid on a suspected chemical and biological weapons site north of Haditha Dam. The raid force engaged numerous gunmen at the site but found no chemical or biological weapons. Scouring Anbar Province, Delta recce operators marked targets for Coalition airstrikes, resulting in the destruction of a large number of Iraqi armored vehicles while the 75th Ranger Regiment's 3rd Battalion conducted a combat drop onto H-1 Air Base, securing the site for future operations in western Iraq.

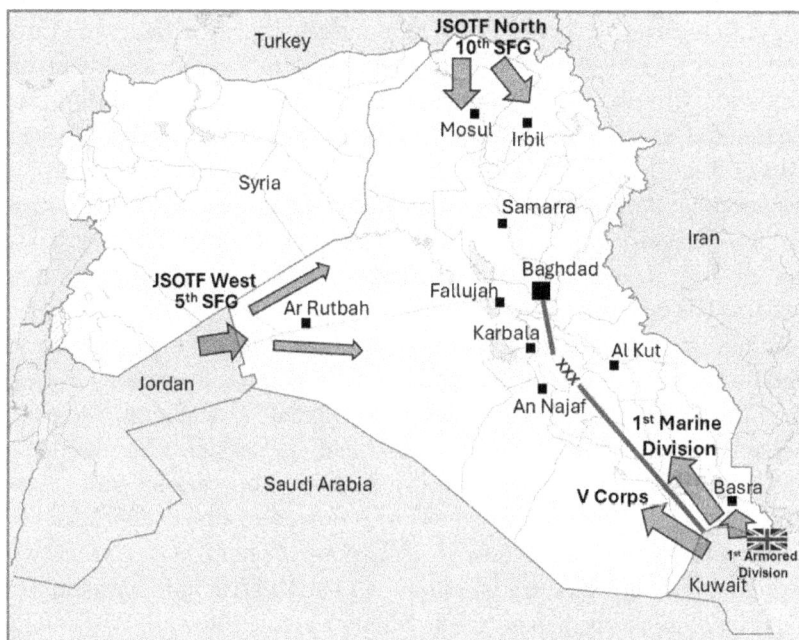

U.S.-led Invasion of Iraq 2003.

In the north, Task Force Viking, comprising 20 ODAs of 10th Special Forces Group, along with a battalion from 3rd Special Forces Group, would thrust southward with their Peshmerga allies. Viking, along with officers of the CIA's Special Activities Division (SAD), was the first to enter Iraq prior to the invasion. Viking fixed the Iraqi divisions stationed along the boundary known as the Green Line and prevented them from reinforcing Saddam's army in Baghdad. Kurdish forces and their 10th SF Group advisors sounded defeated Iraqi forces and uncovered a chemical weapons production plant at Sargat, the only facility of its type discovered in the Iraq war. In a series of battles along the Green Line, the U.S.-led Kurdish forces prevented Saddam's divisions, including thirteen armored divisions, from redeploying to Baghdad to contest the Coalition force coming from the south. The 10th Group teams further assisted their Kurdish allies in recapturing and controlling the key cities of Mosul, Kirkuk and Tikrit.

As for the Iraqi Army, while it was a paper tiger in 1991, it was a tissue paper tiger in 2003. By the end of March, the majority of Saddam's army had either been dispersed or destroyed. Many of Saddam's soldiers and officers just took off their uniforms and ran away. The coalition then occupied Baghdad on April 9, and Hussein and his core leadership went into hiding. Despite being one of the largest in the world, due to poor leadership, outdated equipment, and overwhelming coalition airpower, the Iraqi military quickly crumbled. On April 9, Baghdad fell, then Tikrit on April 15. In less than two months, the U.S. had gained a stunning victory. The war appeared to be over.

When Bush stood on the decks of the U.S.S. Lincoln on May 1, 2003, and announced the victorious end to major combat operations in Iraq, he did so in front of a huge banner that proclaimed, "Mission Accomplished." American-led forces had successfully removed the regime of Saddam Hussein with rapid decisive operations, and yet, as time would tell, the U.S. was wholly unprepared to effectively replace it.

Prior to the invasion, retired Lieutenant General Jay Garner was tapped to head Iraq's transition to a post-Hussein government. His experience assisting displaced Kurds following the 1991 Gulf War and his close ties with Secretary of Defense Donald Rumsfeld made him an obvious choice. Garner believed in a quicker transition to Iraqi governance with early elections, remarking, "I think that what we need to do is set an Iraqi government that represents the freely elected will of the people. It's their country and their oil." However, the White House preferred a more controlled U.S.-led reconstruction strategy. On May 11, 2003, Ambassador Paul Bremer was chosen to take over Garner's role. As events unfolded, Garner's way was the road not taken.

It's one thing to conquer a nation, it's another thing entirely to set up a working administration. The question was what was to take the place of the ousted regime. Named Coalition Provisional Authority (CPA) leader, Bremer would govern Iraq for roughly one year. His first decision, which would have disastrous results, was to purge Baath party members from the Iraqi government. Bremer's premise was the disbanding of the old untrustworthy regime. Once it was gone, the Coalition could start again from a clean slate to create new institutions without the taint of Saddam. This fateful decision was made against the advice of military professionals and without consulting the President's staff. Bremer was warned of the harm this action would have. Garner is reported to have said that none of the ministries would be able to function after this order. "If you put this out, you will put 50,000 people on the street, underground, and mad at Americans." The second decision Bremer made was the disbanding of the Iraqi military and security services. Overnight, 400,000 Iraqi soldiers, mostly Sunnis, were left jobless and without any prospects of assistance or employment from the new government. Bolstered by disenfranchised former Baath soldiers and officials, it was hardly surprising that armed resistance to foreign occupation quickly formed.

The road to hell is said to be paved with good intentions. The net result of Bremer's decisions created a failed state and a power vacuum. As time would tell, without a government, armed militias would soon fill the vacuum and splinter the country along sectarian lines. It's an understatement to say that Bremer's CPA orders 1 and 2 had disastrous effects. The insurgency was created overnight. Almost immediately the capital descended into an orgy of lawlessness and looting. Mobs roamed the city and stole everything that wasn't nailed down.

While the Iraqi Army quickly collapsed in the wake of the invasion, Saddam's Fedayeen remained a force to reckon with. Consisting of some 30,000 paramilitary militia, the Fedayeen, meaning "men of sacrifice," blended into the urban environment and ambushed Coalition supply convoys. The Coalition's fight against this elusive foe foreshadowed the savage counterinsurgency war in which it would soon be embroiled. Characterizing the early insurgency, Russell Crandall offers the following:

With little centralized leadership, the Sunni insurgency started out with small cells that would target the fledgling provincial government politicians or the thinly dispersed Coalition forces, mostly U.S. soldiers. The standard profile of the Sunni insurgent was an 18-year-old with an AK47 and no money. In short, a total of 150,000 American troops were facing an enemy that had no head, no nervous system, and no hierarchical command and control that could be destroyed.[2]

Prior to the invasion, Army Chief of Staff General Shinseki told a Senate hearing that several hundred thousand troops would be needed as an occupation force following the invasion. This did not bode well with Secretary of Defense Donald Rumsfeld, whose figures were nearly half that number.

The strategy the U.S. attempted to employ in Iraq immediately after the invasion was to blanket the entire country with security personnel to protect the population and make it difficult, if not impossible, for insurgents, militias, and criminals to harm the civilian population. Being already committed to the war in Afghanistan, there simply were not enough troops available. Historically, it takes a ratio of roughly 20 security personnel per 1,000 of the population to create the security needed in both counterinsurgency and stability operations. Even if one allowed that the 70,000 Kurdish Peshmerga were more than adequate to secure Kurdistan, the rest of Iraq would still require roughly 450,000 troops to achieve such a ratio.

Making matters worse, the Administration simply had not developed a counterinsurgency strategy prior to the invasion. One would have to be developed. The Administration likewise had no intention of providing the numbers of troops required to actually make a COIN strategy work. This became apparent to American military commanders in late 2003. U.S. planners faced a choice: They could either concentrate the troops they had available in the areas of insurgent activity to try and snuff them out, or they could concentrate those forces in and around Iraqi population centers to try to protect them against insurgents and criminals. Unfortunately, but not unexpectedly, the American military commanders made the wrong decision: They chose the former, adopting an enemy-centric approach. Perhaps seeing the writing on the wall,

---

[2]Russell Crandall, *America's Dirty Wars*, 374.

General Tommy Franks retired from the military on July 7, 2003, a little over a month after the completion of the invasion.

The first suicide bomb occurred in Najaf on March 29, 2003, when a former Iraqi soldier detonated a vehicle-born IED. The blast that killed four U.S. troops in a sense signaled the commencement of a long and savage game. Low-tech and inexpensive, IEDs became the war's signature weapons and by far the greatest killer of American troops, claiming half of all U.S. troops killed in Iraq. On average, a $30 bomb could kill up to four Americans. Sadly, for the price of a used car, a whopping $13,400, insurgents were able to kill 1,790 U.S. troops. This is staggering considering the U.S. spent roughly $45 billion on mine-resistant vehicles to counter the IED threat. As one Iraqi War veteran put it, "We got better at armor as they got better at ordinance." Insurgents filmed IED attacks for recruitment and propaganda purposes. In time, there would essentially be two wings of the Iraqi insurgency, a Sunni wing led by Abu Musab Zarqawi and allied with Al Qaeda, and another led by Shi'a cleric Muqtada al Sadr.

Abu Musab Zarqawi or AMZ grew up in Zarqa, Jordan. As a petty criminal, Zarqawi was in and out of prison. It is even rumored that at one time he worked as a pimp. Following his last stint in prison, he came out a radicalized jihadist. From there, Zarqawi fought in Afghanistan, where he formed Jamaat al-Tawhid wal-Jihad (JTJ) in 2000 in Herat. Then, as the story goes, he tried to meet with bin Laden but was rebuffed, got injured in a U.S. bombing raid, made his way to Baghdad for medical treatment, affiliated himself with the terrorist group Ansar al-Islam, then as post invasion Iraq was descending into chaos, he established JTJ in Iraq with the goal of forcing U.S. forces to withdraw. His ultimate goal was to establish an Islamic caliphate in Iraq.

Zarqawi set the tone of the insurgency by the degree of violence he used. Two attacks in Baghdad would serve to announce the onset of the nascent Sunni insurgency. The first, on August 7, 2003, involved a suicide bombing of the Jordanian Embassy. The driver of a vehicle-born improvised explosive device or VBIED detonated his payload in front of the compound, killing 17 and injuring 40 others. The second VBIED attack targeted the UN facility on Baghdad's Canal Street just over a week later on August 19, 2003. The blast killed 22 people, including the UN's special envoy to Iraq, Vieira de Mello. The United Nations force of about 600 promptly left the combat zone.

Rounding Up the Deck of Cards

By June 2003, with Baghdad and the major cities of Iraq occupied by Coalition troops, CJSOTF-N (Task Force Viking) was deactivated, and CJSOTF-W was re-designated CJSOTF-Arabian Peninsula (AP), having moved to Baghdad in April. CJSOTF-AP began to systematically round up Iraq's former Baathist regime members. The Defense Intelligence Agency developed a set of playing cards to help troops identify the most-wanted members of the regime. For example, Saddam was the ace of spades. General Ali Hassan al-Majid, also known as Chemical Ali, who notoriously used chemical agents to slaughter Kurds and Shi'as, was the king of spades. America's Special Operations Forces now took to task the rounding up process.

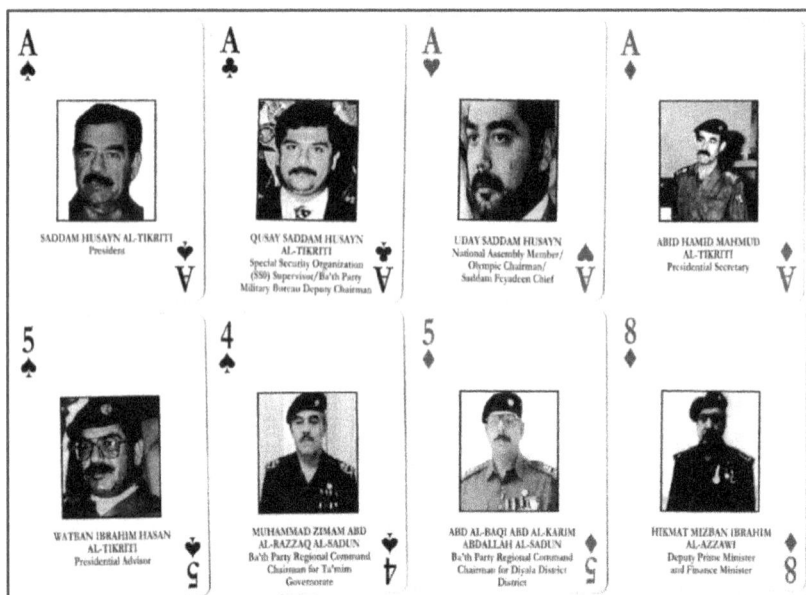

Defense Intelligence Agency Deck of Cards.

During the summer of 2003, hundreds of raids were conducted necessitating the need for a large holding area. In one of his last defiant acts, in a show of amnesty to shore up Iraqi support, Saddam emptied Abu Ghraib prison. The notorious prison was promptly reflagged to hold thousands of orange jumpsuit-clad suspected insurgents. "In reality," as one Iraq War veteran put it, "we went around detaining suspicious looking males." By the end of summer in 2003, Abu Ghraib would be busting at the seams.

Foremost in the fight to round up Iraq's most wanted was Task Force 20. Acting on a tip from an Iraqi informant, Delta commandos and the 101st Airborne Division surrounded the home of Saddam's sons Uday and Qusay in Mosul. Qusay commanded Iraq's intelligence services and Uday organized the paramilitary Fedayeen Saddam. The six-hour gunbattle that eventually killed the two evil sons all began after Uday was coaxed into appearing outside when his Lamborghini was hot-wired. It would take Coalition forces several more months to capture most of the individuals featured on the "deck of cards."

It was one thing to topple a regime, it was another thing entirely to create a working government. At the time, the general sentiment of many in Iraq was that we were building a plane in flight. Bremer's decision to disband the army had serious ramifications. What followed was a series of rush decisions to fill the security vacuum. The first of these was the decision to unrealistically accelerate the training of the New Iraqi Army. When Major General Paul Eaton was given responsibility for setting up a training program for this security force, he was told that his goal was to have nine trained battalions (about 12,000 men) at the end of twelve months. This was a realistic goal, and Eaton's plan was fully capable of achieving it. However, soon after the program started running, the general was suddenly ordered to accelerate his training program to produce twenty-seven battalions in only nine months.

The Administration had come to realize that they were desperately short of troops to fill the security vacuum. Such an accelerated pace would inevitably produce Iraqi soldiers who were neither properly trained nor fully committed to the mission. All things considered, the New Iraqi Army was created without a state to serve. As Bremer saw it, the most pressing need was for full Iraqi sovereignty under an Iraqi government. To that end, Bremer assembled a 25-member Iraqi

Governing Council. Appointed to provide advice and leadership for the country, the council would continue in that role until the June 2004 transfer of sovereignty to the Iraqi Interim Government.

To complement the operations conducted by Coalition military forces, the CPA also created the Iraqi Civil Defense Corps or ICDC. Established on September 3, 2003, the ICDC was tasked with joint patrolling, fixed site and route security. Despite its good intentions, the ICDC was considered a total debacle due to inadequate training, rapid recruitment, and widespread infiltration by insurgents and criminal elements. In Washington's fever pitch to churn out more Iraqi soldiers, it insisted on a breakneck pace that virtually eliminated any ability to vet personnel before they were brought into the defense formation. When it was actually engaged in combat, the ICDC nearly completely collapsed, particularly during the April 2004 uprisings, where nearly half its personnel deserted. Despite best intentions, the ICDC essentially had no combat capability.

By stark contrast, the 36th Iraqi Commando Battalion performed exceptionally well in combat. Organized, trained, and equipped by CJSOTF-AP, the 400-man Commando Battalion was formed from scratch, drawing personnel from various ethnic groups including Arabs, Kurds, Assyrians, and Turkmen. As a combat-proven and experienced unit, the 36th Iraqi Commando Battalion performed particularly well against ISIS during the war against the Islamic State.

As the year 2003 came to an end, Iraq remained largely unstable. Then, on December 13, following a nine-month manhunt, Saddam Hussein was found huddled in a spider hole near a farmhouse close to his hometown of Tikrit. The capture of Saddam was heralded by many as a possible turning point in the war. Washington likewise expressed the hope that rising violence would abate. With the beginning of a new year, the Bush Administration conceded its prewar arguments about extensive stockpiles of chemical, biological weaponry in Saddam's Iraq appeared to have been mistaken. On January 24, 2004, in almost an understatement, the Administration announced that there were in fact no WMDs found in Iraq. This awareness came as the U.S. realized it was now embroiled in its first protracted war since Vietnam.

The First Battle of Fallujah

About 60% of Iraqis live in one of the nation's 15 cities. One of these is Fallujah. Situated on the Euphrates west of Baghdad, Fallujah managed to avoid falling under the control of looters and common criminals in the days following the invasion. The welcome stability was due to a nominally pro-American town council. The U.S. therefore kept a light troop presence in the city. Notwithstanding this easily managed situation, in time, Fallujah would become a veritable petri dish in which the amorphous resistance to the American-led Coalition was cultivated and took shape. This evolved through several stages and was touched off by a series of violent incidents.

The first of these involved the killing of 17 Iraqis and the wounding of 70 more when U.S. troops fired on a crowd who had swarmed a school being used as a headquarters. Then, on April 23, 2003, a crowd of over 200 defied a curfew to demand the reopening of the secondary school. Soldiers from the 82nd Airborne claimed they were responding to gunfire from the crowd, while Iraqi witnesses say that U.S. troops fired indiscriminately. No evidence was ever found that U.S. forces come under attack. Two days later, U.S. forces fired on a protest at the former Ba'ath party headquarters, killing three more people. Following these killings, Fallujah became an insurgency hub for groups like Al Qaeda in Iraq or AQI, providing them with a safe haven to launch attacks against coalition forces.

Fallujah made headlines again on February 12, 2004, after insurgents firing from rooftops attacked a convoy carrying General John Abizaid, who had replaced Franks as commander of U.S. CENTCOM. Then, on February 23, insurgents conducted a coordinated prison break. After diverting Iraqi police to a false emergency, insurgents then raided three police stations and the mayor's office in a coordinated attack. The attack killed 17 Iraqi police officers and freed 87 prisoners. Following the trend of such attacks, Fallujah would become the site of multiple large-scale battles and significant conflict throughout the following years. The first large-scale battle followed the brutal killing of four American contractors.

On March 31, 2004, four Blackwater employees were ambushed while passing through the city. Zarqawi was suspected of being in Fallujah at the time and helped organize the attack. The contractors, Scott Helvenston, Jerry Zovko, Wesley Batalona, and Mike Teague, were killed and dragged from their vehicles. Their bodies were burned and dragged through the city streets and then hung from the King Faisal Bridge. Pictures of the charred, mutilated bodies of the four Americans filled the airwaves and fueled American retribution. The next day, General James N. Mattis, the commander of the 1st Marine Division at the time, devised a plan to respond proportionately to the attack. Albeit, the Bush Administration, embarrassed by publicized images of the contractors' bodies hanging from the bridge, called for an overwhelming response to the American deaths. On April 3, 2004, the 1st Marine Expeditionary Force, under Lieutenant General James Conway received an order to conduct offensive operations against Fallujah.

On the night of April 4, 2004, the 1st Marine Expeditionary Force launched Operation Vigilant Resolve to re-establish security in Fallujah, encircling it with about 2,000 troops. The Marines encountered an estimated insurgent force of over 3,000, armed with machine guns, RPGs, and mortars. It has been estimated that nearly half of Iraq's former Army officers, some 3,000, led insurgent forces after their dismissal. Many of them were drawn to Fallujah's fight. The dense urban environment posed a significant challenge to Marines who navigated through tight alleyways and heavily populated areas while braving intense gunfire. Marines engaged enemy fighters who would fire RPGs and machine guns at them then blend seamlessly into crowds of civilians.

Other U.S. units joined the three-week fight. JSOC's Delta Force conducted raids to capture/kill high-value targets or HVTs, while Green Berets from 5th Special Forces Group deployed in small teams to assist Marines and the Army's 9th Infantry Regiment. The Green Berets provided advanced communications and assault experience. Additionally, the aircraft carrier *George Washington* launched combat sorties dropping laser-guided bombs on insurgent positions, while Tactical Psychological Operations Detachment 910 blared Metallica over their loudspeakers with the aim of demoralizing the enemy.

During the battle, most of the ICDC and Iraqi police abandoned their posts. Some reportedly turned on U.S. forces. The Coalition attack took a toll on civilians as well as insurgents. Some 600 Iraqis were killed, half of which were civilians. As the operation began to come under increasing pressure from the Iraq Governing Council, Bremer called for a ceasefire on April 9. Only a third of the city had come under Coalition control. For many, the situation was dire. Much-needed humanitarian assistance was needed, and hospitals were closed. The ceasefire permitted some 70,000 women and children to leave the city and for government supplies to be delivered. Then, on April 10, U.S. troops pulled back to the outskirts of the city and announced a unilateral truce. The siege continued for the month of April with insurgents conducting hit-and-run attacks on Marine positions. Under intense political pressure, on May 1, 2004, General Conway took a risk and handed over control of Fallujah to the 1,000-man Fallujah Brigade, a CIA-created Sunni paramilitary force. The militia promptly dissolved itself and turned its weapons over to insurgents. The insurgents' acquisition of the brigade's weapons and equipment would eventually lead to the Second Battle of Fallujah in November 2004. Twenty-seven Americans died in the fighting.

Fallujah became a flash in the pan, setting off widespread fighting throughout Central and Southern Iraq. IEDs and suicide bombings increased. Following the battle, Zarqawi transformed the war by organizing the sidelined Sunnis into a multiple-group insurgency. After May 2004, Fallujah became a giant IED-making factory and the point of origin for hundreds of mortar and rocket attacks on the neighboring Coalition forward operating bases or FOBs.

As the fighting intensified, images of Iraqi civilians killed in the violence caused many Iraqis to become resentful of the U.S. presence. A young influential cleric named Muqtada al-Sadr stoked this resentment and became an outspoken critic of the U.S. occupation. The radical anti-American newspaper *al-Hawza* widely propagated Muqtada's calls for resistance and incited violence against occupation authorities. Bremer lost patience and ordered a 60-day closure of the paper on March 28 and called for Muqtada's arrest. On April 3, Bremer sent troops to his home and arrested one of his top lieutenants. The next day, while Marines were launching their attack on Fallujah, the cleric declared a jihad and his 6,000-man Mahdi Army began attacking Coalition forces. His Shi'a militia managed to capture a few Baghdad government buildings and police stations before being driven back.

To the south, hundreds of Mahdi Army militia seized parts of Basra, firing on British troops. To the north of Baghdad, the Mahdi Army took control of Samarra, Latifiya, and Yusufiyah, cutting Highway One between Baghdad and Karbala. While this was underway, a Mahdi coordinated attack was made on Karbala city hall. From April 3–6, 2004, Polish and Bulgarian soldiers along with Iraqi police managed to repulse the attack. Coalition troops sustained six killed to the Mahdi Army's 72. The Madhi Army struck again on May 5 but withdrew after heavy losses. Unfazed by the setbacks, Muqtada continued his calls for resistance.

Several other events around the world and within the theater contributed to Iraq's turmoil. The first of these was the Madrid bombing. On March 11, 2004, 10 bombs exploded almost simultaneously aboard four commuter trains in Madrid, Spain, killing 193 people and wounding another 2,050. The attack that was carried out by an Al Qaeda-affiliated group had a dramatic political impact. By June 30, Spain withdrew its contingent of 1,300 troops from Iraq. Italy and Ukraine soon followed suit.

One of the most notorious events of the Iraq War was perpetrated by U.S. troops. On April 28, 2004, the story of the prisoner abuse scandal of Abu Ghraib broke when photographs showing U.S. Army soldiers abusing Iraqi detainees made their way to the public media. The story made national headlines. Several reserve soldiers from the 327th Military Police battalion were given positions for which they were untrained and unsupervised. Placed in charge of about 200 Iraqi detainees, these reservists were encouraged by a mixed group of CIA and contract interrogators to help elicit information. During the night shift, several reservists, including Staff Sergeant Ivan Frederick, Specialist Charles Graner, and Private First Class Lynndie England, turned their part of the prison into a "little shop of horrors."

To "prepare" them for the following day's questioning, detainees were subjected to all manner of sexual, physical, and psychological abuse. In what has been called an act of "digital documented depravity," photos were taken of the abuse that unfolded. Stripped naked, detainees were bitten by dogs, shot with non-lethal ammunition, forced to form human pyramids, were led around naked by a leash, forced to masturbate, and even perform sexual acts on each other. News of the scandal humiliated the military, and not surprisingly, galvanized the insurgency. It is fair to say, whatever doubt existed in the minds of fence-sitting Iraqis as to how

U.S. troops viewed the Iraqi people, Abu Ghraib seemed to remove all doubt.

Abu Musab al-Zarqawi made headlines again when the body of American Nick Berg was found on a Baghdad overpass by U.S. troops. In a video that was released on May 11, 2004, Berg is seen in an orange jumpsuit screaming as he is beheaded by Zarqawi while others shout "Allahu Akbar." During the video, a masked man reads a statement that the killing was in revenge for the Abu Ghraib torture and prisoner abuse. Hence, the orange jumpsuit.

One of the more controversial events of the Iraq War was the alleged wedding party massacre at Mukaradeeb. The small border town of Mukaradeeb became the scene of a U.S. bombing strike and a follow-on raid on May 19, 2004. The celebration brought two families together, the Rakats and the Sabahs. Like all Iraqi weddings, the party consisted of food, dancing, speeches, and yes, celebratory gunfire. The party broke up around 10:30 pm as neighbors left for home. At about 3am the bombing began and was soon followed up by a raid that killed 42 men, women, and children. Suspecting it as a "foreign fighter safe house," U.S. forces deemed it a legitimate target. In the aftermath, Brigadier General Mark Kimmitt, the coalition deputy chief of staff in Iraq said, "There was no evidence of a wedding: no decorations, no musical instruments found, no large quantities of food or leftover servings one would expect from a wedding celebration. There may have been some kind of celebration. Bad people have celebrations too." Like many other events, the wedding party massacre at Mukaradeeb was merely chalked up to the fog of war.

While it is too convenient to lay all the blame on Paul Bremer, one of the greatest mistakes made by the U.S. was to establish Iraq's Interim Governing Council (IGC) without seeking official consensus from Iraqi leaders. Collaborating exclusively with Iraq's exiled Shi'a leadership, IGC members were hand-picked without in an undemocratic process. Feeling disenfranchised and denied the right to participate in the new government, Sunnis saw the IGC as an American instrument for turning the country over to the Kurds and Shi'as. On June 28, 2004, the U.S.-led Coalition Provisional Authority formally transferred limited sovereignty of Iraqi territory to the Iraqi Interim under Shi'a Ayad Allawi who was named prime minister. Bremer departed Iraq on the same day. Months following the creation of the new Iraqi government, virtually no one outside of Baghdad knew who was running the government.

The Second Battle of Fallujah

By the fall of 2004, the U.S.-led Coalition decided to forcefully confront the growing Iraqi insurgency. As it was decided, this would most notably come about by wrestling the city of Fallujah from insurgent control. All summer long, foreign fighters had poured into the city with estimates putting the number at about 3,000 from virtually every insurgent group in Iraq, including Al Qaeda in Iraq (AQI), National Islamic Army (1920 Revolution Brigade), Ansar al-Sunna, the Islamic Army of Iraq (IAI), along with the Army of Mohammed made up of ex-Fedayeen Saddam fighters. Unlike other battles that would plague Iraq, in Fallujah, Shi'a militants fought alongside their Sunni counterpart. Facing off this sizable contingent of insurgents was assembled a joint force of U.S., British, and Iraqi troops totaling about 13,500.

Code named Operation Phantom Fury, the Coalition aimed to secure the country and prevent insurgent interference in the upcoming Iraqi elections in 2005. On cue, following Bush's reelection, the six-week battle commenced. Urged on by psyop leaflets and broadcasts, most of the city's civilian population of about 300,000 fled prior to the battle. Those measures greatly reduced the potential for noncombatant casualties. Beginning November 7 and lasting until December 23, 2004, intense urban combat was waged over the city's 50,000 buildings. The insurgents divided themselves into platoon-sized units and used standard infantry tactics. U.S. troops grouped in two Regimental Combat Teams consisting of Marines and Soldiers, fought insurgents for each street and every house. Destroying or dispersing the majority of the insurgent formations, the battle was the bloodiest of the Iraq War, killing 2,000 insurgents and capturing an additional 1,500. Coalition forces suffered 110 killed and over 600 wounded, with an estimated loss of 700 civilians. Nearly half of Fallujah was destroyed in the battle. After coalition forces withdrew, however, Fallujah fell back into the hands of militants. Like the first battle of Fallujah, the second battle sent shock waves over the country, breaking insurgent resistance, albeit small-scale attacks continued across Iraq.

In August 2004, with Coalition forces tied up in Fallujah, the Mahdi Army took advantage of the situation to attack Iraqi police stations in Najaf. Marines were sent in to respond. The battle that unfolded took place over the sprawling Wadi-us-Salaam, the world's largest Muslim cemetery. The battle raged down narrow streets and within multi-story buildings as Army troops and Marines contested the Mahdi Army for possession of the city. The battle ended when U.S. troops were ordered to halt, allowing the Mahdi Army to withdraw peacefully.

To the north of Baghdad, American and Iraqi troops fought to retake the city of Samarra, which was overrun by insurgents. Beginning on October 1, 2004, for three days, the joint American Iraqi force of 5,000, supported by M1 Abrams tanks and M2 Bradley armored fighting vehicles, fought a hodgepodge of insurgents for control of the city. The operation code named Baton Rouge, uncovered some 90 insurgent weapons caches, and resulted in 127 insurgents killed, with another 60 wounded, and over 120 captured. Coalition casualties amounted to five, with 20 civilians killed, and 61 wounded.

Following the success of these engagements, American forces began a program to consolidate gains. A campaign was launched to build up the local police forces while tens of millions of dollars were spent on public works projects and hospitals. By this time, the number of FOBs, combat outposts, and other bases in Iraq amounted to 109. Then, in a manner mirroring the Coalition's pause to consolidate, Zarqawi pledged allegiance to Al Qaeda. Posting his allegiance online on October 17, AMZ's forces became Al Qaeda in Iraq or AQI.

Notwithstanding the positive gains of 2004, in October, the Iraqi Intern Government notified the Coalition that nearly 380 tons of mortars, artillery shells, PETN and RDX explosives had been removed from the al Qa'qaa facility. As it was discovered later, back in May of 2003, by the time an Exploratory Task Force arrived at Qa'qaa to search for WMDs, all the explosives, mortars, artillery shells, et al, were gone. As it turned out, in the Coalition's mad rush to take Baghdad, the largest explosives holding facility in the Middle East had been unguarded. The facility which comprised Saddam's largest ammunition and explosives dump had become an insurgent's bonanza. By the end of 2004, 145 Coalition deaths in Iraq were due to IEDs. The question was, how many of the explosives used in those IEDs came from the al Qa'qaa facility?

The Battle of Mosul

In the vacuum of power following the invasion, Kurdish political parties and their militias quickly moved to establish a presence in Mosul and the other cities of the north. As was occurring in the rest of Iraq, former Baathist regime elements in Mosul likewise organized themselves along insurgent lines. The 101st Airborne Division, under then Major General Petraeus, maintained order and balance between Arab nationalists who controlled the west side of Mosul and Kurdish political parties and their militias who controlled the east side. Then on November 8, 2004, as the contest in Fallujah began, insurgent formations under the banner of Ansar al-Islam attacked Iraqi police stations and American military personnel in Mosul. In an effort to gain control of the city, insurgents launched savage coordinated attacks, overwhelming the local Iraqi security forces, leading to a fierce battle that lasted until the 16th.

Due to the overwhelming nature of the insurgent assault, the Iraqi police force in Mosul quickly collapsed, with an estimated 5,000 police officers deserting their posts. The situation was stabilized as Coalition forces, aided by Kurdish Peshmerga, regained control of the city after heavy fighting. The battle resulted in significant casualties. Insurgents killed were estimated at 600, with U.S. losses at 18 killed and 170 wounded. Despite the setback, Ansar al-Islam would remain a force to be reckoned with. A month later, just days before Christmas, a suicide bomber infiltrated FOB Marez, a sprawling U.S. base in Mosul. He walked into a bustling mess hall during the busiest part of the lunch hour and detonated his suicide vest. The deafening blast killed 22 people. Among the dead were 14 troops, four civilian contractors and four Iraqi soldiers. Dozens more were injured.

The Year of Living Dangerously

By 2005, JSOC's counterterrorism (CT) campaign was entering its third year. While JSOC's efforts were playing a crucial role in the war, removing thousands of insurgents from the battlefield, the number of insurgent attacks continued unabated. In fact, insurgent attacks quadrupled from a little more than 200 each week in 2003, to more than

800 a week in 2005. As things turned out, the U.S. invasion of Iraq, along with its twin war in Afghanistan, gave Al Qaeda and its affiliates an even greater stage on which to play out their jihad. Fighters from as far away as Tunisia made their way to Iraq to fight the "Crusaders." As they did, JSOC was more than apt to oblige them in their quest.

JSOC had formed Task Force 20 to conduct covert operations in western Iraq before the invasion. Headquartered at Baghdad International Airport or BIAP, the Task Force focused exclusively on disrupting and dismantling AQI's network. This was done through targeted raids in which the operators captured high-value targets or HVTs. Most of these operations remain classified. Working in concert with CJSOTF-AP, TF 20 hunted former regime elements to round up most of the deck of cards.

One operation demonstrated JSOC's lethality. On the evening of June 11, 2003, 2nd Ranger Battalion's B Company and Task Force Brown's Little Birds attacked and killed over 80 Islamic terrorists at a training camp in a wadi near Rawah, Iraq. Supported by AC-130 and AH-6 gunship fire, the Rangers fought insurgents within the wadi training camp. Achieving complete surprise, the overwhelming maelstrom of JSOC's menacing firepower turned the wadi into hell's half acre. When the dust settled, the Rangers removed 84 insurgents from the battlefield and nabbed quite a haul of weaponry. The plunder amounted to some 2,000 RPGs, 50 RPK machine guns, and 87 SA-7 surface-to-air missiles.[3]

Later that summer, Delta commandos and the 101st Airborne actioned the target that closed out Saddam's sons Uday and Qusay. The Task Force was redesignated TF 714 by the time General Stanley McChrystal took command in October 2003. McChrystal's chief concern was with gaining intelligence over the seemingly amorphous Iraqi insurgency. As one operator observed, "The insurgency was more like a starfish than a spider." In reference to the book by Ori Brafman, *The Starfish and the Spider*, his point was, the decentralized nature of the insurgency we were fighting was much like the multi-headed hydra of Greek mythology. If you cut off a head, two more would grow in its place. As McChrystal soon

[3] Sean Naylor, *Relentless Strike: The Secret History of Joint Special Operations Command* (New York: St. Martin's Griffin, 2015), 229.
3 Ibid., 355.

discovered, AQI was much bigger, faster, and more dynamic than U.S. leaders had anticipated. And it was a starfish. Defeating Al Qaeda in Iraq became the task force's top priority, and Zarqawi became its number one target. During the summer of 2004, JSOC created the "FIND-FIX-FINISH-EXPLOIT-ANALYZE" or F3EA targeting model. F3EA resulted in a cultural change within JSOC (and eventually the U.S. military), with the main effort shifting from the finish phase to the exploit and analyzing phases. By adapting JSOC's intelligence gathering and operational tactics, McChrystal set the task force on the path toward effectively combating AQI's complex networks and significantly degrading their operational capabilities.

By May 2005, JSOC began mounting raids in Anbar Province. The raids targeted foreign fighters coming into Iraq from Syria and Jordan. On May 31, 2005, Delta commandos raided what amounted to a foreign fighter safe house in al-Qaim. Getting wise to the Unit's SOPs, the fighters had strong-pointed portions of the house in preparation for any such contingency. A volley of automatic fire met the assaulters as they entered the compound, killing Sergeant First Class Steven Langmack. Two weeks later, on June 17, 2005, Delta raided another compound in al-Qaim held by enemy fighters loyal to Zarqawi. Master Sergeants Bob Horrigan and Mike McNulty were likewise met by a hail of machine gun fire and were killed in action.

As the war was rounding out its third year, an IED attack occurred every hour, totaling an incredible 1,800 a month. On August 28, 2005, a Task Force convoy traveling along the Syrian border near the town of Husaybah struck an IED consisting of three stacked antitank mines. The weight of one of the vehicles in the convoy detonated the mine and nearly annihilated a Delta team. The blast claimed the lives of Master Sergeant Ivica Jerak, Sergeant First Class Trevor Diesing, Ranger Corporal Timothy Shea, and Sergeant First Class Obediah Kolath. As observed by Sean Naylor,

The fight in Anbar became the bloodiest test of wills that Delta had faced in its history. During one squadron's three-month tour, almost 50% of the entire force had been wounded on that one rotation, an astronomical number.[4]

---

[4] Ibid.

The year 2005 was the year of living dangerously. The general sentiment of most of the military serving in Iraq at the time was, having opened Pandora's box, the U.S. now had a moral obligation to tamp down the violence before heading to the exit ramp.

We Meant Well

One of the greatest challenges to post-invasion Iraq involved reconstruction. While the Bush Administration recognized that reconstruction turned out to be more demanding than they had anticipated, they nonetheless convinced themselves that the problems of the country were simple and straightforward. What was most needed in Iraq from 2003-2006, was basic security and services for the Iraqi people, i.e., electricity, water, sanitation, gas, jobs, medical care, and in many cases food. The reconstruction of Iraq was the largest nation building program in history. In terms of cost size and complexity, it dwarfed even the projects undertaken after World War II to rebuild Germany and Japan. To undertake such a feat, the U.S. State Department contracted Provincial Reconstruction Teams (PRTs). These teams fanned out in Iraq to help improve the country's stability and governance. By 2011, PRTs would spend about $65 billion on various projects. By comparison in today's dollars, the reconstruction of Germany and Japan together cost $49 billion.

Though lavishly funded, as government inspectors later discovered, American "nation building" efforts were characterized by flawed judgment stemming from cultural ignorance, pervasive waste, and inefficiency. Offering once such example, Peter Van Buren, who served as a contractor during reconstruction observes:

In a desert country like Iraq, nothing mattered more than water and its evil twin, sewage. Back in 2004, when the war was still trendy and the coalition of the willing was still in play, the Belgians and the Japanese promised to rebuild the Baghdad sewage plant and even committed a bunch of money. Then Japanese engineering firms drew up plans, produced blueprints, and created a giant three-ring binder of bad English to describe what was to be done. The capacity of the new plant was set to account for the projected population growth in Baghdad. When sectarian hell broke loose, the Belgians got out early and Japanese never visited the plant. The project was eventually awarded to the Bechtel Corporation for $4.6 billion to be completed within one year. Only nothing was ever done.[5]

While many projects managed to get started, fewer were ever finished. Like the half-finished prison in Diyala province which the U.S. spent $40 million to build. Some projects were ill-conceived. For example, U.S. taxpayers spent $88,000 to have certain classics like *Huckleberry Finn* and *Moby Dick* translated into Arabic. The project was intended to encourage Iraqi children to attain literacy skills. Once the truckload of books arrived, they were just dumped behind some school. Many other projects that seemed like success stories ultimately under-delivered. A women's health clinic was started, but then it closed down after six months. There were many other "we meant well" projects.

Then there was corruption. Like the situation in Afghanistan, Coalition forces in Iraq faced enormous challenges. These ranged from public officials taking bribes, to the massive embezzlement of funds meant for development projects. While there is evidence that corruption was already prevalent under the Saddam regime, particularly regarding the UN's oil for food program, there is broad consensus that corruption peaked after 2003. There was just so much money to be had. According to the U.S. Special Inspector General for Iraq Reconstruction, 40% of the reconstruction projects assessed had major deficiencies, including overcharging by subcontractors. Corruption likewise extended into the Iraqi security services where it was common practice for policemen to take bribes. Appallingly, something like 10% of the total $65 billion spent in Iraq on such projects is unaccounted-for.

The Samarra Bombing and Civil War

The year 2005 ended with a new Iraqi constitution. In the next significant milestone, elections for the 275 seats of Iraq's transitional National Assembly were held on January 30, 2005. Shi'a parties won 180 seats, while a Sunni boycott of the elections resulted in Sunnis securing only 17 seats. The boycott was based on the belief that the elections

---

[5] Peter Van Buren, *We Meant Well: How I Helped Lose the Battle for the Hearts and Minds of the Iraqi People* (New York: Metropolitan, 2011), 64.

legitimized the U.S. occupation. In efforts designed to delegitimize the fledgling Iraqi government, Zarqawi used sectarian attacks to gain support within more radical elements of the Sunni community. These actions helped his recruiting efforts and provided safe havens for his organization.[6] By October 2005, Zarqawi joined Al Qaeda to form AQI, but the relationship with Al Qaeda was strained by the degree of Zarqawi's violence. The mass killings of civilians and targeting of Shiite mosques and festivals angered most Muslims, costing AQI legitimacy and support. In a 2005 letter, Al Qaeda deputy Ayman al Zawahiri rebuked AQI's tactics of targeting "civilians, churches, and Shi'a," and encouraged Zarqawi to focus his attacks on U.S. forces and Iraqi security forces.

Undeterred, in a masterstroke on February 22, 2006, AQI turned the tide of the war in Iraq. Early that morning, AQI operatives managed to detonate a bomb that severely damaged the al-Askari Shrine in Samarra, considered one of the holiest sites in Shi'a Islam. The attack triggered Sunni-Shi'a sectarian violence. In retaliation, the newly empowered Shi'a-dominated police force and the paramilitary Mahdi Army launched a series of attacks citywide. One hundred and sixty-eight Sunni mosques were attacked, and 10 imams were murdered, and many others kidnapped.

With the December 2005 parliamentary elections having brought Shi'as into majority rule, the ruling Shi'a coalition selected moderate Shi'a Nouri al-Maliki to be the next Prime Minister. Maliki, a Dawa party old hand, had lived abroad to escape Saddam's death squads. Unlike many other party members, Maliki did not have a base of support or a militia. With Iraq threatened with civil war, in April 2006, Maliki became the first democratically and constitutionally elected prime minister of Iraq.

In a July 26, 2006, address to the U.S. Congress, Maliki said, "Thank you for supporting our people in ousting dictatorship. Iraq will not forget those who stood with her." In their collective memory, the Iraqi Shi'as felt betrayed when the U.S. encouraged an uprising against Saddam's regime in 1991, only to let tens of thousands of Shi'as be slaughtered by

---

[6] Wayne F. Lesperance, *The New Islamic State: Ideology, Religion and Violent Extremism in the 21st Century* (New York: Ashgate Publishing, 2016), 18.

his henchmen. Graciously passing over that, Maliki went on to say, "The people of Iraq will not forget your continued support as we establish a secure, liberal democracy. Let 1991 never be repeated." While Maliki spoke these words to Congress, Iraq was being torn apart by sectarian violence. The cycle of killing was cyclical. Shi'a-led raids against Sunnis in turn produced Sunni car bombs that targeted Shi'a government personnel. Likewise, Sunni and Shi'a insurgent groups targeted families and individuals in Baghdad in retaliatory attacks.

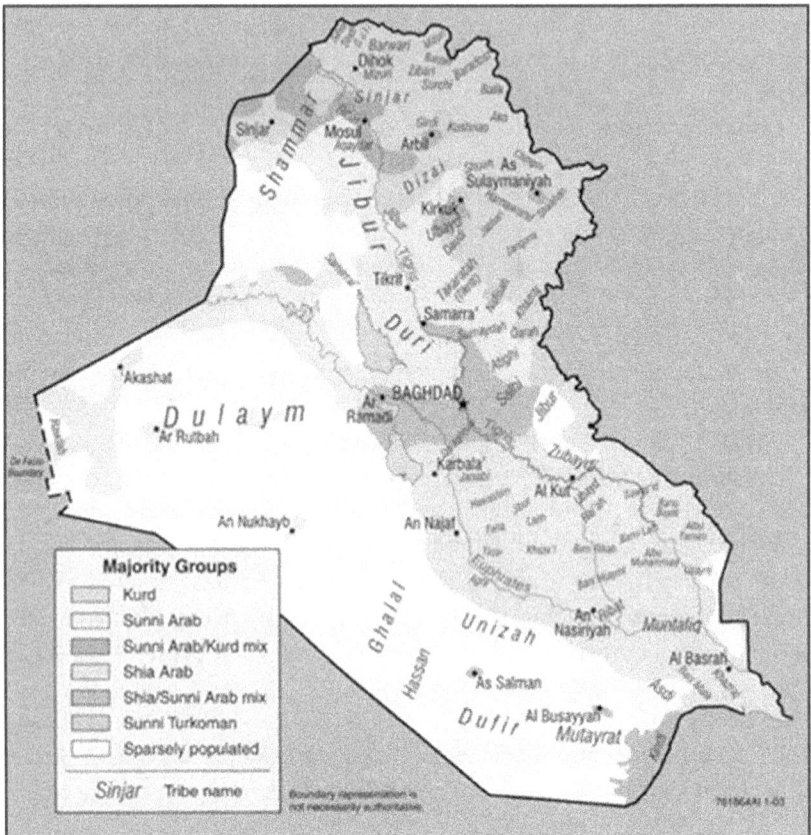

Iraqi Demographics. Courtesy of AP.

During his years as dictator, Saddam pursued the promotion and appointment of Sunni Arabs within his government. This sectarian hierarchy impowered Sunnis over Shi'as and was an effective means to control rebellion within the Iraqi population. This policy was especially important during the Iran-Iraq and first Gulf Wars. Saddam viewed Iraq's Shi'a population as a fifth column backed by the Iranian and American governments. Equally viewing the Kurds as a threat, in an effort to control Iraq's ethno-sectarian demographics, Saddam was notorious for his brutal "Arabization" policies that displaced Kurds in the north. He likewise pursued a violent campaign of persecution against Shi'a Muslims after the uprisings that followed the first Gulf War. Then, in the wake of the invasion, Bremer's order to "de-Ba'athify" Iraqi society, effectively "de-Sunnified" the political landscape, opening the way for Shi'a politicians to dominate the new Iraqi government. As Iraq was coming apart by the seams, and the Coalition struggled to find its footing, Iran ratcheted up its intervention and weaponized its most important instrument in Iraq.

The Badr Brigade

During the Iran-Iraq War of 1980-1988, Tehran organized, trained, and advised a militia of Iraqi Shiite defectors and former prisoners of war. Known as the Badr Corps, this militia proved indispensable for its ability to fight against Saddam. Modeling its approach on the Hezbollah model, the Iranian Revolutionary Guard Corps-Quds Force, also known as the Quds Force, worked to support Shi'a insurgents in Iraq. Under the command of Major General Qasem Soleimani, these Khomeinist militiamen could action targets through sabotage and assassination-style attacks. In 2003, amidst the chaotic days of post-invasion Iraq, the Quds Force infiltrated 10,000 Badr Corps fighters into the country with the goal of ensuring that Iraq could never again threaten Iran.

As the Coalition was distracted by the Sunni insurgency, Iranian and Hezbollah agents in Iraq began to recruit and train Shi'a militia members, including the Mahdi Army of Muqtada al-Sadr. Needing a local partner to fight the growing insurgency, Coalition forces began to rely on those who were the most willing to help. A ready ally was found in the Badr Corps' political arm, the Supreme Council of Islamic

Resistance of Iraq or SCIRI. Through SCIRI, Badrists claimed a place in Baghdad politics and cooperated pragmatically with occupation forces while maintaining ties close with Tehran. By cooperating so publicly with the Coalition, many Sunnis became convinced that Washington was siding with Tehran, leading many Sunnis to see the insurgency as more appealing than the alternative.

Standing in the shadow of its political wing, Badr militiamen were welcomed into the new Iraqi military that the U.S. was building. Led by Hadi al-Amir, who fought in the Iran-Iraq War on the Iranian side, Badr helped to further purge the ranks of the new Iraqi Army of any remaining Sunni Baathists. Badr further saw that members of its militia were placed in the officer corps, strategically taking over those units that could suit its purposes. Now Badrists in the Iraqi army could report to their masters in Tehran about U.S. troop movements and activities. Badr also obtained positions in the Iraqi government, especially the defense and interior ministries. But even as they played the role of helpful local allies to the Americans, Badr ran secret prisons and conducted extrajudicial killings in Baghdad. Directing events from the shadows, one of al-Amiri's preferred methods of killing involved "using a power drill to pierce the skulls of his adversaries." From the perspective of Iraqi Sunnis, the U.S. had ousted a Sunni ruler and installed a pro-Iranian Shiite in his place.

With the aim of getting tough on the insurgents and terrorizing the terrorists, the U.S. created the Wolf Brigade, composed mainly of poor Shi'as from Sadr City. Armed and supported by the U.S., the red beret-clad 2,000-man force wore sunglasses, drove out on raids in convoys of Toyota Landcruisers, and were known to beat and torture prisoners with electric drills. While the Wolf Brigade fought alongside U.S. forces against Sunni insurgents in Mosul and Baghdad, it fell under Iraq's Ministry of the Interior which was led by Badr Corps member Baqir Jabr Al-Zubeidi, who also had some 110,000 policemen under his control. As Coalition forces began to operate increasingly against the Sunni insurgency, the Badrist-infiltrated formations within Iraq's security forces had no problem conducting aggressive military operations in Sunni areas. Likewise, they had zero interest in protecting the Sunni population from Al Qaeda. Under al-Amiri's leadership, the Badr Organization became Iran's most powerful proxy in Iraq. Having risen to preeminence in both the military and political spheres, the Badr

304

Organization further solidified their base by securing 28 seats in the governing council following Iraq's 2005 elections. By the time Maliki came to power, he skillfully managed to stay in the good graces of both the U.S. and the Iraqi Republic, maneuvering himself into a position of strength while solidifying the Shiite militias' control over the security service.

The Anbar Awakening and the Surge

The year 2006 was not a good one for the United States. The U.S. economy experienced a deceleration in employment growth, while inflation increased. In the wake of the Samarra bombing, ethnic and religious violence seemed to be tearing Iraq apart. The American strategy of containment was not able to stem the tide of chaos and civil war. Republicans lost control of the House and Senate in the November midterm elections, and the American public, already weary, began to turn in large numbers against the war. So far, over 2,000 U.S. troops had perished. The general sentiment was that the U.S. had to tamp down the violence before heading to the exit ramp. However, things began to turn for the better in June.

Throughout 2003-2005, the Bush Administration and U.S. planners believed that the problems besetting Iraq were almost entirely the fault of the Iraqi insurgency, which they maintained was largely driven by Al Qaeda in Iraq, especially Zarqawi. The prevalent belief was, once Zarqawi was removed from the battlefield, the insurgency would lose its moxie and die on the vine. JSOC had missed several opportunities to catch the brutal terrorist. Then, on June 18, a task force drone tracked a vehicle containing a man believed to be Zarqawi's spiritual advisor, Abl al-Rahman. After changing vehicles several times to throw off any surveillance, Rahman arrived at a house near the village of Hibhib, in the Baqubah area. After Zarqawi was positively identified inside, an F-16 dropped two laser-guided 500-pound bombs on the house. When a troop of Delta commandos arrived, they found that the Jordanian terrorist had succumbed to internal blast injuries.

Documents found on the scene suggested that JSOC's strategy to dismantle AQI's networks was succeeding. Yet, despite the measure of satisfaction and optimism that came with Zarqawi's death, the violence spiraled ever higher. Al Qaeda quickly promoted Zarqawi's deputy Abu Ayyub al-Masri to assume the late leader's post. As U.S. planners were beginning to realize, they could not raid their way to victory. However, in the latter half of 2006 and early 2007, fortune began to smile on the U.S.-Coalition and the fledgling Iraqi republic.

Notwithstanding their convenient association, by late 2006, relations between Iraq's Sunni leaders and AQI had deteriorated sharply. Throughout 2005 and 2006, the core group of about a thousand AQI operatives systematically alienated itself from its Sunni host. In their orgy of violence, intimidation, and murder, they randomly attacked Iraqi police, many of whom were uniformed tribesmen. They robbed banks, kidnapped journalists, organized suicide bombings (IEDs and VIEDS), and conducted assassinations, along with video-recorded beheadings. Moreover, they flooded tribal areas with foreign fighters, who were ignorant of local customs and values. In perhaps their greatest mistake, Al Qaeda undermined the traditional power of the Anbar tribes by appointing their own local emirs, supplanting the authority of tribal sheikhs.

Recruiting at the lowest level, AQI would say, "Kill this guy. Here's some money." They likewise took control of smuggling routes used by the tribes, virtually putting the Abu Mahal tribe out of business. Moreover, they imposed strict sharia law. And referring to Shi'as as "Persian interlopers," AQI sought to foist Salafism in al Anbar, a radical position they did not sit well with most tribesmen.[7] They even tried to forcibly marry local women to create tribal ties to the organization. Sunni leaders soon began to understand that AQI sought to take control of their land and kill their people, in their quest to establish a Sunni Islamic caliphate.

According to Seth Jones, by the fourth year of the war in Iraq, Al Qaeda had conducted 6,210 attacks that brought about 13,612 deaths. Seventy percent of these victims were civilians and over 50% of victims were Muslims.[8] A tribal revolt against the extremist infiltrators was precisely what was needed to turn the tide of the war. This revolt began after Sheikh Abdul Abu Risha, of the Sunni al-Dulaim tribe, had had enough.

---

[7] Salafism has a history of attacking Shi'a Muslims whom they consider heretics.
[8] Seth G. Jones, *Waging Insurgent Warfare: Lesson s from the Vietcong to the Islamic State* (New York: Oxford, 2017), 52.

His decision was instigated by the murder of his father and brothers at the hands of AQI.

Approaching the Marine guards at Al-Asad Base in September 2006, the sheikh sought to collaborate with U.S. forces to oust AQI from Anbar Province. Wielding significant social, political, and economic force, the Dulaim are the largest and one of the most powerful Sunni Arab tribes in Iraq. Following Sheikh Risha's lead, the Anbar's tribes chose to align themselves with U.S.-led Coalition forces to root out AQI and integrate themselves into the power structure of the new Iraqi government. Risha further encouraged the men of his tribe to join the Iraqi police forces of Anbar. He further established a council that included Sunni leaders from many other tribes in the province.

The movement became known as the Sunni Awakening or the Sons of Iraq. Arguably, there had been limited success in tribal engagement. The Marines had improved things a bit in al-Qaim working with the Abu Mahal tribe. Albeit, up to this point, the U.S. had basically ignored the Sunni tribes. This was due in part to the fact that tribes such as the Dulaimis had formed the very nucleus of the insurgency. The tide had turned. In time the Awakening became a province-wide phenomenon and began to address some of the Sunni political and economic grievances as the tribes reengaged with the political power brokers. From its roots in Ramadi, the movement steadily grew as it expanded westward to encompass the whole of Anbar Province. This was highly consequential. At one time, Ramadi's police were on Al Qaeda's payroll. As Iraqi police recruitment in Anbar province grew from 2,000 to 13,000, one Marine commander described the movement:

One by one, the local tribes are beginning to flip from either hostile to neutral or neutral to friendly. It's probably one of the most decisive aspects of what we've done here. It's bringing those tribes onto our side of the fence.

Anbar's tribal leaders also promoted the work of the Coalition's provisional reconstruction teams. Following up on this success, the governor of Anbar, Mamoon al-Alwani, obtained $40 million in funding from the central government to resuscitate the Anbar economy and fund reconstruction projects, and created jobs. The move undermined support for the insurgency. Al-Alwani was likewise no friend of AQI.

Insurgents had at one time kidnapped his son in 2005, who was later safely returned, and two of his nieces were killed in IED attacks. With the new influx of police recruitment and the backing of the people, inkblots of security, in the form of security checkpoints and stations, soon spread across Anbar. The Awakening effectively denied AQI sources of intelligence and local recruitment, while it gained control of the Al Qaeda dominated cities of Ramadi and Fallujah. The prevailing opinion was that the rule of the tribes was stronger than the state. This proved to be true. In a 2007 speech on the War in Iraq, Bush said plainly:

Anbar province is a good example of how our strategy is working. Together, local sheiks, Iraqi forces, and coalition troops drove the terrorists from the capital of Ramadi and other population centers. Today, a city where Al Qaeda once planted its flag is beginning to return to normal.

By the beginning of 2007, with the blessings of their local sheikhs, the Anbar Awakening gained support from nearly every tribe. The number of car bombings and suicide attacks began to decline. The previous December, Saddam was executed for his crimes against the Iraqi people. To consolidate gains, at a time when most U.S. lawmakers were calling for troop reduction and were anxiously eyeing the exit ramp, Bush decided to bring an additional 30,000 troops into theater. Understanding that stabilization and control was degraded largely by troop rotation, under the working title "The New Way Forward," the president extended the tours of most of 127,000 troops in Iraq. Seeking to encourage his troops, Bush called for "more patience, sacrifice, and resolve" to see the mission through to completion. Part of Bush's new way forward strategy was to bring in a new general to energize events.

After taking command of Multinational Forces Iraq (MNF-I) on February 10, 2007, General David Petraeus inspected U.S. and Iraqi units all over Iraq. Petraeus' first order of business was to place the Sons of Iraq on the payroll. In a bold move, Petraeus paid $400 million for 107,000 tribal members, many of whom were former insurgents. Tribal sheikhs were paid $5,000 a month. As a failsafe, Petraeus ordered the use of biometrics to vet each tribesman and prevent his new ally from falling prey to AQI infestation. In short order, these vetted new allies ousted AQI from Anbar. With minimal U.S. training, the Sons of Iraq were an effective security force and intelligence asset, rapidly spreading beyond Anbar to other western provinces and even into Sunni neighborhoods of Baghdad. As the surge got underway, the security situation across Iraq improved dramatically, giving the Iraqi government and security forces room to consolidate its gains. Down but not out, AQI continued their attacks. After dodging numerous attempts on his life, on September 13, 2007, Sheikh Risha was killed by an IED in Ramadi.

Previously, in January 2006, AQI and seven other Sunni guerrilla groups formed the Mujahideen Shura Council (MSC). Then, following the death of Zarqawi, it disbanded to form the Islamic State of Iraq, led by Abu Omar al-Baghdadi as its first Emir. In October 2006, announcing the dissolution of both AQI and the MSC, al-Baghdadi declared that the previous organizations were replaced by the Islamic State of Iraq. As 2007 came to an end, Coalition operations degraded AQI's operational capabilities, and cleared out several large sanctuaries. AQI's lines of support were nearly severed. Then, by 2008, Al Qaeda was all but defeated in Iraq., and by 2009, the Islamic State of Iraq was all but destroyed. The Islamic State would go underground for three years. In the end, it was the Sons of Iraq who helped the U.S. oust Al Qaeda, not the Iran-backed Shiites.

Besides playing a crucial role in the defeat of Al Qaeda, the Sons of Iraq also protected Sunni civilians from vicious attacks by the Shiite militias. Unsurprisingly, Iran and its Iraqi allies saw the Sons of Iraq as a direct threat to their hegemony. To fight back, the pro-Iranian militias, who were quickly becoming more of a threat to the U.S. mission than Al Qaeda, accelerated their lethal attacks on American servicemen, under the direction of, and with assistance from, Iran's Revolutionary Guard commander Qassem Suleimani. Working in groups of 20 to 60, these Shi'a militias, known as Special Groups, were trained, armed, and funded by Hezbollah and Iran. These Shi'a militias began fielding Iranian-made explosively-formed projectiles (EFPs). These devices had exceptional armor piercing capabilities and began to take a toll on Coalition forces in Diyala province and the Green Zone in Baghdad.

One highly sophisticated raid conducted by Iran's proxy Asa'ib Ahl al-Haqq or AAH signaled the capabilities of such proxy groups. On January 20, 2007, a force of about a dozen AAH posing as an American security team, who spoke English and carried American weapons and traveled in five black GMC Suburban vehicles, were waved through a checkpoint by

Iraqi police onto the Karbala provincial headquarters. Upon entering the compound, the militant team, led by a blond-haired man, breached their way into the provincial headquarters, and in less than 15 minutes were driving away with four U.S. soldiers. The four soldiers were later found executed.

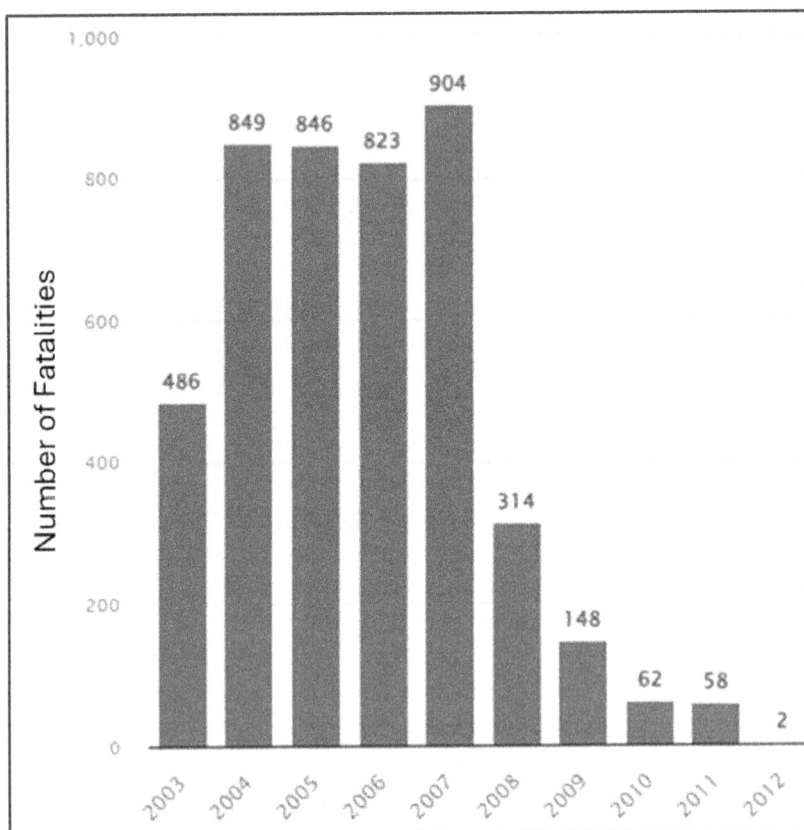

Iraq War Fatalities. Courtesy AP.

In a speech on January 31, 2007, Maliki stated that Iran was supporting attacks against Coalition forces in Iraq. Meanwhile, as the Sons of Iraq were helping to defeat the insurgency in Anbar, Maliki maintained his alliance with these militias, using them to marginalize, isolate, and intimidate his rivals. AAH and other Special Groups in Iraq accounted for over 6,000 attacks against Coalition forces.

Several occurrences throughout 2008 threatened the development of the Iraqi government. On March 24, 2008, Baghdad and the southern port city of Basra erupted in violence as loyalists of Muqtada al-Sadr attacked U.S. and Iraqi security forces. In response, Maliki launched an operation to oust the Mahdi Army. The battle accounted for the first time the Iraqi government had ever addressed the Shi'a militias. With Coalition support, the battle ended inconclusively, though the Iraqi government retained the field. With Bush on his way out, he tapped Petraeus to lead U.S. Central Command, placing him in operational control of both the Iraq and Afghanistan efforts. Lieutenant General Raymond Odierno then succeeded him as the new commanding general in Iraq. Then, in a momentous move, in Anbar, once the country's most restive province, the U.S. military handed over security responsibilities to the Iraqis. This was seen as a symbolic first step toward eventual U.S. withdrawal. Later the same month, Iraq's parliament passed a provincial elections law, clearing the way for voting in most of Iraq's provinces by January 31, 2009.

Obama and the Drawdown

Running on the pledge to end the war in Iraq, incoming President Barack Obama announced, "we would be as careful getting out of Iraq as we were careless getting in." Working with the Maliki Government, the Bush Administration had set a withdrawal date of December 31, 2011. To the consternation of U.S. commanders, in a February 2009 speech, Obama reaffirmed a commitment to this original withdrawal date. By this time, the American people were heartily sick of the war in Iraq. At the beginning of 2009, while the war was stabilizing in favor of the Coalition, by conservative estimates, the war had cost the United States at least $3 trillion and 4,400 troops killed.

Making good on his pledge, in early 2009, Obama announced plans to remove the majority of U.S. troops from Iraq by August 2010. In transition, his plan called for a "modest-sized presence" of some 50,000 troops to train, equip, and advise Iraqi security forces until the end of 2011. By the end of June 2009, U.S. combat troops from more than 150 bases, packed up and turned everything over to Iraqi forces in accordance with a Status of Forces Agreement (SOFA) between Iraq and the United States.

In what may be seen as a step in the wrong direction, between 2008 and the U.S. exit from Iraq in 2011, Maliki succeeded in effectively dismantling the Awakening. Seeing the Sons of Iraq as a threat to his political base, he took steps to disarm and demobilize them. By June 2012, some 70,000 Sons of Iraq were either integrated into the Iraqi Security forces or demobilized, with around 30,000 members continuing to maintain checkpoints. As time would tell, the absence of both the Americans and the moderate Sunni forces created a major power vacuum, one that Islamic State (IS) came rushing in to fill.

The Rise and Fall of the Islamic State

With the coming of the 2011 Arab Spring protests that spread across the Arab world, a civil war in Syria was ignited, and the remnants of the Islamic State was reconstituted. The Islamic State (ISI) formed in Iraq in 2004 when Abu Omar al-Baghdadi founded the Jaish al-Ta'ifa al-Mansurah. The group fought alongside Al Qaeda in Iraq during the 2003–2006 Iraqi insurgency. Built out of the remnants of AQI, in 2013, ISI began seizing territory in Syria and changed its name to Islamic State in Iraq and the Levant (ISIL) or Islamic State of Iraq and Syria (ISIS). As ISIS, known in the Middle East as Daesh, started taking ground in Iraq, they did so in large measure without a fight from the Iraqi Army. Anbar Province was soon thrown back into turmoil. Samarra fell on 4 June 2014, followed by the seizure of Mosul on 10 June, and Tikrit on 11 June.

As ISIS came rolling into Mosul, they were welcomed as liberators by Sunni leaders and ex-Baathists who hoped they would right the wrongs of Maliki's sectarianism. But as the world soon learned, these new ruthless jihadists systematically hunted down those who had worked with America. No one was spared. As the Iraqi Army collapsed, without either the Sons of Iraq or U.S. troops to protect them, Iraq civilians were left defenseless. Following its capture of Mosul, ISIS moved on Baghdad and easily captured Tikrit on June 11, 2014. The next day, in a video-recorded massacre outside Camp Speicher (Tikrit), ISIS summarily executed some 1,700 Iraqi Shi'a cadets, who were captured. From Mosul on July 14, 2014, Baghdadi proclaimed the Islamic Caliphate with himself as caliph. The Islamic State now controlled the lives of some 5 million people. Then, in August, as ISIS overran Sinjar, they threatened to massacre Yazidi civilians trapped on Sinjar Mountain. Coalition airstrikes forced ISIS to retreat, preventing a likely genocide. The U.S. carried out thousands of bombing sorties and cruise missile attacks against ISIS targets in Iraq and Syria.

In what seemed to be a repeat of the Camp Speicher massacre, Kurdish troops identified a compound in Hawijah near Kirkuk in which ISIS was using to house an assortment of political prisoners. The compound was believed to have some 75 prisoners inside. Judging from the fresh graves outside the compound, Kurdish authorities approached U.S. forces with intelligence regarding the execution site. Believing that a mass execution was imminent, JSOC and Kurdish troops conducted a joint raid of the Hawijah compound. It was decided that the commandos of the Kurdish Counter-Terrorism Group or CTG would lead the assault while about 30 Delta Force commandos would assist. In the early morning hours of October 22, 2015, three MH-47 Chinook helicopters carrying the joint force touched down in a field outside the prison compound. After the Kurdish commandos failed to breach through the compounds' outer walls, the Delta commandos took the lead of the assault, climbing over the walls in ladders. The joint force neutralized ISIS fighters and secured all 75 prisoners. The only commando to die was Master Sergeant Joshua Wheeler, a veteran of 14 deployments to Iraq and Afghanistan. Following the departure of the raid force, a pair of F-15E Strike Eagle-leveled the compound to deny the Islamic State their future use.

While it was good it was to pound ISIS back to the stone age, as Michael Pregent suggests:

With the Islamic State in the ascendant, the U.S. repeated the same mistake it had made a decade earlier, partnering with Shiite jihadists to fight Sunni ones, thus empowering Iran while turning potential Sunni allies into enemies. By carrying out airstrikes against IS forces fighting Shiite militias, the American planes were serving, as David Petraeus quipped, as Qassem Suleimani's air force. With the war against Islamic State, Tehran cemented its power in Iraq, making it part of the emerging "Shiite Crescent," a land-bridge from Tehran through Baghdad, Damascus, and Beirut, with Jerusalem as the hoped-for final destination. Suleimani now had a corridor to move troops, rockets, and precision missiles into Syria to be used against Israel.[9]

By December 2017, Iraqi Shiite militias fought alongside the Iraqi Army to oust ISIS from the country. The war would continue unabated in Syria. As some measure of stability returned, Iraq held its elections of 2018. Embellishing their role in ousting ISIS, the Shiite militias created an immensely popular political party known as Fatah. Adil Abdul-Mahdi was elected, replacing Haider al-Abadi, who had replaced Maliki when he stepped down.

As Iraq has now stabilized somewhat, the Badr Organization, and other Shi'a militias continue to play a key role in regional politics. Through such proxies, Iran firmly controls Iraq's military and political landscape. Today, corruption remains rampant, further undermining the Iraqi people's confidence in their leadership. Likewise, economic progress has been challenging as the nation's economy remains dependent on oil, which accounts for over 99% of exports.

In terms of blood and treasure, the Iraq War cost the United States 4,431 killed in action, over 30,000 injured and about $3 trillion dollars. Other casualties of the war include roughly 3,650 contractors killed, and over 200,000 Iraqi civilians. The Iraqi Republic remains an experiment.

Summary and Implications

By way of the five laws of irregular warfare, the following is an analysis of the American COIN effort:

Political objective (s) – Having torn down Saddam Hussein's tyranny, there was nothing to take its place; nothing to fill the military, political, and economic void left by the regime's fall. The result was that the United States created a failed state and a power vacuum, which even as of this writing has not been properly filled. As one veteran put it, "Bremer enacted a one-man performance of the tragedy *Hamlet*. The big difference was Bremer was able to leave 'Denmark' alive."

---

[9] Michael Pregent, How America Helped Iraq Fall under Iran's Grip, Sep 22, 2020, Mosaic Magazine.

Like Afghanistan, the inability of U.S. planners to understand Iraqi culture led to serious miscalculations. The second biggest mistake, after that of CPA orders 1 and 2, was that the U.S. sidelined the Sunni tribes, and formed a lopsided government with an overwhelming Shi'a majority. Equally uninformed was the U.S. proclivity for not only seeing the insurgency as monolithic in nature but lumping Sunni insurgents into the same camp as Al Qaeda. If there is one thing the United States has learned in both Iraq and Afghanistan is that the military can win battles, but it cannot solve problems that are fundamentally political in nature.

Equally as culturally ignorant were the actions of Al Qaeda. Their political objective of an Islamic State that undermined the traditional power of the tribes did not resonate with the Sunnis of al-Anbar. Had AQI not attacked the tribal leaders, it's fair to say things would have turned out vastly different. As events would unfold, Maliki's efforts to dismantle the Sons of Iraq helped to build a new threat, ISIS.

It is fair to say that U.S. objectives in Iraq were flawed from the start. In Afghanistan we started a war to kill one man. In Iraq, we started a war under false pretenses. Moreover, the U.S. entered Iraq with no plan to counter any insurgency, and after nation-building for nearly ten years, having spent $3 trillion, the U.S. withdrew with a misguided plan. In hindsight, it seems the Iraq invasion was a chance to clean house.

In an interview with *Vanity Fair*, former Deputy Secretary of Defense Paul Wolfowitz noted, "the threat of Saddam's weapons of mass destruction was simply the one threat upon which all the senior members of the Bush Administration agreed. We believed that it could be used to justify the war to the public." It may be said, we came for WMDs but stayed for oil and geopolitical gains.

Perhaps we will never know precisely why we invaded Iraq. But if there is one lesson I hoped we learned from the Iraq War (and Afghanistan War) it is this: you cannot occupy a foreign country and impose democracy.

Legitimacy – While AQI clearly delegitimized itself in the eyes of most Iraqis, with its kidnappings, beheadings, and bombings, government corruption and sectarian infighting permitted the insurgency to develop in the Sunni tribal community of Western Iraq. Weak ineffectual government likewise left Shi'a communities to be slowly taken over by vicious sectarian militias. These factors coalesced, along with organized crime rings across the country, stymied the development of governmental institutions capable of providing Iraqis with the most basic services such as clean water, sanitation, electricity, and a minimally functioning economy capable of generating basic employment. The persistence of these problems over time led to the emergence of low-level civil war in Iraq.

Adaptability – The turning point of the war in Iraq came about through the adopting of the Sons of Iraq as partners in counterinsurgency. As the Sunni population watched a barrage of breaking news that appeared to be targeted at their population, the U.S. won their support against AQI. For their part, AQI was busy creating fresh grievances among the Sunni population that made even the Great Satan looked like a reasonable alternative.

The following graph outlines the U.S. IW approach in Iraq:

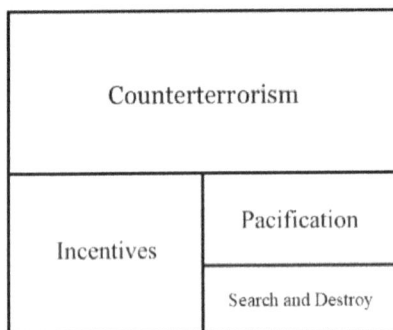

Counterterrorism		
Incentives	Pacification	
	Search and Destroy	

Influence – Success in COIN requires an operational design that stabilizes the affected area by identifying with the people, while working with them to create a better state of the peace, winning their allegiance, and if possible, redress grievances that led to the insurgency. Arguably, through the leadership of commanders like Petraeus, the Coalition tipped the scales in favor of the Iraqi government.

<u>Native Face</u> – In the end, the Sons of Iraq was formed for pragmatic political objectives (defeat of AQI) that lead to a measurable reintroduction of an important sectarian minority into Iraq's political infrastructure. Funded by the U.S. and threatening the power of the Shi'a-dominated national government, its legitimacy on the national level was always questionable at best, but it did restore the legitimacy of the Sheikhs at the tribal level. The Sons of Iraq is an example that supporting a local solution (Native Face) to a regional problem can be effective, especially in the short term.

For Further Reading:

1. *Fiasco* by Thomas E. Ricks.
2. *Endgame* by Michael R. Gordon and Bernard E. Trainor.
3. *By All Means Available* by Michael G. Vickers.
4. *Surprise, Kill, Vanish* by Annie Jacobsen.

Part Five

Great Power Competition

13

U.S. Efforts in African States

"The first, the supreme, the most far-reaching act of judgment that the statesman and commander have to make is to establish by that test the kind of war on which they are embarking; neither mistaking it for, nor trying to turn it into, something that is alien to its nature." – Clausewitz

A Special Forces Sergeant training African soldiers in Niger. Courtesy AP.

Since the U.S. Marines' expedition to Tripoli during the First Barbary War (1801–1805), which may be categorized as the first U.S. counterterrorism campaign, the United States has used irregular warfare in Africa in a variety of ways. Albeit, ever since U.S. Marines stormed the shores of Tripoli, U.S. policies in Africa have largely lacked clear direction and have been reactionary. For the first two hundred years of America's history, her principal relationship with the continent involved the slave trade. During the Cold War and period of decolonization in the

1950s and 60s, the United States began to compete in Africa with Russia and China. Since the end of the GWOT and beginning (or recommencement) of the Great Power Competition, the U.S. military has worked with its allies to counter threats and malignant actors, such as the Islamic State and Boko Haram, while promoting regional security and stability through various IW activities. This chapter will trace out the period of the 1970s to the present and will highlight the PLAIN laws of political objective and adaptability.

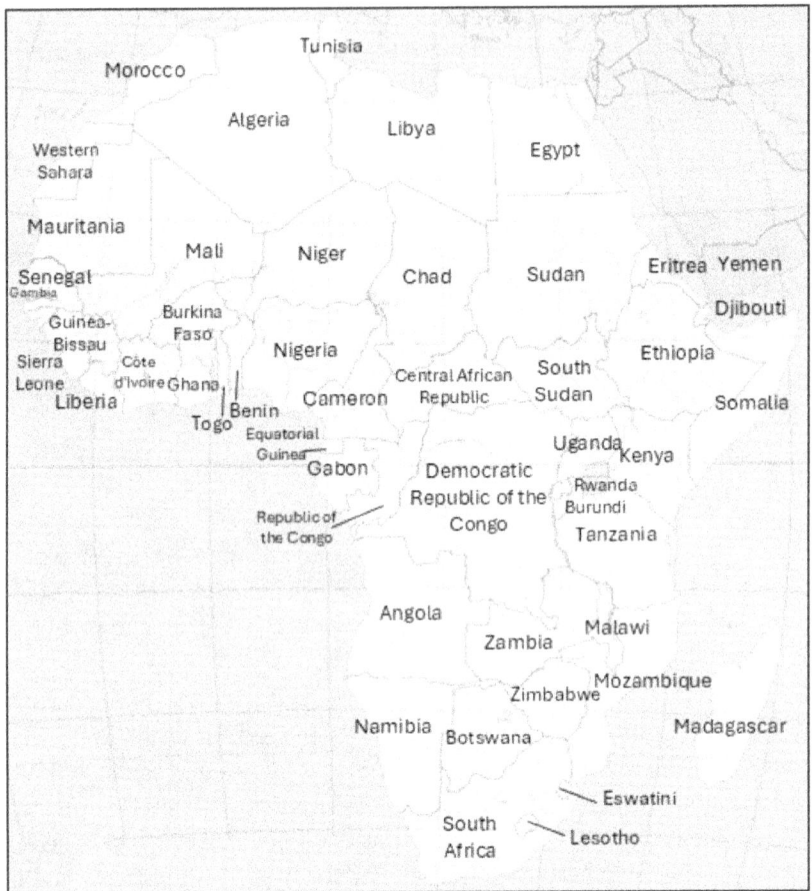

Tunisia
Morocco
Algeria
Libya
Egypt
Western Sahara
Mauritania
Mali
Niger
Senegal
Gambia
Chad
Sudan
Eritrea Yemen
Djibouti
Guinea-Bissau
Burkina Faso
Nigeria
Ethiopia
Sierra Leone
Côte d'Ivoire
Ghana
Central African Republic
South Sudan
Somalia
Liberia
Benin
Togo
Cameroon
Equatorial Guinea
Gabon
Uganda Kenya
Democratic Republic of the Congo
Rwanda
Burundi
Republic of the Congo
Tanzania
Angola
Zambia
Malawi
Zimbabwe
Mozambique
Namibia
Botswana
Madagascar
Eswatini
South Africa
Lesotho

The Nations of Africa.

13 – U.S. Efforts in African States

Proxy Wars

Africa has had its fair share of proxy wars. The decolonization of Sub-Saharan Africa beginning in the late 1950s resulted in several proxy Cold War confrontations between the United States and the Soviet Union. The initial U.S. approach focused on containment, supporting anti-communist regimes, often disregarding democratic principles. Lacking clear direction, this strategy enabled the rise of authoritarian regimes that remain in power today.

The first of these proxy wars occurred in the former Belgian Congo, which gained its independence from Belgium in 1960. The Eisenhower administration believed that the new Republic of the Congo could be a stable, pro-Western government. All that changed within a week after a civil war began. The U.S. backed the duly elected government of Joseph Kasavubu while the Soviet Union and China backed the breakaway Congo provinces of Katanga and South Kasai. Following a series of coups, power was eventually consolidated under U.S.-backed Joseph Mobutu, with an estimated 100,000 casualties throughout the conflict. Mobutu was the anti-communist choice. In time, however, he became the embodiment of a corrupt and brutal dictator. He would go on to reign for the next 32 years.

The next crisis in Africa played out in a similar fashion. Angola had been a Portuguese colony for over 400 years. Following Portugal's withdrawal, a violent power struggle erupted leading to a civil war between three competing former anti-colonial guerrilla movements, the Marxist MPLA (People's Movement for the Liberation of Angola), the FNLA (National Front for the Liberation of Angola), and UNITA (National Union for Total Independence of Angola). Not surprisingly, the Soviet Union backed the Marxist MPLA, and the United States supported the FNLA and UNITA, seeking to prevent Angola from becoming a communist state. The MPLA also had long-established relations with Fidel Castro. The Cubans first engaged the MPLA via Che Guevara in 1965 and provided some military training. The devastating civil war lasted for decades, and eventually Angola became a Soviet satellite. The instability continues today.

In a long-running civil war from 1974-1991, Ethiopia fought to oust its communist Derg regime. In 1974, a military junta called the Derg overthrew Emperor Haile Selassie in a coup, establishing a Marxist-Leninist regime that led to widespread political repression and violence. In the summer of 1977, poverty-stricken Somalia invaded its equally poor neighbor, Ethiopia, in hopes of conquering the Ogaden Desert region, which was populated by ethnic Somalis. What began as a local conflict in a far-flung region of the world grew into a Cold War hotspot as the United States and the Soviet Union took sides. To help him put down various rebel factions and to drive the Somalis out of Ogaden, Mariam enlisted the aid of the Soviet Union who also brought in Cuban troops.

Supported by Cuban troops, Ethiopia pushed the Somalis out of the territory, regaining control of all border posts by March 1978. While Soviet assistance contributed to its victory, it also further complicated the country's internal dynamics, significantly impacting the economy and leading to widespread poverty and famine. Eventually, the Derg regime was defeated following a long and bloody civil war led by a coalition of U.S.-backed rebel forces, the Ethiopian People's Revolutionary Democratic Front or (EPRDF), bringing an end to the regime and the civil war. Some 250,000 people were killed, and another million were displaced during the conflict. Ethiopia's instability continues today. All three of these conflicts were marked by a deep-rooted legacy of colonial division as well as strategic American missteps.

United Nations Operation in Somalia I and II

After the end of the Cold War, several nations collapsed primarily due to the removal of external support. One of these nations was Somalia. The authoritarian regime of Siad Barre collapsed. This led to a power vacuum which was exploited by clan-based militias and resulted in widespread civil war, famine, and a protracted state of instability.

The first major combat operation in Africa after WWII, and America's biggest combat operation since the Vietnam War occurred in Somalia in the horn of Africa. In 1991, Somali warlords overthrew brutal president Mohamed Siad Barre. Immediately, a civil war broke out as the two most powerful warlords, Ali Mahdi Muhammad and Muhammed Farah Aideed, began fighting each other. In the wake of the civil war, and the famine that ensued, an estimated 250,000 Somalis died. To alleviate the famine, the United Nations began a peacekeeping mission. While managing to negotiate a ceasefire, the UN was unable to stop the violence or address the famine, as the goal of distributing food and medical aid was largely unsuccessful. In December 1992, with security deteriorating and thousands of tons of food stranded in portside warehouses, President George H.W. Bush proposed the sending of U.S. troops to Somalia to protect aid workers and get the needed food to Somalia's starving people.

In a televised address, Bush said that America must act to save more than a million Somali lives, but reassured Americans that the operation was not open-ended, and that "we will not stay one day longer than is absolutely necessary." The well-intentioned decision of the president notwithstanding, America's troops would become entangled in Somalia's civil war, and the mission would stretch out for nearly two years before being called off by President Bill Clinton in 1993.

Beginning in December 1992, the U.S. committed about 25,000 troops to war-torn Somalia. The first U.S. troops to arrive were U.S. Marines from the 7th and 11th Marine Regiments. With the aid of U.S. troops and forces from other nations, such as those from Pakistan, the UN succeeded in distributing desperately needed food to many starving Somalis. The stability operations that commenced also sent Special Forces ODAs from 5th and 10th Groups. However, as the sectarian fighting continued, due to the lack of a central government, the UN went far beyond the limits of traditional neutral peacekeeping missions. To better stabilize the environment, the UN moved to disarm Somali civilians, and then, incentivized the humanitarian aid as a reward for support to the mission. In this way, the UN mission threatened the traditional power of the Somali clans as well as the political balance in the civil war. When President Bill Clinton took office in January 1993, he immediately reduced the U.S. footprint to just over a thousand troops.

Transitioning to a second UNOSOM action, the $1.6 billion UNOSOM II began in March 1993. By mid-1993, only 1,200 American combat soldiers remained in Somalia. This force was aided by troops from 28 other countries acting under the authority of the UN. The already unstable situation took a turn for the worse when 24 Pakistani soldiers were ambushed and killed by Aideed's militia. The Pakistani troops were gunned down while inspecting a weapons-storage facility. Forty-four other Pakistani soldiers were wounded. As the leader of the Habr Gidr clan with several thousand fighters, Aideed had previously been responsible for the killing of several soldiers using improvised bombs. Aideed's men were armed with assault rifles, RPGs, and machine guns, with many technical – heavy weapons mounted on vehicles.

The UN passed a resolution calling for the immediate apprehension of the warlord. In response, Clinton approved the deployment of Task Force Ranger, composed of 400 Rangers from 3rd Battalion's Bravo Company of the 75th Ranger Regiment, and commandos from Delta's C Squadron, who were supported by helicopters from the 160th Special Operations Aviation Regiment. The task force was commanded by Major General William Garrison.

From August to October 1993, TF Ranger conducted Operation Gothic Serpent with the primary mission of capturing Mohamed Farrah Aidid and his lieutenants. During the month of September 1993, the task force conducted six missions against Aideed's militia, capturing some of the warlord's associates. Then, on the afternoon of October 3, 1993, the task force was informed that two of Aidid's advisors were meeting in a building a block away from the Olympic Hotel in central Mogadishu. The building was in the epicenter for the Habr Gidr clan, the Bakara market. Like its previous six missions, the task force would infiltrate via helicopters then exfiltrate by ground convoy. The previous raids were conducted primarily at night, as this provided a tactical advantage for the task force who was able to operate using night vision optics. However, the time-sensitive nature of the target required the October 3 raid to be launched during the day.

Just after 3:30 PM, the assault force and ground convoy consisting of 19 aircraft, 12 HMMWV vehicles, and 160 men launched from the task force's base at an airfield three miles from the target. C Squadron commandos would raid the target building to capture Aidid's men while Rangers in blocking positions would surround the target building to isolate the target. After a 10-minute flight, four Little Bird helicopters inserted one element of the raid force, while six Black Hawks inserted the remainder of the commandos onto the target and the Rangers into their blocking positions around the target building. Within 20 minutes, TF Ranger had secured the target building, having captured 24 of Aidid's militia, including one of his top lieutenants. The assault force commander then called for the ground convoy to his location to load the prisoners and the assault force.

While the raid was underway, armed militiamen and civilians converged on the target area from all over the city. Summoned to the fight by cellphones, an estimated 4,000 militiamen converged on the American raid force. When two MH-60 Black Hawk helicopters were shot down, ground forces converged to their locations to recover the personnel. The ensuing battle to get to the downed helicopters turned out to be the most intensive close combat that U.S. troops had engaged in since the Vietnam War. After loading the detainees onto the vehicles, the ground convoy set off for the crash site but never made it as they were under a heavy barrage of small-arms and RPG fire. Forced to return to the airfield, roughly half of the 75-man ground force were injured.

During the raid, Master Sergeant Gary Gordon and Sergeant First Class Randall Shughart, a C Squadron sniper/observer team, provided precision and suppressive fires from Black Hawks above the two helicopter crash sites. Learning that no ground forces were available to rescue one of the downed air crews, and aware that a growing number of armed militiamen were closing in on the site, Shughart and Gordon volunteered to be inserted to protect their critically wounded comrades. After their third request, they were inserted one hundred meters south of the downed Black hawk. Armed with their personal weapons, the two commandos fought their way to the downed fliers through intense small arms fire, a maze of shanties and shacks, as the enemy converged on the site. After pulling the wounded pilots from the wreckage, they established a perimeter and fought off a series of attacks. The two commandos continued to protect their comrades until they had depleted their ammunition and were themselves fatally wounded. Their actions saved the life of Chief Warrant Officer Three Michael Durant. The two Delta snipers were posthumously awarded the Medal of Honor.

As night set in, the assault force became split up into six separate positions. The surrounded task force soldiers fought off thousands of militiamen and may very well have been overrun had it not been for the task force's helicopter gunships. Gun running throughout the night, the TF 160 pilots held the clan fighters at bay. Just before daylight, a cobbled quick reaction force finally reached the beleaguered task force. The wounded were loaded into vehicles while the uninjured ran back to the base. The one-hour raid had taken over 14 hours. In the aftermath, 18 U.S. troops were killed with another 84 wounded. Estimates of enemy fatalities are about 1,000. Then, on October 14, after a brokered deal, Somali captors released Durant, along with the remains of the other Americans killed during the battle.

The failure of UNOSOM II to restore order in Somalia had substantial repercussions for the country. Somalia continued to be mired in internal conflict and the international community became reluctant to intervene in other civil conflicts, such as the genocide in Rwanda in 1994.

In a U.S. Army War College publication, Hamid Lellou made the following analysis:

American strategy during the Cold War shifted from direct military engagement to more nuanced forms of influence, such as economic aid and covert operations. The lack of a coherent, long-term plan often led to short-term alliances with authoritarian regimes, which undermined efforts to promote democracy and stability. This inconsistency has had lasting repercussions, with American strategies enabling the rise of authoritarianism and allowing Russia and China to capitalize on the resulting instability to expand their influence in the region.[1]

Enduring U.S. Missions in Africa

Since the start of the Global War on Terrorism, the U.S. military has since been involved in a variety of military operations in Africa. These include Operation Enduring Freedom – Trans Sahara (OEF-TS) and Operation Enduring Freedom – Horn of Africa (OEF-HOA). Both operations work to promote regional stability and security, provide humanitarian relief, stop violent extremist organizations or VEOs, and eradicate terrorist organizations using every form of irregular warfare.

Launched in February 2007, Operation Enduring Freedom – Trans Sahara, redesignated Operation Juniper Shield in 2013, is a U.S.-led military operation that focuses on counterterrorism and policing in the Sahel region of Africa. Employing its Special Operation Forces, the United States works to train, advise, and assist local forces in missions against terrorist groups like Al Qaeda, Boko Haram, and the Islamic State.

---

[1] Hamid Lellou, *U.S. Relations with Africa and the New Cold War.*

The problems within Africa's Sahel are complex. Instability in the region is primarily driven by a convoluting mixture of factors. These include desertification which results in the loss of topsoil, reduced plant growth, and decreased ability to sustain agriculture. This in turn leads to food insecurity and increasing competition for scarce resources, which fuels tensions between communities, particularly between those in the cities and the nomadic tribes. Other major problems include weak political institutions, corruption, and a lack of basic services; all of which create fertile ground for extremist groups to operate. To aid rebel groups, the region has seen a proliferation of weapons from other conflicts. The Sahel is also susceptible to trafficking activities like drug smuggling, which further destabilizes the region. The composite sum of these factors contributes to a cycle of violence and displacement across the region, particularly in countries like Mali, Burkina Faso, and Niger, Operation Juniper Shield's primary operating location.

One of the biggest contributors to the instability of the Sahel was the collapse of the Libyan state in 2011. In the morning of September 11, 2012, a culmination of bureaucratic failures materialized as members of a terrorist organization known as Ansar al Sharia attacked the U.S. diplomatic compound in Benghazi, Libya. That attack resulted in the deaths of U.S. Ambassador to Libya John Christopher Stevens and U.S. Foreign Service Information Management Officer Sean Smith. As the attack continued, the CIA Annex in Benghazi also fell under attack and Global Response Staff (GRS) employees, Tyronc Woods and Glen Doherty, subsequently lost their lives in the defense of the compound.

Following the collapse of the Gaddafi regime in Libya, estimates suggest that hundreds of thousands, potentially up to a million, small arms and weapons, including heavy weaponry, were looted from government stockpiles, significantly contributing to regional arms proliferation, with many finding their way to various armed groups in neighboring countries.

Sahel Region of Africa.

While precise numbers are difficult to determine, estimates suggest that between 250,000 and 700,000 firearms were accessible, including significant quantities of heavy weapons like anti-aircraft systems. These weapons spread beyond Libya, reaching armed groups in countries like Mali, Chad, Niger, Nigeria, Central African Republic, and even further afield. These nations are home to the world's fastest growing and most-deadly terrorist groups.

Jama'at Nusrat al-Islam wal-Muslimin (JNIM) is a terrorist group based in Mali. It has been active across much of West Africa, including parts of Burkina Faso and Niger. The Islamic State West of Africa Province (ISWAP) and Boko Haram have been active in Nigeria since 2009. In 2015, Boko Haram was rebranded as the Islamic State in the West African Province (ISWAP). A splinter faction of the original Boko Haram was active until 2021, when ISWAP killed its leader, absorbed its territory, and relegated its members to remote islands in Lake Chad. ISWAP has since established control of northeastern Nigeria and parts of Niger. Niger has the largest increase in deaths from terrorism, with deaths more than doubling since 2007. This is the highest terrorism death toll attributed to attacks by the Islamic State in the Greater Sahara (ISGS).

Battle of Tongo Tongo

At the request of the government of Niger, the U.S. deployed a special operations task force to conduct a variety of counterterrorism and security force assistance missions. Working in concert with Niger's military, these activities were designed to build the capacity of those forces to conduct operations to counter Boko Haram, Al Qaeda in the Islamic Maghreb, and the Islamic State in the Greater Sahara (ISGS).

On October 4, 2017, a joint US-Nigerien patrol was ambushed by militants from the Islamic State in the Greater Sahara near the village of Tongo Tongo, Niger. The 3rd Special Forces Group team involved in the October 4 attack, Operational Detachment Alpha 3212, had previously deployed to Niger in 2016. However, personnel turnover during the previous year prevented the ODA from conducting key pre-deployment collective training as a complete team. When they arrived in Niger, only half of the team had conducted any collective training together. When the Team arrived back in Niger, they concentrated primarily on training the new CT company and conducted two partnered operations with their Nigerien counterpart.

On October 3, 2017, ODA 3212 left Camp Ouallam, situated 50 miles north of Niamey, with Nigerien forces on a CT operation targeting a key member of ISGS. Before departing, the team did not conduct pre-mission rehearsals or battle drills with their partner force. Once at their objective in Tiloa, unable to locate the target, the Team and their partner force conducted a key leader engagement (KLE) with a partner force commander. During the mission, no command higher than the company level was aware that the team's mission sought to capture a key member of ISGS. The acting company commander (a captain) was the most senior officer aware of the true nature of the mission. Becoming aware of a high-value target (HVT) near the team's location, the Special Operations Command and Control Element (SOCCE) commander, in N'Djamena, Chad, directed a multi-team raid. A second ODA and their partner force would converge on the HVTs location. Weather denied the second team's ability to take part in the operation. ODA 3212 went on to execute the raid but had a dry hole. Following the raid, the team and their partner force began their movement back to Ouallam.

While on their trek back to base, the team's partner force needed water necessitating the convoy's stopping near the village of Tongo Tongo to resupply. While there, the ODA commander conducted an impromptu KLE with village leaders. After the completion of the KLE, the U.S.-Nigerien force left Tongo Tongo and came under intense enemy fire immediately south of the village. The U.S.-Nigerien force dismounted and returned fire and attempted to counterattack the enemy in a flanking maneuver, killing four ISGS fighters. However, the enemy massed and began to envelop the friendly force. Realizing this, the detachment commander ordered a return to the vehicles and initiated a break contact to the south. The team acknowledged the order. Then, under increasing enemy fire, the force began to withdraw, throwing smoke canisters to mask their movement. However, two Nigerien vehicles and one U.S. vehicle, including three U.S. Soldiers (Staff Sergeant Black, Staff Sergeant Wright, and Staff Sergeant Johnson) did not withdraw from the initial ambush site. They were last seen by team members actively engaging the enemy from defensive positions near their vehicle and preparing to withdraw with the rest of the team.

Meanwhile, the rest of the team advanced to a point about 700 meters away and began to consolidate and attempted to make contact with the enveloped third vehicle. As the enemy continued to press the attack, the team was forced to continue its withdrawal. As this occurred, Sergeant La David Johnson was in a prone position to the rear of his vehicle when the order to break contact was given. Having acknowledged the order, he and two partner Nigeriens were unable to get back into their vehicle. Under heavy enemy fire, they were forced to evade on foot and were killed in action while engaging the threat. As the third U.S. vehicle became quickly overrun by enemy fighters, Staff Sergeant Black was killed, and the other two Green Beret staff sergeants, Wright and Johnson, continued to engage the advancing enemy until each was shot and killed by enemy fire.

Receiving the report of troops in contact, Nigerien and French partner forces responded immediately. A Nigerien ground QRF departed, and a French Mirage aircraft conducted several low-level shows of force, unable to engage enemy fighters because they did not have communications with the team on the ground. The Nigerien QRF located the remains of the three Green Berets along with one Nigerien Soldier as they swept through the ambush site. Following a larger search, the remains of Sergeant La David Johnson were found. The enemy ambush resulted in the deaths of four American soldiers, several Nigerien troops, and an estimated 100 plus ISGS fighters.

Following the Tongo Tongo ambush, the U.S. military implemented several changes aimed at better protecting troops and mitigating similar situations in the future. These changes include providing more armored vehicles to troops, increasing air support with armed drones for overwatch, improving medical evacuation response times, reviewing mission planning procedures, and emphasizing better training and intelligence gathering before deploying to high-risk areas in Africa.

In recent years, counter insurgency has significantly decreased the activities of the Islamic State in the West African Province in Nigeria, with deaths dropping over 90% from 2,131 in 2015 to 178 in 2021. The remainder of the Sahel region, however, has witnessed an increase in violence and terrorist attacks. Adding to the instability, several military coup scenarios have played out in countries like Mali, Burkina Faso, Chad, and Niger. Some have blamed the United States for military coups in West Africa, siting at least 14 coups involving U.S.-trained African officers, such as the January 2022 coup in Burkina Faso. Since 1950, there have been 492 attempted or successful coups carried out around the world, Africa has seen 220, with 109 of them successful.[2] According to one Special Warfare Journal article, "The number of violent episodes in the Sahel region, centered around Mali, Burkina Faso, and Niger, has quadrupled from 700 incidents in 2019 to over 2,800 incidents in 2022."[3] Sadly, as the article goes on to say, these same countries' armed forces have overthrown their governments "and turned to Russia for diplomatic support and military aid."

Wagner Group in Africa

With the declining U.S. presence in the Sahel, and the withdrawal of French forces in 2017, Mali, Burkina Faso, and Niger strengthened their ties with Russia. The organization that these nations encounter the most is the Wagner Group. Since it rose to prominence in the Syrian Civil War, Wagner Group has expanded its footprint into Africa. Wagner Group is a Russian state-funded paramilitary organization, also described as a private military company (PMC). Since 2017, Wagner's network of mercenaries has provided military support, security and protection for several governments in Africa including the Central African Republic, Sudan, Libya, Mali, Burkina Faso, Niger, and Mozambique, among other African countries. In these African nations, many of which have a limited U.S. presence, Wagner has immersed itself. In nations like Mali, Mozambique, Central African Republic (CAR), and Sudan, Wagner does everything from leading training exercises, fighting anti-government forces, and brutally quelling protests. In return, Russian and Wagner-linked companies have been given access to these countries' natural resources, such as rights to gold and diamond mines. The Russian military has likewise been given access to strategic locations such as airbases and ports. One of Wagner's strategies includes providing local

---

[2] Megan Duzor and Brian Williamson, VOA News.
[3] Special Warfare, Oct. 24, 2024.

leaders, journalists and social media influencers with training in Russia, after which they distribute pro-Russian propaganda with a local touch upon their return.

In September 2020, Mali's transitional government agreed to accept 1,000 Wagner group mercenaries "to conduct training, close protection, and counterterrorism operations." As part of a Mali-Russia military cooperation, Mali's government received four MI-17 Russian attack helicopters, along with weapons and ammunition. Given the chronological proximity of these two actions, Wagner appears intimately involved in the growing military-to-military relationship between Mali and Russia. Albeit all has not gone well for the group. On July 25, 2024, Tuareg rebels allied with Jama'at Nasr al-Islam wal-Muslimin ambushed a military convoy transporting Malian and Wagner personnel to Tinzaouaten. Nearly a hundred Russian contractors and scores of Malian troops were killed.

The presence of the Wagner Group in Africa presents a serious problem for the U.S. and its allies. Wagner continues to expand its foothold in the Sahel. Analysts predict that Burkina Faso could soon hire Wagner to help counter a growing jihadi insurgency in the wake of the withdrawal of French troops.

The Lord's Resistance Army

Over the years, the United States has provided security assistance and training to various African nations. In October 2011, Obama announced that he had ordered the deployment of 100 U.S. military advisors to Uganda with a mandate to train, assist, and provide intelligence to help combat the Lord's Resistance Army or LRA. At least 30,000 people died and another two million were displaced as the LRA spread terror in northern Uganda for more than 20 years. Elements of the U.S. Army Special Forces were deployed at a cost of approximately $4.5 million per month. Eventually the LRA was ousted from Uganda and is now believed to number between 200 and 300 fighters. In recent years the group has carried out attacks in the DRC, South Sudan, and CAR. Although the group is reportedly in decline, the LRA abducted over a hundred people and killed at least three others in 2015.

Operation Enduring Freedom – Horn of Africa (OEF-HOA)

One of America's enduring efforts in Africa has been Operation Enduring Freedom – Horn of Africa. Beginning in October 2005, the objectives of this military operation include defeating Al Qaeda and its affiliates along with strengthening the militaries of East African partner nations. The operation's primary military component is the Combined Joint Task Force – Horn of Africa (CJTF-HOA), which includes the naval components of Combined Task Force 150 (CTF-150) and Combined Task Force 151 (CTF-151), which operate under the United States Fifth Fleet.

The Horn of Africa is a politically volatile region with multiple actors, requiring careful navigation and coordination to maintain positive relationships. In 2016, a 10-year plan was implemented that focused on the following lines of effort: First, neutralize Al-Shabaab and transition the African Union mission in Somalia to the Somali National Army. Second, degrade violent extremist organizations in the Sahel-Maghreb and contain instability in Libya. Third, contain Boko Haram (now Islamic State in Western Africa Province). Fourth, interdict illicit activity in the Gulf of Guinea and Central Africa. And fifth, build peacekeeping and humanitarian assistance disaster response capacity for African partners.

CJTF-HOA has actively participated in community development projects like building schools, clinics, and water wells, conducting medical and dental care initiatives (MEDCAPs, DENTCAPs), and providing humanitarian assistance during natural disasters. The task force has demonstrated flexibility by adjusting its operations to respond to changing situations and emerging threats in the Horn of Africa.

In addition to various SFA, COIN, and humanitarian activities, OEF-HOA has conducted numerous CT missions against Al Qaeda linked groups such as al-Shabaab and has launched an anti-piracy initiative in the region. In one such anti-piracy operation, an American merchant mariner captain named Richard Phillips was taken hostage by Somali pirates but was rescued by U.S. Navy SEAL snipers who fatally shot his captors. Many OEF-HOA missions in the region have centered on disrupting terrorist activities, capturing high-value targets, and gathering intelligence. In January 2012, one SEAL mission recovered two hostages taken captive by Somali pirates. Dubbed Operation Octave

Fusion, the U.S. military rescue mission was launched on January 25, 2012, to rescue American citizen Jessica Buchanan and Danish citizen Poul Hagen Thisted, who had been kidnapped and held hostage by pirates.

In the early morning hours of January 25, 2012, operators of SEAL Team 6 (ST-6) conducted a HALO infiltration a few miles from where the two hostages were being held. Three months earlier, Somali pirates had kidnapped the two after they had given a workshop on the topic of landmines in the city of Galkayo, situated about 350 miles northeast of Mogadishu. Using drones, JSOC soon found out where the pirates were holding them and were able to track the hostage movements and daily routines.[4] With the negotiations with the kidnappers dragging on, Obama authorized JSOC to attempt a rescue mission.

The SEALs realized in freefall that they couldn't make it to the primary drop zone. There was too much fog. Using microphones and earpieces, they managed to discuss the change in plans and managed to land at their alternate drop zone, far enough away that the pirates never heard or saw them.[5] Having offset infilled, the operators closed the distance to the camp, then surreptitiously entered it. The nine kidnappers didn't have a clue they were under attack until it was far too late. Within minutes, the SEALs neutralized all nine of the captures, then moved on foot to a pickup zone where Task Force Brown helicopters flew them to Galkayo's airport where an MC-130 flew them to Djibouti.

The Great Power Competition is in full swing in Africa. However, the U.S. can stand to step up its game. In his article entitled *America Ignores Africa at Its Own Peril*, Joe Bruhl offers the following critique of U.S. dealings in Africa:

Throughout its history, the United States lacked clear objectives on the continent and, as a result, its policies were largely reactionary, vacillating between exploitation, benign neglect, and half-hearted attempts at democratization and humanitarian assistance. Today, Washington still primarily views the continent as a problem to be managed rather than as a partner in shaping the next century. Instead of a problem to be solved, China and Russia view Africa as an opportunity to be seized. From 2007 to 2017, U.S. trade with Africa dropped by 54% as

---

4 Sean Naylor, *Relentless Strike: The Secret History of Joint Special Operations Command* (New York: St. Martin's Griffin, 2015), 425.
5 Ibid.

China's grew by 220%. While Russia's total investment in Africa pales in comparison to the United States and China, it has grown by 40% since 2015. China supports 46 port projects in Africa — financing more than half and operating 11. The United States supports zero.

The United States has deep and meaningful military relationships across the continent. The U.S. military has trained African armies, helped to fight violent extremists, and improved the effectiveness of troops deploying to U.N. operations. A successful U.S. strategy would shift the current downward trajectory by increasing military investment in Africa. The U.S. Army's employment of a Security Force Assistance Brigade in Africa is a good step, but to sufficiently support U.S. policy objectives, the Department of Defense needs to reverse its drawdown and include more than one paragraph about Africa in the next National Defense Strategy.[6]

The United States must pay more attention to Africa. It's time for the United States to compete like a great power in Africa.

When I was training the Mauritanian army in 2018, the U.S. had just opened a gleaming new embassy in the capital to the tune of hundreds of millions of dollars (as a replacement for the slightly less gleaming old but still massive one.) Touring the city, we passed by the Chinese embassy, which was indistinguishable from a Radio Shack in a strip mall. Meanwhile, Chinese had constructed Mauritania's major port and supervised its local maintenance and administration. They own the economy that pays for the military we were training. Who has loyalty, despite whatever personal inroads we make with the military's (or country's) future leaders. Competition on a political level must be more than military, similar to building out an area complex in UW.

Summary and Implications

By way of the five laws of irregular warfare, the following is an analysis of American IW efforts in Africa:

Political objective (s) – According to the U.S. Department of Defense, the U.S. strategy for Africa is to continue to partner with African governments to help shape the global security environment, preempting and deterring conflict and building regional mechanisms for security.

---

[6] Joe Bruhl, *America Ignores Africa at Its Own Peril.*

This will prove challenging as Africa has seen more political unrest, violent extremism, and democratic reversals than any other region in the world. With the exception of Nigeria, in which the U.S. provided substantial resources and training to help fight terrorism, instability in the Sahel, combined with worsening insurgent violence and frequent coups, exposes a lack of strategic clarity in U.S. Africa policy.

Legitimacy – Since the Wagner Group has emerged as an armed actor in Africa, the private military company has established ties in several African countries, where they provide security services and paramilitary assistance. Wagner is most active in the Central African Republic (CAR), Libya, Mali, and Sudan – nations that are at odds with the West. Likewise, China's belt and road initiative (BRI) is exploiting African states, and in fact now own many airports and port facilities. In this new scramble for Africa, Russia and China are competing for the control of the abundant natural resources of the African continent. The U.S. political and military interest in Africa has increased significantly in recent years. Albeit, a lot more needs to be done to stem the tide of Russian and Chinese exertions. Whatever is to be done, the U.S. would do best to retain the moral high ground.

Adaptability – It stands to reason that the United States should evaluate and rethink the way it employs IW in Africa. Violent extremist organizations thrive in regions where there is a general lack of essential services. If the U.S. military were to focus more on stability operations, in particular civil affairs in support of a larger interagency effort, it could provide greater regional security and stability.

Influence – The United States needs to counter Russia and China in Africa. In May of 2024, following a military coup that ousted the elected government, Russian forces moved onto a $100 million airbase in Niger that U.S. troops had at one time used for counter-terrorism missions. Moreover, once a key U.S. partner in counterterrorism efforts in the Sahel, Chad is now turning to Russia for security assistance. If left unchecked, China and Russia's growing influence in Africa threatens regional stability and American interests.

<u>Native Face</u> – The efficiency of a unit conducting IW depends largely on its knowledge of the people and the terrain. The United States needs to develop far better cultural intelligence in Africa.

For Further Reading:

1. *Relentless Strike* by Sean Naylor.
2. *In the Company of Heroes: The Personal Story Behind Black Hawk Down* by Michael J. Durant, Michael Durant, Steven Hartov.
3. *Black Hawk Down* by Mark Bowden.

14

The Syrian Civil War

"The first, the supreme, the most far-reaching act of judgment that the statesman and commander have to make is to establish by that test the kind of war on which they are embarking; neither mistaking it for, nor trying to turn it into, something that is alien to its nature." – Clausewitz

U.S. soldiers stand along a road across from Russian military armored personnel carriers (APCs), near the village of Tannuriyah, Syria, May 2020. Courtesy AP.

The Syrian Civil War is one of the most complex of conflicts as well as the greatest human disasters of the twenty-first century. What began as a peaceful protest movement challenging an autocratic government has evolved into a complex multi-layered war. At the time of this writing, an estimated 500,000 Syrians have been killed and another 1.9 million wounded. A further 5 million have fled the country and an additional 6 million are internally displaced, all of which are out of a prewar population of 21 million. This chapter outlines the American involvement in the conflict, demonstrating the PLAIN laws of political objective and adaptability.

In December 2010, a wave of pro-democracy protests and uprisings known as the Arab Spring spread through the Middle East. Beginning in Tunisia with the self-immolation of a fruit vendor decrying corruption, pro-democracy protestors overthrew the El Abidine Ben Ali regime on 14 January 2011. The wave spread eastward across the Arab world, coming to Egypt, overthrowing the regime of Hosni Mubarak on 25 January. Taking great inspiration, in March 2011, 15 boys in the southwestern city of Deraa, Syria, spray-painted on a school wall: "The people want the fall of the regime." After they were arrested and tortured, protests broke out against the rule of Syrian dictator Bashar al-Assad. When three protestors were killed by Assad's goons, anti-regime protests soon spread from Deraa to major cities such as Damascus, Hama, and Homs where they were likewise met with regime violence.

Events in Deraa offered a preview of what was to come: the Syrian Army fired on unarmed protesters and carried out mass arrests. Torture and extrajudicial executions were frequently reported at detention centers. As large-scale unrest spread over Syria, President Obama issued a statement on Syria, condemning in the strongest possible terms the use of force by the Syrian government against demonstrators. The general sentiment in the U.S. was that the Syrian protests would naturally follow the path of Tunisia and Egypt eventuating in the downfall of the Assad regime. In late April 2011, the Syrian army brought in attack helicopters, tanks, and artillery, laying siege to Deraa. Cut off from food, water, and medicine, the civilian death toll mounted. Amid international condemnation, Assad repeated the Deraa response in other cities where there were protests. An armed resistance to Assad began to take form.

Appalled by the indiscriminate violence, a group of Syrian Army officers who defected from Assad, formed the Free Syrian Army or FSA in July 2011. Their goals were to protect unarmed protesters and bring down the Assad government. The FSA's civilian counterpart, the Syrian National Coalition (SNC), was likewise established in summer 2011, in

Istanbul. The United States, Turkey, and among others, soon recognized the SNC as "the legitimate representative of the Syrian people." By the late summer of 2011, rebels had formed over a thousand local militias. Skirmishes between the regime and rebels became commonplace.

Syria. Courtesy of AP.

As it appeared Assad might be overthrown, Iran and its proxies rushed to his aid. Tehran began airlifting men, weapons, and supplies to Damascus. Qassem Suleiman organized loyal militias in Syria along the same model it had used in Iraq. The Quds Force supplemented these Shi'a militias with Shiite fighters from Iraq, Lebanon, Afghanistan, and Pakistan. After a year of bloodshed, Assad's forces hammered rebel enclaves, killing an estimated 8,000 rebels. Camps in Jordan, Lebanon and Turkey began to overflow with refugees. The initial American response was to send humanitarian aid. Reeling from the aftermath of Libya's chaos in the wake of toppling Qaddafi, the Obama Administration was leery of providing kinetic aid to the rebels. Watching the genocide take place, those in Congress, like Senator John McCain, were calling for U.S. air strikes to create safe havens for Syrian rebels. In the estimation of General Martin Dempsey, the Chairman of the Joint Chiefs of Staff, it would take something like 70,000 U.S. personnel to carve out safe havens for the rebels. This was a step Obama was not willing to take. However, extremist groups connected to Al Qaeda were willing. Funded by an array of wealthy Sunnis from countries like Saudi Arabia, as early as the end of 2011 they began to arrive. Among these groups was the nucleus of what would eventually become ISIS.

For the U.S., the real conundrum was to identify who could be trusted among these rebel groups who could mount an effective resistance against the Syrian regime. From Damascus, Assad's rhetoric aimed at delegitimizing the rebels: "They call it a revolution, but it has nothing whatsoever to do with revolutions. A revolution needs thinkers. A revolution is built on thought. Where are their thinkers? They are a bunch of criminals." With mounting political pressure, CIA Director David Petraeus proposed arming a group of rebels. These would be organized and trained in Jordan. Having just pulled out of Iraq, Obama demurred. Over the next two years, the civil war intensified. More than 60,000 Syrians were killed, and hundreds of thousands were forced to flee their homes, while Iran continued to send arms to the regime. Emboldened by the West's inaction, Assad initiated a new phase in the war, the deployment of chemical weapons.

The first chemical attack was carried out on August 21, 2013, in a rebel-held suburb east of Damascus. Calculating that use of chemical weapons would strike terror in the hearts of the civilians and eliminate rebel support, the Syrian Army fired twelve rockets with a sarin gas chemical payload at the opposition held district of Eastern Ghouta. Sarin gas is a nerve agent that causes lung paralysis and results in death by suffocation. The attack killed an estimated 1,400 people. This became a red line for the Obama Administration. Obama vowed that there would be enormous consequences if there were more chemical weapons used. Despite the threat, Assad remained defiant.

Plans were drawn up to launch strikes on August 31, 2013. However, the next day Obama had second thoughts. Wanting to shore up support, the president then sought authorization from Congress for the use of force. While this was being deliberated, as Rebekah Koffler relates, Russian reshaped the civil war. In an op-ed, Vladimir Putin appealed to the American people that they refuse to support Obama's potential action against Assad. Putin reminded Americans:

That Russia and the United States, despite the fact that they stood against each other during the Cold War, they were also allies once and defeated the Nazis together. Putin invoked the images of innocent victims, inevitable civilian casualties, including the elderly and children, escalation, potentially spreading the conflict far beyond Syria's border, unleashing terrorism, and further destabilizing the Middle East and North Africa. He tried to sow doubt that Assad was responsible for gassing his own people by putting the blame on opposition forces, knowing that, like in Russia, Americans believe that you are innocent until proven guilty. In this way Putin weaponized information to achieve a psychological effect on Obama and the American people and prevented the enforcement of the red line to undermine support for punishing Assad.[1]

A few days later, Russian foreign minister Sergei Lavrov convinced Secretary of State John Kerry that Russia would remove the chemical weapons arsenal from Syria. There would be no need for a punishing strike. The Syrian people looked to the skies for America's help in vain. As these events unfold, the Kurds, who long dreamed of autonomy, took up arms and succeeded from the government. With U.S. inaction, many moderate rebels began to join with extremist groups. ISIS emerged as the strongest of the factions, wresting government control away of Raqqa in northeastern Syria. Six months later, ISIS crossed back into Iraq. U.S. officials were in shock as the Iraqi security forces that the U.S. had spent billions on folded. Facing little or no resistance, ISIS moved deeper inside Iraq, capturing Iraq's military bases and seizing everything from small arms to light-armored vehicles and tanks.

The U.S. response was robust. On September 22, 2014, Obama launched a U.S. air campaign against the Islamic State, targeting ISIS positions across a huge swath of northern and eastern Syria. These

---

[1] Rebekah Koffler, *Putin's Playbook: Russia's Secret Plan to Defeat America* (Washington, D.C.: Regnery gateway, 2022), 160.

strikes were in support of the Free Syrian Army (FSA) and the Kurdish People's Protection Units (YPG). Operation Inherent Resolve, with the aim of defeating the Islamic State in Iraq and Syria, was now underway. Beyond defeating ISIS, working with various international partners, the U.S. goal was to establish conditions for post-conflict stability in the region. Coalition air power took to task the annihilation of ISIS, conducting over 8,000 air strikes, dropping over four thousand 2,000-pound Joint Direct Attack Munitions or JDAMs. Lethal support to Syrian Kurdish forces who were besieging ISIL-controlled Kobani then began in October 2014.

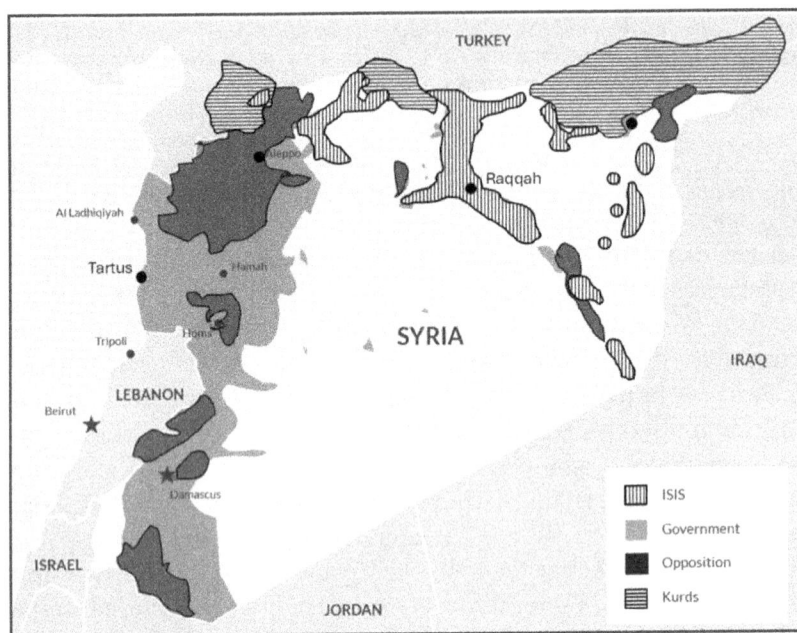

Syrian Civil War. Situation Summer of 2014.

Realizing that bombs alone would not defeat ISIS, though he initially rejected the proposal, Obama relented and approved Timber Sycamore, the CIA program to arm and train Syrian rebels fighting the forces of Assad. Pursuant to the arming of these rebels, the U.S. shipped 994 tons of Soviet-type weapons and ammunition from Eastern Europe to Jordan.

However, these weapons were stolen by Jordanian intelligence operatives and then sold to arms merchants on the black market. As things turned out, the program "significantly augmented the quantity and quality of weapons" of the Islamic State. Eventually, the U.S. began supporting rebels, but only those who would fight ISIS.

From Civil War to Proxy War

In time, the Syrian Civil War morphed into an international proxy war as an immensely complicated global power struggle played out within the nation's borders. To best understand the war, what is needed is to see it from the broader picture of a regional proxy war. In what is referred to as the Saudi-Iran Proxy War, the two main actors, Saudi Arabia and Iran vie on multiple levels over geopolitical, economic, and sectarian influence in pursuit of regional hegemony. Having many comparisons to the cold war, the cause of the rivalry is the age-old schism between Sunni and Shi'a Islam. The present rivalry has developed in stages. The first stage occurred in the wake of Iran's Islamic Revolution in 1979. Saudi Arabia began to promote the fundamentalist Salafist strain of Sunni Islam to counter Tehran's Shi'a ideology. The second stage gathered force in the power vacuum resulting from Saddam Hussein's ouster from Iraq in 2003. This led to the rise of Sunni jihadism, along with a renewed assertiveness of Iran to counter it.

In January 2015, the United States began to train opposition forces. Assembling a coalition of nations including Jordan, Qatar, Saudi Arabia, and Turkey, thousands of rebels were trained and equipped for the fight. The Pentagon's goal was to have 3,000 fighters to have completed training by the end of 2015. There was some static involved in the training of the rebel groups. While the U.S. trained and armed the Syrian Free Army as well as the Syrian Democratic Forces, Turkey viewed the Kurdish SDF formation as a terrorist group.

Following appeals from Assad for help, Russia entered the war in September 2015. As his closest ally in the region, Russia had much to lose. Putin's only port of access in the Mediterranean, the port of Tartus, stood in danger of being lost. To bolster Assad, Putin deployed an impressive array of aircraft consisting of 30 fighter planes, six Su-34 bombers, and 15 Mi-24 attack helicopters. In addition to air forces, the

Russians brought just under 2,000 personnel, along with six T-90 tanks, 15 artillery pieces, and 35 armored personnel carriers. On September 30, 2015, Russia launched its first airstrikes against targets in Rastan, Talbiseh, and Zafaraniya in Homs Province.

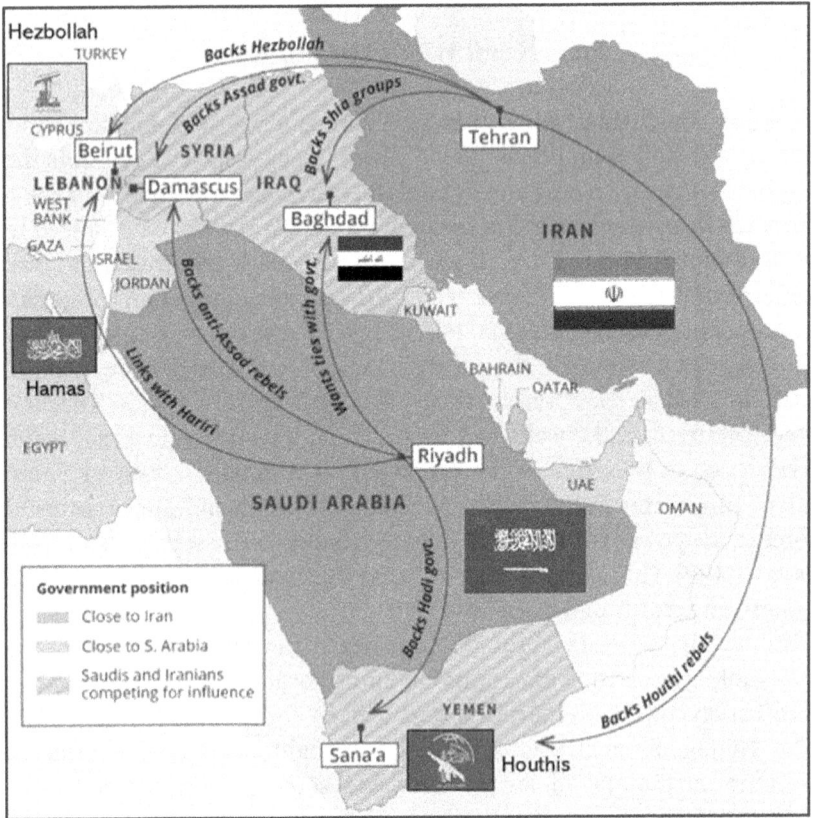

Saudi-Iran Proxy War. Courtesy AP.

For Iran's part, the war was all about countering its regional rival, Saudi Arabia. At the height of Iran's intervention, from 2015 to 2018, an estimated 10,000 Islamic Revolutionary Guard Corps or IRGC forces and 5,000 Iranian Army soldiers were stationed in Syria. Russian airpower and the force of Iran's Hezbollah and other Shi'a militias were brought in to prop up their ailing ally. Essentially, the conflict in Syria drew up four sides in the fight: the Assad regime backed by Russia and Iran; the Kurdish forces backed by the U.S.; the opposition forces backed by Saudi Arabia, Turkey, Jordan, the U.S., and other Gulf States; and the Islamic State backed by no one. To complicate matters further, Turkey began to bomb Kurdish forces while these U.S.-backed Kurdish formations are battling ISIS and the Assad regime, while Russian aircraft bomb opposition forces.

The Rise and Fall of ISIS

In 2015, the Obama administration authorized U.S. boots on the ground with an initial deployment of fifty U.S. Army Special Forces soldiers from Fort Campbell's 5th Special Forces Group. The American force would train and advise elements of the Free Syrian Army. The aim of the program was to train an army of 15,000 rebels to fight ISIS. The nucleus of what was to be Division 30 of the new Syrian Forces was trained in Qatar. After three weeks of training, the first group of Division 30 rebels set up a headquarters in Syria. Before the group could get off to a start it was attacked from the sky by Assad's warplanes and eventually eliminated the Al Qaeda-affiliated group al-Nusra front. Having handed over their brand-new trucks, weapons and ammunition to the al-Nusra Front, the surviving members fled to join other rebel groups. Additional groups of infiltrating Division 30 were captured as they attempted to enter Syria. Following the failed $500 million program to train moderate Syrian rebels, the program was abandoned. U.S. efforts then pivoted to placing all their stock in the SDF.

In support, U.S. Air Force A-10 Warthog and F-15E fighter planes were positioned in Incirlik Airbase in Turkey.

In the chaotic days of 2011, the remnants of AQI in Syria rebranded themselves ISIS and began seizing territory in remote areas of western Syria. As ISIS forces were advancing into Iraq, beginning in September 2014 to January 2015, they laid siege to the Kurdish-held city of Kobani, in northern Syria. ISIS fighters were held back by relentless U.S. airstrikes. With the battle for Kobani ongoing, Combined Joint Task Force – Operation Inherent Resolve (CJTF-OIR) was established on October 17, 2014. Its mission was to work with partners to defeat ISIS throughout Iraq and Syria and set conditions for post-conflict regional stability. Over time, CJTF-OIR encompassed the contributions of more than 60 partners.

After Donald Trump won the 2016 presidential election, he announced that the U.S. would withdrawal from the conflict. Then, in one of the largest urban battles in history, with the influx of Russian and Iranian support, Assad was able to recapture the city of Aleppo that had fallen into rebel hands. The four years of savage fighting left over 31,000 people dead, including the destruction of an estimated 33,500 buildings. Following up on this victory, in the Spring of 2017, Assad again used chemical weapons on his people, killing 85 people, including 20 children. Trump vowed a response. And on April 7, 2017, for the first time in the war, the U.S. deliberately attacked Syrian government forces in a missile strike on the regime's Shayrat Airbase.

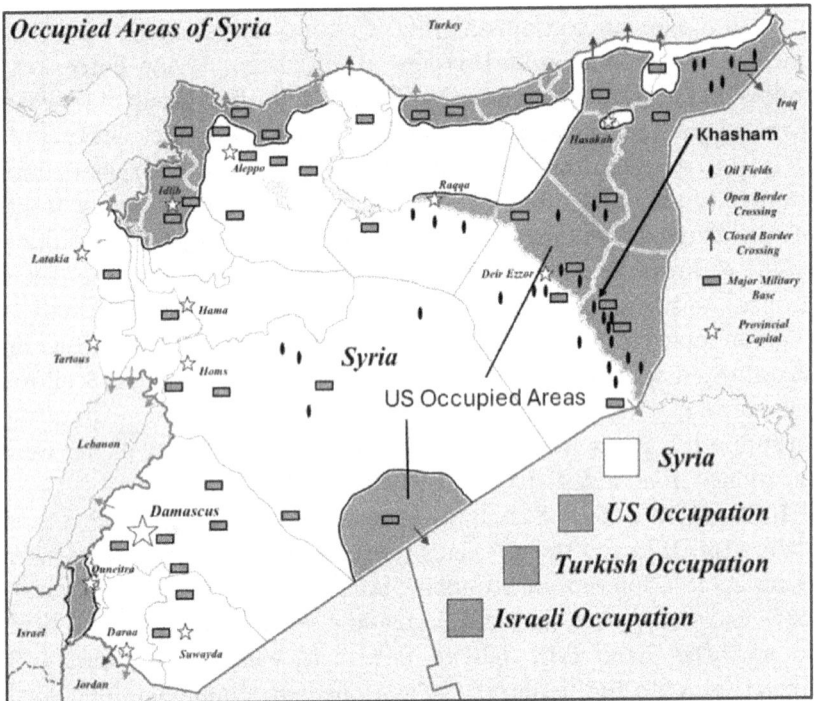

Syrian Civil War. Situation as of 2017. Courtesy AP.

By mid-2017, U.S. partner forces consisted of about 50,000 Syrian Opposition soldiers. Throughout 2017, JSOC struck ISIS in a variety of raids designed to decapitate its network. On January 8, 2017, JSOC commandos carried out a raid in Deir Ezzor, capturing ISIS fighters and killing at least 25. Then, on March 21, 2017, JSOC and Kurdish SDF fighters launched a heliborne assault on ISIL around the Tabqa Dam. In the fighting that lasted for three weeks, senior leader of ISIS, Abu Sayyaf, was killed.

Battling ISIS for the eastern side of the Euphrates, the combined U.S.-Syrian force liberated the Syrian cities of Tabqah, Manbij, and Shaddadi. In concert with these efforts, Iraqi Security Forces and their American advisors liberated approximately 70 cities, including Tikrit, Haditha, Ramadi, Fallujah, Hit, and Mosul. Then, in June 2017, springboarding from Kobani, the SDF, supported by SOJTF-OIR and Coalition airpower, launched their attack on the ISIS capital of Raqqa, one of the last remaining ISIS strongholds. Despite heavy resistance from ISIS defenders, the SDF secured the city by October 2017 and wrenched the al-Omar oilfield from ISIS fighters.

The Battle of Khasham

Following the fall of Raqqa, the Euphrates River became the de facto border between U.S.-backed rebels and the Russian-backed Syrian government. In support of the Assad regime, Russia deployed an array of surface-to-air missile systems such as the S-300 and S-400 along the new de fact border. While Coalition forces and their partners consolidated control of these liberated areas they faced a new threat, Wagner Group. The now infamous Wagner Private Military Company had been established one year prior in 2014, and by now had amassed a litany of war crimes. Wagner had an estimated 5,000 mercenaries in Syria. Giving Putin a measure of deniability, Wagner was sent to train, advise, and coordinate Assad's forces. By this point in the war, Wagner PMCs had been involved in earlier offensives in 2016 and 2017.

As the war was not getting any cheaper, their central task was to seize oil and gas fields in Syria, protecting them for the Assad government. As an incentive, any oil field Wagner liberated would yield the private military company 25% of the profits. Just over the Euphrates River from

Wagner's location was the Conoco gas plant in Khasham. Before long, it became clear to the Coalition that Wagner's chief Yevgeny Prigozhin planned to seize it. To take it, they would have to contend with a team of about 30 soldiers from Joint Special Operations Command who were stationed there.

Monitoring drone feeds, the JSOC troops began to note the arrival of dozens of fighters just across the Euphrates. Then, on February 7, 2018, a force of Iranian backed militia groups, and the Wagner PMC crossed the Euphrates River into SDF territory. The size of the force numbered was about 500 strong and was armed with machine guns, artillery, armored personnel carriers, and 27 Russian T72 tanks. The enemy force far outnumbered the small contingent of U.S. forces at the gas plant. The nearest friendly forces were a Special Forces ODA and a platoon of Marines were located at a mission support site about 20 minutes away.

The enemy force massed around the gas plant throughout the day. By nightfall, the JSOC team watched as Russian T72 tanks moved within a mile of the refinery. As one veteran of the battle recounted, "I think part of the tell was that Russian doctrine says that they're going to do things that look like exercises right up to the point of attack." At 10pm, the Wagner-led mixed force mounted their tanks and armored personnel carriers, closed the distance towards the oil refinery, and opened fire. The enemy tank, artillery, and mortar fire were heavy in volume but largely ineffective. Fearing the situation could escalate into open war with Russia, the Trump Administration notified the Russians to recall the attacking force. However, the Russians denied any involvement, claiming the forces arrayed in that area were only Assad's. As the small JSOC team came under fire, they accurately destroyed several advancing tanks. This would be the Russians' first taste of the javelin. Likewise, the Marines and the ODA took off in a race for the plant.

While the Russian tanks continued to fire, enemy ground troops made a frontal assault over open ground. The predator drones conducting surveillance on the situation launched all their Hellfire missiles to destroy as many enemy tanks and artillery positions as possible while the JSOC team engaged the assaulting force. The drones, now out of ammo, continued to remain on station to provide surveillance of the fighting. Soon the U.S. Air Force arrived. The United States did not know how Russia would react once it began to launch a full-scale air operation.

Once on station, F-22 raptors delivered punishing precision strikes against Russian artillery and mortar positions. Not long afterward, U.S. Army Apache helicopters arrived delivering a withering barrage of missiles and rockets onto the advancing tanks and armored vehicles. Once the QRF finally reached American defenders, they provided some much-needed relief to the woefully outnumbered force unleashing hell on the advancing troops. While Russian and Syrian forces were shooting blindly at night, U.S. forces were able to engage ground targets accurately up to 1,000 meters with night vision, infrared lasers, and array of heat and light sensors. Technology also allowed U.S. ground troops to mark targets for aircraft. Even an AC130 gunship arrived overhead, as Air Force combat controllers began to direct the next wave of American firepower. The 500-strong enemy force was shattered. An estimated 100 pro-Syrian and 200 plus Wagner troops were killed. The only casualty on the U.S. side was an allied Syrian fighter who had been lightly wounded.

Afterward Russia denied all responsibility for the attack. Months later however, Russia news confirmed that an unusually high number of dead Wagner troops arrived in Russia after the battle. Additionally, Iran's Supreme Leader, Ali Khamenei, condemned the confrontation and stated "Today, the U.S. government is the cruelest and most merciless system in the world, which is even worse than the savage ISIS members." Assad's foreign minister called it a "brutal massacre and a crime against humanity."

Coalition forces and their partners spent most of 2018 consolidating control of liberated areas and rooting out remaining pockets of resistance in the Euphrates River Valley. Sensing that the U.S. mission in Syria had concluded, U.S. President Donald J. Trump announced his intention to withdraw the remaining 2,000–2,500 U.S. ground troops in Syria on December 19, 2018. The SDF struck a deal with the Assad government after Trump announced a troop withdrawal from the country.

On March 23, 2019, the SDF pronounced the "destruction of the so-called Islamic State organization" after clearing the last ISIS stronghold in Baghouz in southern Syria. In the west, with the assistance of their foreign backers, Assad's forces besieged and bombarded the rebels' final redoubts in Syria's northwest in late 2019, imperiling hundreds of thousands of civilians. By December, the regime and its allies advanced into Idlib, where Russia-backed forces launched a devastating air campaign and clashes resumed between the regime and Turkish forces seeking to protect their opposition posts in the area. Then, on October 26, 2019, 100 JSOC commandos conducted a heliborne raid, launching from western Iraq into the Idlib province to kill or capture Abu Bakr al-Baghdadi, the leader of ISIS. After being cornered inside a tunnel, Baghdadi died by detonating a suicide vest, killing two of his sons alongside him.

Meanwhile, hostilities between the regime and the Turks intensified in February 2020, when Syrian government forces killed Turkish troops in direct combat for the first time, spurring Turkey to retaliate with strikes against dozens of regime targets. The fighting endangered Idlib's population, which ballooned to three million as government authorities offered rebel fighters and civilians the choice of surrendering—risking conscription or arrest—or being bused north to the province. The heightened violence resulted in the war's largest mass displacement to date, some nine hundred thousand people being forced from their homes.

Syria's civil war took a surprising turn in late 2024 when rebel forces in partnership with the Turkish government, Hay'at Tahrir al-Sham or HTS, launched a rapid assault that toppled Assad's regime in a matter of days. The rebel offensive sent Assad fleeing to Moscow. The new leader of Syria is Abu Mohammed al-Jolani. He was sent to Syria by Abu Bakr al-Baghdadi to establish Al-Nusra Front, an affiliate of Al Qaeda. In 2016, Jolani rebranded al-Nusra Front as Jabhat Fatah al-Sham. The following year it became Hayat Tahrir al-Sham (HTS). In the wake of Assad's fall, the U.S. and Israel carried out numerous air strikes on the former dictator's arsenal of weapons. The Israeli military estimates it destroyed up to 80% of Syria's military capabilities. Meanwhile, America's mission to Defeat ISIS remains ongoing in Syria.

As Christopher Phillips suggests, "the Syrian civil war cannot be explained without a detailed understanding of the international dimension."[2] While the U.S. long had aims for regime change in Syria, the immediate need in 2014 was the destruction of ISIS, denying Al

---

[2] Christopher Phillips, *The Battle for Syria: International Rivalry in the New Middle East* (New Haven, CT: Yale University Press, 2016), 3.

Qaeda Syria as a base of operations. As for Iran, when Syria, which has long been a satellite within the Iranian orbit, began to have problems, Iran, its more powerful Shi'a neighbor came to the rescue. Likewise, Saudi Arabia, who has long sought to remove Assad and make room for a Sunni government, began to assist the fledgling Sunni rebels. As seen, Russia's aim was to prop up its failing ally.

With the collapse of the Assad regime, Russia lost a crucial ally in the Middle East. Likewise, the seismic shift weakened Russia's core allies, namely, Iran, and her proxies, such as Hezbollah. An Israeli campaign in southern Lebanon in late 2024 likewise severely weakened Hezbollah leadership. Since the fall of Assad, the Russian military has been focused on grounding it out in Ukraine. By far, it seems that Turkey is the greatest benefactor in Assad's fall. While the Turkish government has cared for thousands of refugees, they have also backed the Free Syria Army, and the Hay'at Tahrir al-Sham or HTS, which is Syria's new master.

While many see Jolani and HTS as "moderates" who could signal the decline of jihad movements like the Islamic State and Al Qaeda, others see a radical Islamic group in control of a state with access to chemical weapons. What will become of Syria is yet to be seen.

Summary and Implications

By way of the five laws of irregular warfare, the following is an analysis of the American IW effort in Syria:

Political objective (s) – Unlike the many miscalculations of the U.S. in Afghanistan and Iraq, from the beginning, the U.S.-led Task Force OIR worked by, with, and through regional partners to defeat the Islamic State. In another way that was a vast improvement over OEF and OIF was the manner in with the U.S.-led Coalition partnered with numerous rebel groups to bring about this end goal.

Legitimacy – While the Assad legacy was one of horror with hundreds of thousands dead and millions displaced or in exile, the justification America had for intervening militarily in Syria was predicated on international law, particularly the right to self-defense, which is enshrined in Article 51 of the UN Charter.

<u>Adaptability</u> – The U.S. "by-with-through" approach to fighting the Islamic State precluded the need for a large footprint, reduced the financial cost of the campaign, and kept coalition casualties to a minimum. Following an initial failed train and arm program, the U.S. pivoted to backing the SDF with great success in ousting ISIS from the country. Then, following the end of the Assad regime and takeover by the SDF and Syrian opposition, subsequent air strikes ensured that ISIS did not take advantage of the circumstances.

<u>Influence</u> – The Bashar al-Assad regime's rapid collapse demonstrates the crucial importance of both legitimacy and influence. As numerous desperate forces aimed at shoring up political power, Syria's future remains uncertain.

<u>Native Face</u> – Operation Inherent Resolve worked by, with and through regional partners to militarily defeat the Islamic State of Iraq. The U.S. strategy of backing partnered forces such as the SDF, and to a lesser degree the Free Syrian Army, is an IW formula for success.

For Further Reading:

1. *The Battle for Syria: International Rivalry in the New Middle East* by Christopher Phillips.
2. *Destroying a Nation: The Civil War in Syria* by Nikolaos van Dam.
3. *Burning Country: Syrians in Revolution and War* by Robin Yassin-Kassab and Leila Al-Shami.

15

The War in Ukraine

"With two thousand years of examples behind us, we have no excuse for not fighting well." – B.H. Liddell Hart

U.S. Army veteran trains Ukrainian soldiers outside Kyiv. Courtesy AP.

The war in Ukraine is entering its fourth year. While neither Ukraine or Russia will likely win nor lose the war solely based on what happens at the front, both Kyiv and Moscow now seek to bring the war to each other's backyard. Each has attacked the other with bombs and drones, but Ukraine is also taking the war behind the Russian lines by organizing partisan groups in the occupied territories and within Russia. This chapter highlights what U.S. special operations can learn from the way

Ukraine is employing its irregular warfare potential to degrade Russia's warfighting ability.

While much attention has been paid in the media to the World War One-style trench warfare at the forward line of troops or FLOT, and the real-time revolutionary applications of drones to modern warfare, Ukraine is also an extended case study in the effectiveness of irregular warfare, from the importance of cultural and societal trends to the applications of guerrilla warfare, sabotage and subversion. From the outset of Russia's invasion, Ukraine has made the most of its highly tech-literate society to democratize the conflict to best enable citizen participation in the war effort. Ukraine has been masterful in its use of the information space to broadcast narratives legitimizing its government and war efforts both internally to maintain morale and externally to garner external support, while delegitimizing Russia through fact-checking Russian disinformation and highlighting Russian war crimes as well as instances of logistical and tactical incompetence.

Under the auspices of 'freedom,' Moscow sent 'little green men' into the Crimea to 'protect' the citizens of the Russkiy Mir (Russian World). In February 2014, heavily armed masked Russian troops invaded Ukraine's Crimea Peninsula to "defend" Crimean residents. Moscow followed up by formally annexing the peninsula. The previous year, Euromaidan protests erupted in Ukraine's capital, Kyiv, against Ukrainian President Viktor Yanukovych's decision to reject a deal for greater economic integration with the European Union (EU). Before fleeing the country, Yanukovych attempted to brutally suppress these protests with state security forces. By May 2014, pro-Russian separatists in the eastern Ukrainian regions of Donetsk and Luhansk held their own independence referendums. Armed conflict in the regions quickly broke out between Russian-backed forces and the Ukrainian military. Russia denied any military involvement.

The Minsk Accords, in February 2015, failed to negotiate an end to the violence. Then, in April 2016, to deter possible future Russian aggression elsewhere on the continent, NATO announced the deployment of four battalions to Eastern Europe, rotating troops through Estonia, Latvia, Lithuania, and Poland. Then, during the fall of 2021, months of intelligence gathering and observations of Russian troop movements of armor, missiles, and other heavy weaponry moving toward Ukraine, all indicated that a Russian invasion was imminent. On February 24, 2022, President Putin announced the beginning of a full-scale land, sea, and air invasion of Ukraine with the goal of demilitarizing and denazifying Ukraine. U.S. President Joe Biden declared the attack "unprovoked and unjustified."

Russia's initial attack consisted of long-range missile strikes designed to wreak havoc on Ukrainian military assets, communication and transportation infrastructure. Hospitals and residential complexes were also shelled in these bombing attacks. Also on the morning of February 24, 2022, a Russian assault force consisting of approximately 34 helicopters and 200 to 300 Russian airborne soldiers began an attack on Hostomel Airport. Their objective was to decapitate the Ukrainian government. The close proximity of the airfield at just 12 miles from the capital's center offered the combined force of the 31st Guards Air Assault Brigade and 45th Separate Guards Spetsnaz Brigade a great opportunity to do just that.

Having captured the airport, an air bridge could be made to bring in additional airborne battalions who would follow on transport planes. They could then rapidly take control of Kyiv and overthrow the government. Having gained an initial toehold of the airport, Russian forces failed to achieve the objective of the assault. As related by John Spencer, "Ukrainian National Guard conscripts, backed by artillery units, were able to delay the elite Russian airborne troops long enough to prevent the Russian military from using the airfield as an airbridge to support a rapid seizure of Ukraine's capital."[1] Several Russian helicopters were shot down and the elite Russian troops suffered several hundred casualties.

The world then watched in wonder as a forty-mile-long Russian military convoy, composed of hundreds of tanks, armored vehicles, towed artillery, and supply trucks stalled out on its drive to Kyiv. While the reasons for the stall are disputed, the most likely explanations are that the convoy simply ran out of fuel. With the initial Russian invasion blunted, Russia withdrew all troops from Ukraine's capital region and redeployed its forces to the Donbas region.

---

[1] John Spencer, *The Battle of Hostomel Airport: A Key Moment in Russia's Defeat in Kyiv.*

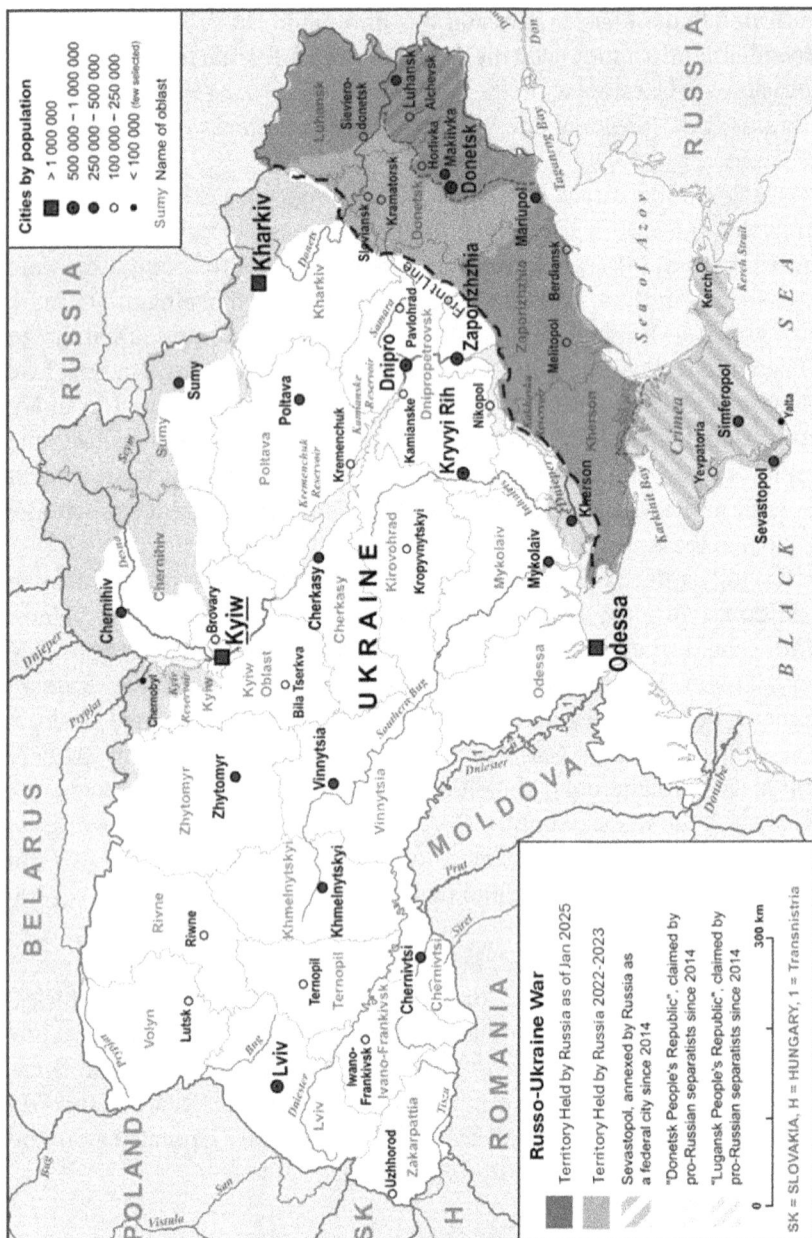

Russo-Ukraine War. Courtesy AP.

In the aftermath of the Russian withdrawal from Kyiv's surrounding areas, Ukrainian civilians described apparent war crimes committed by Russian forces, including accounts of summary executions, torture, and rape. On March 2, 141 of 193 UN member states voted to condemn Russia's invasion in an emergency UN General Assembly session, demanding that Russia immediately withdraw from Ukraine.

Undeterred, on April 18, Russia launched a new major offensive in eastern Ukraine, seizing Mariupol, a port city that had been under siege since late February. Indiscriminate and targeted attacks against civilians in the city, including an air strike on a theater and the bombing of a maternity hospital. Then, in August, Russian forces seized the Zaporizhzhia facility, the largest nuclear plant in Europe. Fighting in the territory surrounding the facility sparked international fears of a nuclear disaster. Representatives of the International Atomic Energy Agency (IAEA) called for an immediate cease fire.

In early September, the Ukrainian military accomplished a major feat of arms with deception at its foundation. A much-publicized Ukrainian southern offensive aimed at Kherson was in fact a disinformation campaign to distract Russia from the real one being prepared in the Kharkiv region. While massing troops in the Kherson region, on September 6, 2022, Ukrainian troops surprised Russian troops along the front at Kharkiv. Surprising Russian forces, Kyiv witnessed a hasty Russian withdrawal across the Dnipro River as Ukrainian forces retook the city of Kherson and all territory west of the river. The ruse enabled Ukrainian forces to retake significant territory, making strong advances in the northeast and mounted a southern counteroffensive in the Kharkiv region. Following Ukraine's Kherson rouse, Russia then redeployed forces eastward to Donetsk, in addition to sending tens of thousands of reinforcements to the area in advance of a February 2023 offensive. Russia also annexed four occupied territories: Luhansk, Donetsk, Kherson, and Zaporizhzhia. In a speech, Putin also hinted at the possibility of nuclear escalation.

Then, in March 2023, following a winter stalemate, Russia launched an offensive surge. However, the attack made little progress and devolved into a costly siege of Bakhmut in which Russia suffered an estimated 100,000 casualties. Unwilling to give up the town, Ukraine likewise took heavy casualties. By late May, Russia claimed to have taken the city. Then, in June 2023, Ukraine launched a counteroffensive, attempting to break through Russian defenses eastward in Donetsk province, including around Bakhmut, and southward in Zaporizhzhia province. Ukrainian forces met stiff resistance and suffered heavy losses against hardened Russian defensive positions.

Then, on June 23, 2023, Putin faced a major internal challenge when Wagner Group's Yevgeniy Prigozhin claimed the Russian Ministry of Defense shelled his forces. Wagner convoys advanced more than halfway to Moscow before Belarussian President Alexander Lukashenko negotiated for Wagner soldiers to stand down. Two months after the revolt, Prigozhin died in a private plane crash outside of Moscow. Putin denied any involvement. Since 2024, the war in Ukraine has stalemated.

Ukraine's War in the Shadows

As the war on the frontline has ravaged hundreds of thousands of Ukrainians and Russians, Ukraine's special operation forces have been steadily taking the fight behind the Russian lines. Using guerrilla tactics, including commando raids, targeted assassinations, and sabotage, pro-Ukrainian partisans are creating havoc in Ukraine's Russian-occupied territory and beyond.

Similar to the British MI5, the Security Service of Ukraine or SBU conducts everything from counterintelligence and combating organized crime, to counterterrorism, covert action, and irregular warfare. Kherson was one of the first major Ukrainian cities to fall to the advancing Russians. With the Ukrainian Army unable to enter the city, and as the Russians became more entrenched, the SBU organized a resistance to fight back. Kherson partisans left a banner with a message on a pole in the city, which read: "Russian occupier and everyone who supports their regime. We are close—we are already working in Kherson. Death awaits you all! Kherson is Ukraine!" The town's resistance passed along information regarding Russian dispositions that eventually helped the Ukrainians in their push to retake the city.

Recognizing the vulnerability of Russia's rail lines, the SBU set out to compromise the vital network. On October 8, 2022, a remote-controlled truck bomb detonated on the Kerch Bridge, a 19-kilometer-long bridge that serves as the main supply artery of Russia's support to the Donbas.

Repairs on the damaged roadway and rail line took several months. Then, in July 2023, and again in December 2024, Ukrainian maritime drones struck the bridge again further weakening this critical infrastructure. Sea drones have also been used to sink Russia's modern warships in the Black Sea. Each attack demonstrates Ukraine's ability to exploit vulnerabilities in Russian defenses using innovative tactics and advanced technology, and to capitalize on these successes through information operations and subversion.

Attacking Russia's railway infrastructure degrades Moscow's ability to bring additional men and material to the front. Hoping to compromise a vital conduit for weapons being shipped to Russia from North Korea, the SBU went after one of Russia's main rail arteries. Thousands of miles east of the front, saboteurs managed to place explosives on a Russian 50-car freight train carrying diesel and jet fuel. Saboteurs timed the blast to obstruct the nine-mile-long Severomuysky Tunnel, a vital piece of infrastructure in Russia's transportation system that is part of the Baikal–Amur Mainline (BAM). There are only two rail lines that cover the vast expanse of Russia: the trans-Siberian, which stretches 5,772 miles from Vladivostok to Moscow, and the newer BAM, which links China and Russia. The attack came at a moment when Ukrainian forces on the front were struggling to stave off relentless Russian assaults. Russian-language Telegram channel Baza, which is linked to Russia's security services, said that a fuel tank caught fire while moving through the tunnel in Buryatia in the early hours of Thursday. It took months to restore the mountain pass to full working order.

Ukraine's partisan forces are becoming more organized and effective. In September 2022, a pro-Ukrainian partisan unit named Atesh, meaning fire in Crimean Tatar, was formed and began to operate in Russian-occupied Crimea. On 20 October, Atesh partisans sabotaged a railway line near Novooleksiivka in Kherson Oblast. The attack denied Russian forces needed supplies in Southern Ukraine. Atesh partisans have expanded their sabotage operations to include the destruction of Russian bunkers, trenches, and weapon arsenals.

All told, since the Russian invasion of February 2022, some 76 cases of railway sabotage have been carried out by pro-Ukrainian partisans. Ukrainian sabotage efforts go beyond trains. Over 500 miles from the front, aerial drones evaded Russia's best air defenses to strike the Nevsky oil refinery in the middle of St. Petersburg. Russian media claimed that three empty fuel tanks caught fire. In an April 21, 2022, television interview, the mayor of Russian-occupied Melitopol, Ivan Fedorov, said Ukrainian partisans had killed about a hundred Russian soldiers in the city in nighttime ambushes. Later that summer, Melitopol-based partisans sabotaged a section of rail that ran through the town. The mayor also claimed that the Russian army was struggling to deal with these unseen attacks. Throughout 2022, numerous pro-Russian officials were kidnapped and later found shot. In one high vis attack, the Russian-appointed deputy head of the occupied Luhansk region, Oleg Popov, was killed in a car bombing.

Ukrainian sabotage has taken aim at Russia's bombers. On December 26, 2002, a Ukrainian drone struck a Russian bomber base in Engels, about 400 miles northeast of the Ukrainian border. Russian media reported the "liquidation" of four saboteurs. In the wake of the attack, the Russian military tightened its security around its installations, surprised at how far the partisan attacks had reached into Russia.

Ukraine is not alone in using guerrilla tactics. Russia is also employing spies, saboteurs and collaborators, and it targets trains, as well. According to Polish authorities, dozens of saboteurs under the direction of Russian intelligence have been detained. Their main targets, the ministry said, were "trains transporting military and humanitarian aid to Ukraine and preparing for train derailments."

Perhaps the most sophisticated and highly publicized sabotage of the war so far has been the wrecking of the Nord Stream 2 pipeline in the Baltic Sea. Nord Stream 1 and Nord Stream 2, stretch over 700 miles from the northwest coast of Russia to northeast Germany. The pipelines which are intended to supply Western Europe with Russian gas have reportedly cost over $12 billion. On September 26, 2022, massive pressure loss and signs of large gas were reported. The pipelines, which at the point of attack are about 80 meters below the surface, were ripped apart by deep sea explosions. Investigators reported that the explosives used in the sabotage were most likely planted with the help of experienced divers. One estimate puts the cost of repairing the pipelines at about $500 million. While details are fuzzy at best, it is believed that the sabotage operation involved a small sailing boat and a team of six Ukrainian soldiers and civilians with relevant expertise.

Since the pipeline sabotage, there has been much speculation about what transpired on the seafloor. Ukraine immediately accused Russia of planting the explosives but offered no evidence. Russia, in turn, accused Britain of carrying out the operation. Investigative journalist Seymour Hersh published an article in *Substack* concluding that the United States carried out the operation at the direction of Biden. It's no secret that the U.S. has opposed the pipeline as it gives Russia energy dominance over Europe and America's allies. In 2021, after a meeting with German Chancellor Olaf Scholz at the White House, President Biden said that Putin's decision about whether to attack Ukraine would determine the fate of Nord Stream 2. Biden said:

If Russia invades, that means tanks and troops crossing the border of Ukraine again, then there will be no longer a Nord Stream 2. We will bring an end to it. When asked exactly how that would be accomplished, Mr. Biden cryptically said, "I promise you we'll be able to do it."

Allegedly, Ukrainian President Volodymyr Zelenskyy approved the plan, but later backtracked after U.S. intelligence found out about it and asked Kyiv to call it off. A senior official from the SBU said that Zelenskiy "did not approve the implementation of any such actions." As one credible narrative goes, Zelenskyy ordered the pipeline destroyed but after canceling the order, his directions failed to reach the saboteurs. The pipeline remains inoperable.

In its war with Russia, Ukraine has achieved considerable strategic depth. On July 27, 2024, a convoy of the Malian army and its Russian auxiliaries from the Wagner Group was ambushed by pro-independence Tuareg rebels. The death toll was eighty-four Russians and forty-seven Malians. Two days after the battle, a spokesman for Ukraine's military intelligence service (GUR) made a statement that did not go unnoticed. He suggested that his services were collaborating with the Tuareg rebels, and having received useful information, "it allowed them to carry out a successful military operation against Russian war criminals."

Having the advantage of interior lines, Ukraine has been ability to capitalize on Russia's blunders. Yet, having a numerical inferiority in manpower and warfighting capacity, a war of attrition is not to Ukraine's advantage. At the commencement of hostilities, Ukraine's military prospects seemed slim. It seemed the Zelenskyy Administration would collapse or at least escape to become a government in exile with a subsequent guerilla war. Instead of the government in exile option, famously, Zelenskyy asked for ammo. As the war grinds on, Ukrainian partisans continue to undermine the combat potential of Russian forces in Russian-occupied areas as well as within Russia. As Seth Jones has aptly said, "The war in Ukraine is changing right now, as Ukraine increases the number of guerrilla operations against Russian forces and decreases conventional operations. The goal is to deliver death by a thousand cuts." In this sense, the war in Ukraine offers a test case to one of T. E. Lawrence's axioms. In a conventional war between industrial powers, irregular warfare operations can be a significant supporting element to the main strategy. As it is said, every war begins like a vibrant young man yet eventually dies like a tired old pensioner. While it seems that Russia plans to slowly ground out a victory, as Ukraine's IW efforts have demonstrated, the game can change.

Technology, Information and Narrative

The conflict in Ukraine has been an extended showcase of the importance of adaptability in warfare, with the Ukrainian government displaying an ability to harden its durability through the application of technology and democratizing the war effort to its populace, an instance of "centralized planning and decentralized execution." Ukraine enjoys a highly tech-literate populace. This has been a double-edged sword, as it has allowed the Ukrainian government to rapidly modernize its interface with its civilian populace but has also made Ukraine an international hive for cyber criminality. As we will see, even this has sometimes worked to the government's advantage.

In 2020, the Ukrainian government formed its Ministry of Digital Transformation, which created a digital civilian interface. Named Diia and referred to by the Ukrainians as "state on a smartphone," it allows citizens to conduct almost all governmental transactions from a smart phone. Accessed through a smart phone's biometrics, each citizen's account allows them to do everything from access their birth certificate, renew their passport or driver's license, or pay their taxes. With the advent of the invasion, the Ukrainian government added multiple apps allowing citizen participation in the war effort. The first was EVorog (E Enemy in Ukrainian,) which allows Ukrainian citizens to report Russian troop or vehicle locations and movements utilizing an AI enabled chatbot to answer questions resulting in a S.A.L.T. report.[2] Later came the ePPO app. ePPO stands for "electronic air defense." The app utilizes AI to enable Ukrainian citizens to report the location of Russian drones to enable mobile ADA teams in their destruction. The app utilizes the phone's GPS for location and the user points their phone in the direction of the drone and presses a single button to register its direction and flight trajectory. The more user reports generated, the more accurately the app triangulates the drone's position and trajectory. The app has resulted in thousands of Russian drones shot down. Diia also has platforms for citizens to post videos, such as acts as defiance against the Russian war effort as well as evidence of Russian atrocities, often boosted from government accounts.

Diia's success led to the creation of Delta, the military interface enabling rapid technology adoption and inter-unit communication. On of the apps used is GIS Arta. It allows Ukrainian artillery batteries to conduct fires missions using data from forward observers as well as drones in the air. Nicknamed "Uber for Artillery," it enables artillery units not currently engaged to place themselves in an "on" status, inviting calls for fire from adjacent units. Some of these batteries have claimed rounds in the air from calls for fire in as little as 30 seconds.

Social media has been a crucial tool for the Ukrainian war effort since the earliest hours of the invasion. Ground-level videos of Ukrainian soldiers choosing to detonate themselves along with the bridges they stood on in last-ditch efforts to slow the Russian advance, farmers using their tractors to tow away Russian tanks that ran out of fuel, and a grandmother handing an invading Russian soldier sunflower seeds and telling him it was so when he died in Ukraine "something beautiful will grow from your body" went viral internationally in the early hours of the conflict, boosting Ukraine's communal commitment to resistance and rallying the international community in its support. Today multiple ground-level soldiers operate YouTube and Telegram channels showing

---

[2] S.A.L.T. stands for size, activity, location, and time.

the realities of the conflict and broadcasting patriotic commitments to the ongoing war efforts.

With its tech-savvy population, the Ukrainian government has benefitted from what it calls its "IT Army," an international confederation of fact-checkers and hackers that both spoil Russian disinformation and deface pro-Russian websites and conduct cyber-attacks. Concurrently, the Ukrainian government has used mis-and-disinformation to its advantage. One early viral story was of a group of Ukrainian soldiers stationed at a signals site on Snake Island in the Black Sea. Faced with radio calls for their surrender from a Russian war ship, the soldiers replied "Russian warship, go f..k yourself" in a clip that instantly went viral world-wide, along with the story that the Ukrainians then heroically died to a man. Later, it came to light that the soldiers had actually surrendered and were exchanged in a prisoner swap, but neither stopped the Ukrainian government from making an image of a soldier flipping off a Russian battleship into a postage stamp nor dampened the legendary character of the exchange. Similarly for the first nine months of the conflict, the Ukrainian government boosted stories of a Ukrainian fighter ace flying an aged MIG warplane blasting Russian pilots from the sky by night, fully fabricated but still fondly remembered as "The Ghost of Kiev."

The state has utilized social media effectively for its own aims directly in support of operations. Following the VBIED attack on the Kerch bridge (a multi-billion dollar project which featured Putin personally driving the first truck across), multiple government accounts quickly piled on to thumb their noses at Russia, with the Tourism Ministry's Twitter account simply stating "sick burn" below images of the burning bridge, while the Defense Minister juxtaposed video of the burning bridge with video of Marilyn Monroe singing "Happy Birthday Mister President." The attack was timed to coincide with Putin's birthday. The Ukrainian war effort has also included the "I Want to Live Campaign" designed to provide conscripted Russian soldiers a way to surrender their way out of the conflict. The Ministry of Tourism's Twitter account hosts instructions, not just for how to safely surrender to Ukrainian forces, but how to avoid conscription altogether, how to go AWOL, and how to conduct simple sabotage against the Russian war effort, all examples of subversion with a sabotaging intent.

Ukraine has utilized effective deception techniques throughout the conflict, most conspicuously in the Kherson Ruse mentioned earlier in the chapter. But its deception practices are ubiquitous. Ukraine maintains a cottage industry in the production of decoy equipment (inflatable tanks and troop carriers but also hyper-realistic towable pieces such as ADA batteries made from plywood) intended to waste Russian fires assets. Two recent tactical-level deceptions illustrate how insidious and malicious their efforts can be. In the first, they identified that many ground-level Russian troops are poorly supplied, and the Russian war effort is reliant on donations from patriotic citizens. They also knew enough about Russian drone operator TTPs to anticipate that drone pilots supplied with new first-person-view (FPV) drone goggles would place them on their heads prior to turning them on. They then donated multiple sets of FPV goggles to the Russian war effort. The googles were packed with 10-15 oz of explosives rigged to explode when turned on. The googles made it to front-line troops. As of this writing, the count was eight Russian drone pilots who popped their own grapes.

The second example is a rash of arsons targeting ATM machines in Russia from December 2024 to December 2025. On the face of it, this is an act of sabotage with the intent of adding additional hardship to a Russian population already experiencing severe economic hardship due to international sanctions. The second and third order effects are more insidious. When the Russian security forces arrested the arsonists, they learned two surprising things. First, most of them were geriatrics pensioners. Second, they didn't commit these acts of arson for ideological reasons. Instead, they were victims of phone scammers, who had likely targeted geriatrics due to their fear of authority and vulnerability to social engineering techniques. The Russian government's overriding internal narrative is strength, law and order, and consequences for bad behavior, and as a result the majority of these old people were indicted on charges of domestic terrorism. Once indicted, Russian citizens enjoy an approximately 0.26% chance of not being found guilty at trial. If we look at the societal implications of that, the effect is not just on gramps spending his final days in prison, but friends and acquaintances painted with the same brush and likely receiving additional scrutiny or surveillance. The Ukrainians created a cycle of grievance inside Russia from thin air.

Incidentally, the Russians immediately accused the Ukrainian government of being behind the attacks, who responded with two different narratives. Some government figures declared that the attacks were in fact orchestrated by the SBU, while others stated that the attacks were perpetrated by Ukraine's virulent cyber-criminal elements, who "sometimes choose to be patriotic." The truth could be a cross of the two, with the SBU beginning the scam, and the criminal element noticing and then doing it better.

Summary and Implications

By way of the five laws of irregular warfare, the following is an analysis of the Ukrainian IW effort:

Political objective (s) – Ukraine's war is a defensive one aimed at expelling its invaders and the restoration of its territorial integrity, including Crimea and the Donbas. For Russia, the war is to "denazify" Ukraine and to bring the Russians outside of Russia back into the fold.

Legitimacy – Russia's invasion of Ukraine has been qualified by legal experts as a crime of aggression which constitutes a manifest violation of the Charter of the United Nations. Moreover, since the invasion, the Russian military has committed numerous war crimes, such as deliberate attacks against civilian targets, including forced deportations, and the killing and torture of Ukrainian prisoners of war. According to Ukrainian authorities, over 120,000 Ukrainian children have been abducted from their murdered parents and deported to Russia's eastern provinces.

Adaptability – According to a recent RAND study, using these unorthodox and imaginative tactics, Ukraine's irregular warriors are notching strategic gains against Russia. With Western support and technology, these gains could become even more potent. This strategy to attrit Russian army logistics and undermine morale is optimal for indirect, irregular warfare. It will support the more substantial conventional offensive operations needed to expel Russian forces from Ukraine.

<u>Influence</u> – Ukraine's strategic communication is carefully crafted. The Kyiv narrative emphasizes Ukraine's moral superiority over Russia, characterizing its aggressor as a terrorist state as it lauds the audacity of Ukrainians on the battlefield. As the war has surpassed its one thousandth day, Russia is losing influence in its former backyard of the Soviet Union as well as globally.

<u>Native Face</u> – Recently, Putin brought in some 12,000 North Korean soldiers to the fight. The inexperience of these soldiers as well as the language barrier between them and their Russian counterparts are likely to minimize their impact. Russia's bringing in foreign troops such as North Koreans soldiers signals at least two things. First, Putin has no plans to concede anything for peace. And secondly, it shows his resolve to employ alternative force generation to support the war. When Putin called for a partial mobilization wave in September 2022, hundreds of thousands of Russians fled the country to avoid being enlisted to the fight.

As the war in Ukraine enters its fourth year, and as Trump has been elected, the question is how the conflict will play out. In mid-November 2024, Biden authorized Ukraine to use US-supplied missiles to strike deeper inside Russia. In response, Putin doubled down on Russia's ability to use tactical nukes. The West could help by providing greater support to Ukrainian operations behind Russian lines. However, backing a Ukrainian insurgency could draw NATO and Russian forces into conflict. Only time will tell.

For Further Reading:

1. *The Gates of Europe: A History of Ukraine* by Serhii Plokhy.
2. *Borderland: A Journey Through the History of Ukraine* by Anna Reid.
3. *Subversion: The Strategic Weaponization of Narratives* by Andreas Krieg.

16

Winning the War of the Flea

"With two thousand years of examples behind us, we have no excuse for not fighting well." – B.H. Liddell Hart

Special Forces during a reconnaissance mission in Kunar Province to identify the site for a future Village Stability Platform, February 2012.

As is so often the case, strands of thought become hard to keep together, fall apart during development, and can become overly difficult to tie together at the end. The focus of this chapter will be to tie the loose ends of the argument together. Having surveyed over a hundred and twenty-five years of American irregular warfare, the question is, can there be said to be a formula for success? Let the answer be framed in light of the global rivalry America now finds herself in, the Great Power Competition (GPC). A great power is a sovereign state that is recognized as having the ability to exert its influence on a global scale. The GPC is a global rivalry among three dominant states: the United States, Russia, and China. In the Great Power Competition, "third states" will be the irregular warfare battleground. The 2022 National Defense Strategy (NDS) defines the principal threats to the United States as Russia and China. Other persistent threats are identified as Iran, North Korea, and Violent Extremist Organizations (VEOs). VEOs consist of bad actors such as Hezbollah, the Islamic State (IS) and its affiliates, Al Qaeda (AQ), Abu Sayaf Group (ASG), along with other similar groups who threaten the United States and its interests abroad. Prioritizing the threats of Russia and China and a return to great power competition, in the words of the National Defense Strategy, "The United States and its Allies and partners will increasingly face the challenge of deterring two major powers with modern and diverse nuclear capabilities – the People's Republic of China and Russia – creating new stresses on strategic stability.[1]

China continues to use predatory economics to intimidate its neighbors while militarizing features in the South China Sea. Iran continues to sow violence and remains the most significant challenge to Middle East stability. Russia has violated the borders of nearby nations and pursues veto power over the economic, diplomatic, and security decisions of its neighbors." Arguably, Russia's seizure and annexation of Crimea in March 2014 signaled the transition from the Cold War era to the return to Great Power Competition. In his book, *Three Dangerous Men*, Seth Jones rightly argues that Russia, China, and Iran have built ruthless irregular warfare campaigns that are eroding American power. The following is a discussion on just how these three dangerous men are doing just that.

---

[1] Secretary of Defense (SecDef), *Summary of the 2022 National Defense Strategy of the United States of America* (Washington, DC: Department of Defense, 2018), 4.

Russia

The difficulty in understanding Russia's motivations and actions once led Winston Churchill to observe, "Russia is a riddle wrapped in a mystery inside an enigma." Churchill said this in a radio broadcast, shortly after the Nazi-Soviet Pact was signed in October 1939. It's fair to say that the West has been trying to decipher Russia ever since. Following WWII, the United States and Russia began viewing each other as a chief adversary, each fearing that the other would unleash a surprise nuclear attack. That fear led to an arms race to stockpile nuclear warheads. Since the start of the arms race, Russia has the most nuclear missiles with 5,889 warheads, while the U.S. has 5,224. Just how many of those warheads would actually work today we are not sure. During the GWOT, the nuclear aspect of this rivalry faded to the background of geopolitics. Now, after twenty years of chasing Al Qaeda and ISIS, the United States has recently placed Russia back at the top of its threat list. Be that as it may, for Russia, however, the United States has always posed a serious and persistent threat.

By consensus, Russia is a threat to the United States for three main reasons: First, Russian leaders, principally Putin, define their country's national interests on a geopolitical collision course with those of the United States and her allies. A former KGB officer well-versed in the Soviet practices of subversion, termed "active measures" in Soviet doctrine, Putin blames the fall of the Soviet Union and Russia's geopolitical decline under Yeltsin on ideological contamination of democratic and liberal ideas from the West. Heading a heavily authoritarian government, clamping down on freedom of speech and sources of information that differ from official Russian channels, Putin has successfully created a national narrative where the state dictates the populace's worldview. While deplorable from the standpoint of human freedom that we as Americans hold dear, such a system has real advantages in countering subversion and dissent from both internal and external enemies in comparison to a free society. From this essentially defensive bastion of informational control, Russia has launched myriad offensive disinformation campaigns against its neighbors and around the globe, with the goal of subverting and coercing its way into a favorable position in global competition.

Russia continues to view the United States and NATO as the main enemy. Before invading, Putin declared Ukraine a red line where he would stop western expansion. The core strategy for Russia under Putin, is the reclamation of "Mother Russia" via the establishment of dominance over the post-soviet countries and the consolidation of the Russian diaspora. The problem for U.S.-Russian relations is that achieving this requires Moscow to undermine U.S. foreign policy goals. Since 2008, Putin has invaded two of his neighbors, Georgia and Ukraine, and has illegally annexed portions from both. The end goal for Putin is to end American dominance and reestablish Russia as a global power.

The second way Russia is a threat to the United States is because it uses hybrid warfare. In addition to its modernized weapons arsenal, Russia can mount significant combat power using the "little green men" approach, or by employing novel electronic warfare, anti-satellite, elaborate disinformation techniques, and cyber capabilities against U.S. forces, including the threat of nuclear weapons. As we have seen since 2014, Russia has threatened to use nuclear weapons, even in regional and local conflicts to coerce concessions from its adversaries. To date, Russia has not followed through and has had mixed success in exacting such concessions. For example, in September 2023, Elon Musk personally disabled Starlink access to Ukrainian special operators attempting to conduct a drone attack against Russian naval vessels near Crimea, announcing that he had interfered to avoid an escalation of the conflict to a third world war or nuclear conflict. Ukrainian forces have since hit a multiplicity of Russian naval vessels with drones, causing major damage to the Russian Black Sea fleet, even driving it from the Black Sea. As always, the success of coercive influence depends on the nature of the target audience.

Regarding disinformation, the two-year, million-dollar investigation by special prosecutor Robert Mueller concluded that not a single American, let alone the U.S. president himself, conspired with the Russians to steal the 2016 election. But just because the Trump-Putin collusion turned out to be a hoax, doesn't mean that Russia did not have a hand in the disruption of the 2016 presidential election, and ensuing upheaval.[2]

Third, Russia is a threat to the U.S. because it's less constrained to operate morally. While those who live in glass houses should not throw stones, Putin gave cover to Bashar al Assad's use of chemical weapons which he readily used to murder thousands of his own people, including children. Likewise, GRU cyber units hacked the computers and e-mail accounts of various members of the Clinton campaign. In both conventional and hybrid operations, Russia shows a disregard for civilian casualties and a propensity for committing a multitude of war crimes.

One thing is for sure, Putin wants Russia's geopolitical footprint back, and subversion is one of the key tenets of his strategy. Russian subversion strategy is devoted to ensuring Russia's continued relevance on the world stage, with concessions from foreign nations enabling it (including ours.) Its subversion operations target both sides of America's political spectrum. It follows a six-step process detailed in Andreas Krieg's book *Subversion*, paraphrased below.[3]

Step 1: Orientation. Russia pollutes the information space with weaponized narratives to disrupt fact-finding or relativize facts, making it harder for society, media or policymakers to differentiate from fact and fiction, creating information paralysis. You can think of this step as preparation of the environment for the information space. It prepares for the introduction of "new facts." An example would be Tucker Carlson's interview with Putin, where Putin turned the interview into a "history lesson" that was cherry-picked to legitimize Russian grievances with the West. Another example would be encouraging to ramp up the heat of political discourse and encouraging the vilification of opposing political ideologies. Both sides of the political aisle now routinely partake in "cancellation" of members of their own echo chambers that do not show sufficient orthodoxy or loyalty to the narrative or "facts" of the moment.

---

[2] Rebekah Koffler, *Putin's Playbook: Russia's Secret Plan to Defeat America* (Washington, D.C.: Regnery Gateway, 2022).
[3] Andreas Krieg, *Subversion: The Strategic Weaponization of Narratives* (Washington, D.C.: Georgetown University Press, 2023).

Step 2: Identification. Russia identifies key social, psychological, infrastructural and physical vulnerabilities in the adversary's information environment. Target audiences chosen on both sides of the U.S. political spectrum tend to have an antiestablishment worldview, with concurrent sociopolitical and socioeconomic grievances regarding alienation from mainstream society and disenfranchisement from the establishment's failure to meet their needs. An example from the left is Russia's identification of *Black Lives Matter* as a vehicle to cause physical chaos in the streets. On the other side of the aisle, Russia identified the far right's' propensity for entertaining conspiracy theories (given a nod by President Reagan in his "nine most terrifying words in the English language") as a vehicle to sow distrust in the U.S.

Step 3: Formulation. Russia has formulated a major narrative describing its nature, goals and actions both at home and abroad. At home, it emphasizes nationalism, strength and revived greatness. In the former Soviet satellite states, this is pushed as a shared identity as the "Russian World." For non-Russian-speaking countries, Russia has offered an insurgent narrative in which it offers an alternative to U.S. preeminence, liberal values and globalization. In formulation, Russia legitimizes itself by pushing the above narrative while delegitimizing convenient targets through the introduction of disinformation to promote social discord. For example, in 2014 Russia fabricated stories that an Islamic State training facility had been built in Ukraine, a story that not only resonated with disaffected Russian speakers in eastern Ukraine but made its way into the U.S. Congress's debates on aid to Ukraine. Similarly, Russia targeted African Americans with fake stories of the Klu Klux Klan infiltrating the U.S. police departments during the 2016 election cycle.

Step 4: Dissemination. At this point in the process, Russia is able to utilize a large network of "indirect and coincidental surrogates" in addition to its official channels to broadcast its narratives. Another term for "coincidental surrogates" would be useful idiots, influencers who have swallowed the cumulative confusion and disinformation that they independently advocate in accordance with Russia's major narrative. While the narrative, because it is not directly under Russian direction, is no longer homogenous (like playing a game of telephone with an abstract thought,) it discourages attribution, sows confusion, and allows the target audience to triangulate "the truth" to their own tastes and biases. An example would be the significant airtime conservative media apologists spent describing the defensive nature of Russia's invasion of Ukraine in 2022, with the true culprits being NATO expansion and western anti-Russia policies. To be clear, sometimes the morning's talking points (the commentary to comment on,) regardless of which side's morning show you're watching, are Russia's talking points.

Step 5: Verification. Russia utilizes co-opted or fake experts in order to provide legitimacy for its narratives, often through the use of think-tanks. An example would be the Canada-based Centre for Research on Globalization, which in the aftermath of Malaysian Flight 17 downed over Ukraine published fictitious evidence supporting Russia's denials of responsibility.

Step 6: Implementation. Finally, Russia ensures that its operations in the digital sphere translate into effects in the physical realm, with the true test of effectiveness being effects at the policy making level in foreign governments. Perhaps the most high-profile example was the Kremlin's Internet Research Agency, led by Yevgeny Prigozhin, shaped debates about Hillary Clinton's email scandal. In hindsight, former FBI director James Comey admits his decision to reopen an investigation into Clinton days before election day 2016 was prompted by disinformation claiming a coverup by the Obama Justice Department.

While the Kremlin targets audiences on both sides of the political spectrum, the right has overlapping values with those espoused in Putin's metanarrative-nationalism, the importance of the image of strength and masculinity, anti-globalization, conservative social values, and a historic propensity for entertaining conspiracy theories. In reality, the state of American governance that would seem most likely to fulfill Putin's territorial ambitions would be one that prioritizes control of domestic narratives and a rejuvenated imperialist mindset, as well as hyper partisan legislative deadlock (where the incentive is a sound bite "owning" the other side, rather than functional governance) resulting in an executive branch that constantly and incrementally assumes greater legislative function. Such a platform enables global competitors to negotiate the carving up of territories to suit their mutual interests.

What Putin needs now is the war to end, with him able to maintain the appearance of strength abroad but more importantly at home. The Russian economy is near collapse, the population does not support further conscriptions, and if your best friend in the world is Kim Jung Un, things have gone awry. The best possible scenario vs U.S. governance for him to be able to make that deal is the one that he currently confronts.

Given Russia's strategy of reclamation, it seems likely that Russia will at some point attempt to recover the Baltic States of Estonia, Latvia, and Lithuania – all republics in the former Soviet Union. The U.S. is currently working with these Baltic States to blunt any such attempt by Russia to absorb them once again into the Russian orbit.[4] The big question, however, is how will the war in Ukraine end? With the Ukrainian War now beginning its fourth year, with U.S. support, Ukraine is making gains against Russia via irregular warfare means. As has been argued, with continued support, these gains could become even more potent as these IW efforts support Ukraine's conventional fight needed to expel Russian forces from Ukraine.

While we wait to see how things will unfold in Eastern Europe, Russia has since reestablished ties with the nations it supported during the Cold War, including Angola, the Central African Republic, and Mozambique. Exchanging military and security support for economic, geopolitical, and military gains, Russia targets these nations as they are rich in resources and weak in governance. Between 2016 and 2021, Wagner Group deployed to fourteen African nations. As this number grows, and as Russia continues to build out its capabilities in Africa, it seems more than likely that America will continue to have confrontations with Russia's proxies, like the paramilitary company Wagner Group.

Russia is likewise a threat to America's interest as well as the regional stability of Central and South America. Russia's key vehicles for its expansionist agenda in the Western Hemisphere are the anti-U.S. authoritarian regimes, including Cuba, Nicaragua, and Venezuela. Russia has a long history of supporting Cuba as well as several central American regimes. Russian troops regularly conduct training missions

---

4 See *Resistance Operating Concept* by Otto C Fiala.

in Nicaragua and give other forms of support. Similar military support has been given to Venezuela, Russia's most important trading and military ally in the region. Russian support to its politically unstable and economically impoverished South American ally has included military trainers, technicians, along with Wagner Group mercenaries. Russia heightened the threat considerably in September 2008, when it flew Tupolev Tu-160 nuclear-capable heavy strategic bombers to Venezuela.

On the world stage, in the opinions of many analysts, Russia's conflicts in Ukraine and Syria have been the testing grounds for some of its new weapon systems and fighting concepts. Russian military planners are forecasting that a kinetic war with the United States is only a matter of time. While that waits to be seen, Russia's close ally in the conflicts in Syria and Iraq, and chief recipient of Russian weaponry continues to be Iran.

Iran

Iran is an authoritarian theocratic republic of 90 million with a Shi'a Islamic political system. Iran's destabilizing actions threaten vital U.S. interests in the Middle East and beyond. How the United States should approach Iran will remain one of Washington's primary foreign policy challenges for years to come. In the words of the National Defense Strategy,

Iran is taking actions that would improve its ability to produce a nuclear weapon should it make the decision to do so, even as it builds and exports extensive missile forces, uncrewed aircraft systems, and advanced maritime capabilities that threaten choke points for the free flow of energy resources and international commerce. Iran further undermines Middle East stability by supporting terrorist groups and military proxies, employing its own paramilitary forces engaging in military provocations, and conducting malicious cyber and information operations.[5]

In light of what we have experienced in the wars in Iraq and Syria, the question is, what can the United States expect in its competition with Iran in the years ahead? It seems more than likely we can expect more of

---

[5] Secretary of Defense (SecDef), *Summary of the 2022 National Defense Strategy of the United States of America*, 5.

the same. Iran has several proxies across the Middle East at its beck and call. These include Hezbollah in Lebanon, Hamas in Gaza, Shi'ite militias in Syria and Iraq, and the Houthis in Yemen. Through its intelligence service and Quds Force, Iran leverages these proxies to meddle abroad.

The origins of Shi'a Islam go back to the death of Islam's charismatic prophet leader around 632 AD. Following Mohammed's death, the community was plunged into a crisis as to who would succeed him. Should the successor be the most pious Muslim or a direct descendant of the prophet? The seeds of dissent in the Islamic community were planted when the companions of the prophet moved quickly to select Abu Bakar, Mohammed's father-in-law. The caliph was to be the political leader of the community. As the protector of Islam, the caliph led the jihad and was to govern the community by the Sharia Islamic law. Those who accepted the choice of Abu Bakar, the majority of the community, became known as Sunnis.

A minority of the community, the Shi'a Meaning the party or followers of Ali, took strong exception to the selection of Abu Bakar. They believed that before his death Muhammad had designated the senior male of his family, Ali, the prophet's cousin and son-in-law, to be leader, or imam, of the community. Ali was eventually chosen as the fourth in a succession of caliphs but was assassinated after five years of rule in 661 AD. Ali's son, the brave and charismatic Hussain, and his army battled to try to regain power and reinstate what they believed to be the true values of Islam. Hussein's forces were defeated by the Sunni army of the Khalif Yazid in 680 AD. The death or martyrdom of Hussein in the battle of Kabbalah became a defining symbol for Shi'a Muslims.

Since Iran's Islamic Revolution, Saudi Arabia and Iran have been at loggerheads with each side promoting their own fundamentalist strains of Islam to counter the other. However, as the conflicts in Iraq and Syria have shown, in an invasion or perceived one, such fundamental distinctions are set aside. Sunnis and Shi'as often pull together when faced by a common external threat but then later fall back into intra-religious conflict.

Terror in the Name of Islam

The Soviet invasion and occupation of Afghanistan galvanized Muslims from around the globe to jihad. A decade later, the subsequent September 11th attacks and the U.S. invasions of Afghanistan and Iraq reignited new jihads against the occupying "Crusaders." For the foreseeable future, America will face Islamic terrorists. From America's warfighting standpoint, it is important to note that Islam itself is not the enemy. Like all the world's major religious traditions, Islam has its own extremist fringe. There represents a new form of terrorism, born of transnationalism and globalization. It is transnational in its identity and recruitment and global in its ideology, strategy, targets, network of organizations, and economic transactions.

Many have proposed it is just an inevitable clash of civilizations. The attacks of September 11th and the global threat of Osama bin Laden and Al Qaeda resurrected a knee jerk response of the clash of civilizations approach. But Islam is not monolithic. The challenge is to recognize radicalized versions of Islam. There can be no excuse for terrorism in the name of Islam. Suicide attacks, bombings, assassinations in the name of any cause, whether justified in the name of God, justice, or state security, are still terrorism.

A recent phenomenon has been the suicide attack. However glorious this is made to appear by jihadist propaganda, suicide is forbidden in Islam. But militant Muslims do not see this as suicide. For them, it is self-sacrifice for the cause. What drives young Muslims to become suicide bombers? The use of concepts like jihad and martyrdom to justify suicide bombing provides a powerful incentive: the prospect of being a glorified hero in this life and enjoying paradise in the next.

Over against terror in the name of Islam, Islamic scholars and religious leaders across the Muslim world have made strong, authoritative declarations against the initiatives of those like Al Qaeda:

Islam provides clear rules and ethical norms that forbid the killing of noncombatants, as well as women, children, and the elderly, and also forbids the pursuit of the enemy and defeat, the execution of those who surrender, infliction of harm on prisoners of war, and the destruction of property that is not being used in the hostilities.[6]

In short, bin Laden and his successors, hijacked Islam, using Islamic doctrine and law to legitimize terrorism.

Let it be accepted as a fact that attacks by Islamic terrorists will only continue to rise. It's worth noting, as Seth Jones has observed, since the end of the Cold War, there has been an increase in the number of insurgencies involving extremist Islamic groups, while the percentage of insurgent groups motivated by communist ideology has dramatically declined.[7] Though defeated regionally, the Islamic State and Al Qaeda continue to be highly active in their quest to spread insurgency.

How does this pertain to Iran? While Iran wants to appear to the world as a respectable Islamic republic, they nonetheless undermine any degree of propriety by continuing to operate through terrorist-related activities. As details have emerged from the October 7 Hamas-led attack on Israel, it is clear to see that Hamas could not have been able to plan and carry out such a brutal attack on Israeli civilians had it not been for Iran's support. The vicious unprovoked attack claimed the lives of 1,180 civilians, including 36 children. A further 200 civilians were taken captive. A week later, Israel Defense Forces began offensive weeps in Gaza to destroy Hamas and rescue the hostages. While a truce in November 2023 freed more than 100 hostages, the fight goes on. Iran is directly behind the October 7 Attack as it lies within their mandate to destroy Israel.

What we can expect from Iran is likely more of the same. It's reasonable to expect Iran to continue in its desire to build out its capabilities in Iraq, Syria, and other parts of the Middle East. The chief way this is done is through her proxies – terrorist organizations such as Hezbollah, Hamas, and the Houthis rebels. Two big questions emerge from any discussion regarding Iran. Will the Middle East become immersed in a regional war? On October 1, 2024, in a retaliatory strike, Iran launched a barrage of about 180 ballistic missiles into Israel. The

---

[6] John L. Esposito, *Unholy War: Terror in the Name of Islam* (New York: Oxford, 2002), 158.
[7] Seth G. Jones, *Waging Insurgent Warfare: Lessons from the Vietcong to the Islamic State* (New York: Oxford, 2017), 12.

strike was in retaliation for Israel's assassination of Hezbollah terror chief Hassan Nasrallah who was killed the previous month in Tehran. Thwarted by Israel's aerial defense system and global allies, most of the missiles were destroyed before reaching their targets. The ones that got through caused minimal damage. Tensions in the region remain high.

The second question is, how will the post-war situation in Syria develop? The situation in Syria is likely to remain highly unstable and fraught with challenges. Given the fragile political landscape, the devastating economic crisis, widespread displacement, and continued sectarian tensions, the potential for further conflict due to the influence of external actors like Russia, Iran, and Turkey is highly probable.

China

While the last full-scale war China was involved in was back in its 1979 invasion of Vietnam, in which it performed poorly, the United States is right to take a cautious stance toward China. As Russia's ally, China is led by an aggressive and expansionist President Xi Jinping. China has hundreds of nuclear weapons, the world's second largest population, and a force of three million active-duty, paramilitary, and reserve personnel to do its bidding. China's ally North Korea could have as many as eight million troops. At present, with the exception of Taiwan, while China is not on a war footing, its low-intensity conflict approach is quite a different story.

China's Belt and Road Initiative or BRI, commenced in 2013, is its major strategy to dominate the globe economically through an extensive network of land and maritime routes logistically connecting China with the rest of Asia, Pacific Oceania, Africa, and Europe. The BRI is Xi Jinping's means to further the Chinese Communist party's (CCP) state interests. In countries like Zambia and Tanzania, China has built railways and invested in mining. By means of what is called debt-trap diplomacy, China extends debt to borrowing nations, who in the course of time default on the loan and lose various concessions. Such is the predicament of Uganda. China built its new Entebbe Airport for a loan of $200 million. Using such tactics, China saddles countries with unsustainable debt and sets the conditions for the CCP to garner control of foreign assets, including unrestricted military access. On August 1,

2017, China opened its one and only base in Djibouti. However, the Chinese military stands to gain bases everywhere as the Chinese military strategy works hand-in-glove with the Belt and Road Initiative to gain global dominance.

While the BRI in many respects is China's long game, China has long been exporting its support to fledgling Maoist insurgent groups. Since 1960, China has supported the Naxalite insurgency in India. This long-running insurgency that has enveloped several Indian states has claimed tens of thousands of casualties and destabilized India. A similar story has evolved in Myanmar's Civil War. Though Islamic insurgency is on the rise, something America cannot afford to forget is that Maoist insurgency is not a thing of the past. It remains a global challenge.

Unambiguously, the greatest question regarding China is what can be expected from the situation in Taiwan. China considers the island nation of 24 million people a breakaway territory that must be brought under its control. In a delicate diplomatic strategy, the United States views Taiwan's status as undetermined. The United States has not formally said it would intervene if China were to attack Taiwan, but U.S. presidents have repeatedly suggested they would deploy U.S. troops to defend the island. It goes without saying that such an attack by China could precipitate an unraveling of U.S. power in the Pacific and a broader global destabilization.

In the event China does invade Taiwan, a likely scenario would be one in which all Taiwanese political space would be taken but the CCP would further have to pacify the populace who, it is hoped, would carry on the resistance via partisan warfare. As an irregular war would ensue, Taiwanese resistance forces would likely fight it out from mountainous and urban sanctuaries. As such a resistance would require external support, foreseeably U.S. special operations forces would be available to ply their trade. In such a protracted fight, it would not only be incumbent on the Taiwanese people to remain vigilant despite the great losses they would incur, but it would require the willing resolve of America and her allies to stay the course, insisting on full liberty.

Today, America faces many domestic and foreign threats. Regarding the latter, to avoid catastrophe, it is imperative that America's irregular warfighters understand our strategic opponents. Our enemies are counting on our ignorance to win.

The Way Forward

Many argue that the U.S. military was never designed for or able to become effective at irregular warfare, nor is it likely to get much better at it with practice. For many critics, this appears to be so. However, in view of the ground that has been covered, such an assessment is wrong on its face. There appears to be a formula for success in irregular warfare. However, as we have considered the many complexities involved in IW, we are mindful that we are dealing with certain realities that cannot be objectified. And so, while any quest to find a gold-plated formula for success may forever remain elusive and ephemeral, the following summary of the PLAIN laws constitutes a balanced approach:

P – Political Objective: Unambiguously, political objectives are the goals that a war is to be fought to achieve. This would seem to be axiomatic. However, history shows that America has launched campaigns with scant justification and without so much as a well-defined plan, while strapping the American people with huge debt, as well as a loss of American lives.

Having surveyed twelve case studies involving the American way of IW, one thing is certain. Where U.S. policy directives were vague, with the exception of the Philippine Insurrection, the wars that followed were fraught with ambiguity, mission creep, and failure. This typifies the wars in Afghanistan and Iraq to a tee. Had it not been for exceptional operational and tactical level adaptivity, as in the case of village stability operations in Afghanistan, and Petraeus' enlistment of the Sons of Iraq, both conflicts might have imploded. Granted, Afghanistan did not end well, and it remains to be seen if Iraq's democracy will survive. Yet, as has been argued, if it had been given time to mature, the VSO program could have worked.

According to the Powell Doctrine, named for General Colin Powell, the U.S. should only fight wars: with overwhelming force, clear goals, and a decisive exit strategy. In both Afghanistan and Iraq, our reach exceeded our grasp. Both nations were invaded with only one of Powell's doctrinal tenets – overwhelming force. Arguably, the U.S. followed the wisdom of Powell in Syria. It stands to reason that our next war should square with it fully.

384

U.S. planners must also bear in mind that the U.S. center of gravity is its political culture. Likewise, time and resource limitations are dominant factors, given the tendency of the American people to grow weary and impatient of protracted and unproductive conflicts in distant regions. Huge commitments, with unclear objectives, run the risk of limited prospects of success. The way forward should be to have an actionable policy that squares with the Powell doctrine that is undergirded by a doable goal.

L – Legitimacy: Legitimacy is the qualitative condition of the governing authority, as is perceived by the people who give their consent. David Galula has rightly said, "The counterinsurgent reaches a position of strength when his power is embedded in a political organization issuing from and firmly supported by the population." The same could be said for insurgents. The U.S.-sponsored insurgencies in enemy-occupied Europe and Asia achieved such positions of strength from their perspective affected populations. Regarding COIN, while this positional strength was attained by General Bell during the Philippine Insurrection, as well as Magsaysay and Lansdale during the Huk rebellion, such support has seldom since been achieved by U.S. IW practitioners since.

Arguably, this has most often been the case due to leadership failure at the host national level. President Diem's corrupt, weak, and culturally incompetent governance foredoomed American COIN efforts in Vietnam. To a similar extent, the same could be said of America's experiences in Afghanistan, and to a lesser degree Iraq. It is fair to say that despite significant U.S. support, the host nation governments during those wars never really worked to develop state legitimacy. The way forward should be to have competent leaders in place and then carefully maintain the condition of perceived legitimacy through various means. Easier said than done. Perhaps the best example of this comes from America's experience with the Huk rebellion and the reform policies that help curb its popularity.

A – Adaptability: Irregular warfare is a learning competition that requires adaptively modifying approaches to complex and unpredictable threats. America prides itself on ingenuity. In America's most successful wars, American leaders have modified and adapted their tactics as much as possible to win. This was especially seen Bell's use of circulars and resettlement in the Philippine Insurrection, Lansdale's scheme to undermine the Huks, the CORDS program in Vietnam, McChrystal's change of approach in Afghanistan, and the co-opting of the Sons of Iraq by Petraeus. The only thing we learn from history is we don't learn from history. American leaders must learn from our IW experiences. With over 120 years' experience of asymmetric warfighting, there is simply no excuse.

Regarding adaptability, the way forward should be one that takes into account the words of David Galula: "The soldier must be prepared to become a propagandist, a social worker, a civil engineer, a schoolteacher, a nurse, a Boy Scout. But only as long as he cannot be replaced, what's better is to entrust civilian tasks to civilians."[8]

I – Influence: Influence is the capacity to affect perception. As we consider irregular warfare, we are dealing with certain realities that cannot be objectified. The importance of influence in this endeavor may be expressed in the form of a metaphor. As has been mentioned, as all true carpenters know, a sharp nail splits the wood, but a blunted tip keeps it whole. Like a tiny axe, a sharp nail acts like a wedge, splitting the wood. Yet, blunting the nail serves to punch a hole. The blunted nail keeps the wood intact. Early on, the U.S. strategy for the war in Afghanistan may be said to have epitomized the sharp nail, as the search and destroy mentality often counteracted other efforts, such as SFA, SO, and COIN. Arguably, McChrystal's blunted tip strategy, leveled the American approach, targeting Taliban infrastructure, while generally producing a more stabilized Afghanistan.

In Afghanistan, Al Qaeda likewise sought to exert a tremendous amount of influence. Bin Laden skillfully cultivated and developed his relationship with Mullah Omar and the Taliban, providing financial support, building roads and other construction projects, and sending his Afghan Arabs to fight alongside the Taliban in critical battles. Conversely, for the first eight years of the war, while there were stability efforts, the primary aim of the U.S. was to search and destroy insurgents

---

[8] David Galula, *Counterinsurgency Warfare: Theory and Practice* (Westport, CT: Praeger, 2006).

who were all lobbed into the monolithic category of Al Qaeda. It wasn't until the eighth year of the war that the U.S. evaluated its war performance and made a course correction. The U.S. COIN tactic of VSO was working to win over fence-sitters and neutralizing nay-sayers. However, like CORDS in Vietnam, VSO came at the eleventh hour. By then America had grown ever so tired of the conflict which had been sidelined by the Iraq invasion. There is one thing that's clear from America's IW experience. The way forward calls for a holistic approach, as victory in IW runs in packs.

N – Native Face: If there is one thing history shows it is that irregular warfare practitioners cannot win as an outside power. One RAND study offers sound advice: Even if tactically successful, a unilateral operation by external forces may ultimately lead to failure by undermining and delegitimizing the very indigenous capability the external actor is trying to build. Illustrating precisely that point, Russell Crandall remarks that "an army of liberation can have the appearance of an army of occupation."[9] This was precisely part of the problem that plagued the U.S. operation in Afghanistan, and to a lesser degree in Iraq. By contrast, the light footprint approach during OEF-Philippines, and Operation Inherent Resolve, working by, with, and through regional partners was undoubtedly the secret sauce for eliminating any such unfavorable perception. The challenge for this to work is for American IW practitioners to find local solutions to local problems.

Define-Isolate-Disintegrate-Consolidate-Resource

Regarding counterinsurgency, with all its complexities, it's fair to say we cannot unravel this Gordian knot with abstract reasoning. The riddle of counterinsurgency is not revealed at the conclusion of a syllogism as a perfect answer lies beyond both our cognitive reach and understanding. Before we capitulate, it may be said, however, there appears to be an American formula for success. Our adventures in counterinsurgency have taught us several weighty lessons we dare not disregard.

---

[9] Russell Crandall, *America's Dirty Wars: Irregular Warfare from 1776 to the War on Terror* (New York: Cambridge, 2014), 467.

According to the current U.S. doctrine, counterinsurgency is to follow a Shape-Clear-Hold-Build-Transfer methodology. The COIN force begins by *shaping* the environment to their advantage, concentrating on population centers and the reduction of insurgent influence over them. In *clearing*, the COIN force destroys, captures, or forces the withdrawal of insurgents to secure a physical and psychological environment that establishes or re-establishes government control. The COIN force then *holds* the contested areas and *builds* defensive networks, radiating security and influence from this cleared area. Build tasks include improving infrastructure, roads, reestablishing services, schools, and facilities. Finally, in *transition*, the COIN force hands over operations to host nation security forces.

As a nuance to Shape-Clear-Hold-Build-Transfer, we suggest a Define-Isolate-Disintegrate-Consolidate-Resource methodology.

The COIN force begins by defining the problem, <u>define</u>. Normally, this consists of a real or perceived grievance, as in the Huk's slogan, "land for the landless." Underscoring the simplicity and importance of this, Seth Jones observes,

Insurgencies are caused, in part, by grievances that motivate rebels. If rebels didn't have these motivations, insurgencies wouldn't happen. Charismatic leaders that use hyper nationalist, religious, or other types of rhetoric are generally important to mobilize individuals and organize rebellion. Insurgent leaders can use grievances instrumentally like an advertising campaign: to persuade individuals to join and to retain those loyalties that already support the cause. After all, insurgents generally view themselves as agents of change. As Che Guevara wrote, "we must come to the inevitable conclusion that the guerrilla fighter is a social reformer and that he fights in order to change the social system that keeps all his unarmed brothers and ignominy and misery."[10]

<u>Define</u> takes place before shaping by understanding the nature of the problem, identifying the root cause(s) of the insurgency and its leadership, while devising a realistic solution.

In <u>isolate</u>, applying a discriminate use of force, insurgents are physically and psychologically removed from the affected populace, thereby depriving the insurgency of tangible support, intelligence, etc. In

---

[10] Seth G. Jones, *Waging Insurgent Warfare: Lesson s from the Vietcong to the Islamic State* (New York: Oxford, 2017), 19-20.

addition to merely clearing insurgent strongholds, the isolate step may include such tactics as population resettlement or village stability efforts.

In disintegrate, the COIN force reduces the tangible support the insurgents may receive from both the populace and external actors while reducing insurgent sanctuaries, thereby limiting their will to fight, and operational reach. Lines of operations in tandem with this step are border control, strikes into insurgent sanctuaries, etc. Arguably, the U.S. failed in Afghanistan primarily because of the Taliban's cross-border sanctuary. The Taliban could always slip over the border into Pakistan's FATA.

Similar to hold-build, in consolidate, the COIN force consolidates gains by stabilizing the affected area, through building out security and defensive networks, improving infrastructure, roads, reestablishing services, schools, and facilities, etc. Finding local solutions and employing a native face with a light U.S. footprint, consolidating aims at identifying with the people, working with them to create a better state of the peace, and winning their allegiance.

Finally, in resource, the COIN force continues to resource the people to achieve their needs, working to redress the grievances that led to the insurgency. As in any counterinsurgency approach, the host national government leadership is crucial to success.

Arguing that effective COIN practices run in packs, in their analysis, a RAND panel of authors recommended seven key ingredients to successful counterinsurgency. These dovetail with our findings and include:[11]

1. Plan to pursue multiple mutually supporting lines of operation in COIN.
2. Build and maintain forces that can engage in these lines of operation simultaneously.
3. Ensure the positive involvement of the host-nation government.
4. Strike a balance in the approach used against insurgents.
5. Be adaptable.
6. Avoid collective punishment.

---

[11] Christopher Paul, *Victory Has a Thousand Fathers* (Santa Monica, CA: RAND, 2010), 93-99.

7. Ascertain the specific support needs of and sources of support for insurgent adversaries and target them.

Further, given the American proclivity to grow impatient with long wars in faraway places, and the fact that no nation ever benefits from long wars, time and resources are therefore the dominant factors.

For Further Reading:

1. *Three Dangerous Men* by Seth Jones.
2. *Unholy War: Terror in the Name of Islam* by John Esposito.
3. *Unrestricted Warfare* by Qiao Liang and Wang Xiangsui.

Conclusion

"Insurgency and counterinsurgency will remain alive and well for the foreseeable future. The challenge is to better understand this type of warfare."
– Seth G. Jones

As this book has tried to demonstrate, America has a long "irregular" or "asymmetric" war-fighting capacity. Notwithstanding this fact, irregular war is as old as organized war and is used by everyone, some better than others. China and Russia have conventional armies, navies, and an air force. For them, irregular warfare is merely one component of a larger strategy. Lacking a large, modernized military, Iran relies exclusively on its proxies and irregular warfare. Given this state of things, predictably, the United States will most likely continue to face, who Seth Jones refers to as, the "Three Dangerous Men" (Russia, China, and Iran), in a great power rivalry within third states, as we continue to compete for influence and control.

Echoing this sentiment, retired General Cleveland suggests,

Competition with other nations is the primary national security concern of the United States. This includes competition with the revisionist powers of China and Russia, the "central challenge to U.S. prosperity and security," and rogue regimes (e.g., Iran, North Korea) that threaten to destabilize regions through "their pursuit of nuclear weapons or sponsorship of terrorism." Deterring the aggression of these adversaries will certainly require that the United States expand the lethality of its conventional and nuclear capabilities to sustain its military supremacy. However, actual competition and conflict with these nation-states will most likely be irregular. This reflects a deliberate calculus by our adversaries, who understand that conventional or nuclear war with the United States would be risky and prohibitively costly, but that the United States is vulnerable to irregular approaches. Rather than risk conflicts with casualty numbers that are virtually unthinkable, these adversaries will continue to deploy a blend of information, legal, and proxy warfare in challenging the United States and its allies. Maturing the American way of irregular war is critical for the United States to remain competitive against these threats.[1]

---

[1] Charles T. Cleveland, *The American Way of Irregular War: An Analytical Memoir* (Santa Monica, CA: RAND, 2020), 224-226.

Cleveland goes on to say that there is a growing desire for those in government to want to get back to the type of warfare we know how to fight. However, as he presciently states,

With the growing threats from rival powers demanding that U.S. military capabilities be rededicated to their traditional missions, it is likely that the United States will return to conventional approaches to conflict without outside pressure to do otherwise. As such, there is a real risk that we repeat the mistakes that we made in the wake of Vietnam—specifically, that we again fail to take seriously the lessons of past conflicts to adapt to a changing threat landscape. For many, the return to traditional war and the pivot from these contests against adversaries wielding nonconventional means are much welcomed. It is their comfort zone, after all, and where the United States has dominated for a hundred years. Our adversaries have moved to dominate in the space below the threshold of war. It will be a strategy built around an American way of irregular war that defeats them.[2]

Let it be accepted then that irregular warfare will only increase for the foreseeable future. This needs no further argument. In conclusion, let it be said that the golden rule of irregular warfare is that the people are the contested *land*. Success depends not necessarily on occupying political space, but rather on the abstract psychological metrics of legitimacy and influence. It certainly cannot be overstated that an IW-fighting force cannot kill its way to success. All the key terrain could be captured, yet without at least the passive support of the affected population, one will always fail. Granted, battles must be won, but the decisive battle in irregular warfare is always in the hearts and minds of the people.

As an American warfighter, you may be called upon to execute any number of missions like the ones described in this volume, or others, we have yet to dream of. As you do, remember the immortal words of Clausewitz and Sun Tzu. The Prussian rightly said, "In war many roads lead to success, and they do not all involve the opponent's outright defeat." Likewise, Master Sun said, "To subdue the enemy without fighting is the acme of skill." These two axioms sum up the essence of irregular warfare genius.

De Oppresso Liber

---

[2] Ibid.

Name of Conflict	Dates	Duration	IW Activity Employed	Tactics Used	Result	Forces Involved	U.S. Cost
Philippine Insurrection	1899-1902	3 Years	COIN	Search & Destroy Resettlement	U.S. Win	125,000 U.S. 40,000 Insurrectos	$400 million 4,200 killed
WWII, Japanese Occupation of the Philippines	1941-1945	4 Years	UW	GW, Sabotage, Subversion	U.S. Win	260,000 Allied Guerrillas 500,000 Japanese	$10 billion +30,000 killed
WWII, Nazi Occupation of France	1944-1945	1 Year	UW	Sabotage, GW	U.S. Win	300,000 Maquis 93 Jed Teams	$43 billion
Hukbalahap Rebellion in the Philippines	1950-1954	5 Years	COIN/FID	Search & Destroy, Reforms, Resettlement	U.S. Win	250,000 Filipino Troops 100,000 Huks	$520 million
Vietnam War	1954-1972	18 Years	COIN/FID/ SFA/SO/CT	Search & Destroy, Reforms, Amnesty, Resettlement	U.S. Loss	2.7 million U.S. 400,000 Vietcong	$1 trillion 58,220 killed
Cuban Revolution	1953-1959	6 Years	COIN/FID	Search & Destroy	U.S. Loss	30,000 National troops 800 guerrillas	$8 million
Nicaraguan Civil War	1978-1990	12 Years	UW	GW, Subversion, Sabotage	U.S. Loss	120,000 Sandinistas 125,000 Contras	$100 million
Salvadoran Civil War	1980-1992	12 Years	COIN/FID	Search & Destroy	U.S. Loss	48,000 National troops	$1 billion
Somalian Civil War UNNOSOM II	1992-1995	3 Years	FID/SFA/CT COIN	Search & Destroy Amnesty, Reforms	U.S. Loss	25,000 UN troops +5,000 Militia fighters	$44 million
Afghanistan War	2001-2021	20 Years	CT/COIN/FID SFA/SO	Search & Destroy, Village Stability	U.S. Loss	800,000 U.S. 20,000 Taliban	$3 trillion 3,590 killed
Iraq War	2003-2011	9 Years	CT/COIN/FID SFA/SO	Search & Destroy, Reforms, Amnesty	U.S. Win	1.5 million U.S. 20,000 Insurgents	$3 trillion
OEF-Philippines	2002-2017	15 Years	COIN/CT/FID	Search & Destroy, Amnesty, Reforms	U.S. Win	1,200 U.S. 1,000 Abu Sayyaf	$588 million 17 deaths
Syrian Civil War	2014-	10 Years	CT/COIN/FID	Search & Destroy	Ongoing	2,400 U.S.	$14 million 118 killed

Annex B: Select List of Axioms

The following axioms are a compilation of arguably the world's best wisdom regarding irregular warfare.

1. The golden rule of IW is the people themselves are the contested *land*.

2. Irregular warfare is a learning competition. Who adapts wins.

3. Control of the populace is the master weapon of irregular warfare (Sepp's law).

4. Counterinsurgents can't want it more than the host nation does.

5. The counterinsurgent reaches a position of strength when his power is embedded within and supported by the host nation's government and populace (David Galula).

6. UW is not about the U.S. soldier; it's about leveraging indigenous partners (Brodie's first law).

7. UW is not about U.S. resources; it's about leveraging indigenous resources (Brodie's second law).

8. If a Soldier cannot articulate how he plans to win "through and with" indigenous partners, he has failed to demonstrate that he is proficient in UW (Brodie's third law).

9. IW favors the indirect approach (B. H. Liddell Hart).

10. Find local solutions to local problems (Robin Sage axiom).

Annex C: The Twenty-Seven Articles of T.E. Lawrence

T.E. Lawrence from *The Arab Bulletin*, 20 August 1917.

The following notes have been expressed in commandment form for greater clarity and to save words. They are, however, only my personal conclusions, arrived at gradually while I worked in the Hejaz and now put on paper as stalking horses for beginners in the Arab armies. They are meant to apply only to Bedu; townspeople or Syrians require totally different treatment. They are of course not suitable to any other person's need, or applicable unchanged in any particular situation. Handling Hejaz Arabs is an art, not a science, with exceptions and no obvious rules. At the same time we have a great chance there; the Sherif trusts us, and has given us the position (towards his Government) which the Germans wanted to win in Turkey. If we are tactful, we can at once retain his goodwill and carry out our job, but to succeed we have got to put into it all the interest and skill we possess.

1. Go easy for the first few weeks. A bad start is difficult to atone for, and the Arabs form their judgments on externals that we ignore. When you have reached the inner circle in a tribe, you can do as you please with yourself and them.

2. Learn all you can about your Ashraf and Bedu. Get to know their families, clans and tribes, friends and enemies, wells, hills and roads. Do all this by listening and by indirect inquiry. Do not ask questions. Get to speak their dialect of Arabic, not yours. Until you can understand their allusions, avoid getting deep into conversation or you will drop bricks. Be a little stiff at first.

3. In matters of business deal only with the commander of the army, column, or party in which you serve. Never give orders to anyone at all, and reserve your directions or advice for the C.O., however great the temptation (for efficiency's sake) of dealing with his underlings. Your place is advisory, and your advice is due to the commander alone. Let him see that this is your conception of your duty, and that his is to be the sole executive of your joint plans.

4. Win and keep the confidence of your leader. Strengthen his prestige at your expense before others when you can. Never refuse or quash schemes he may put forward; but ensure that they are put forward in the first instance privately to you. Always approve them, and after praise modify them insensibly, causing the suggestions to come from him, until they are in accord with your own opinion. When you attain this point, hold him to it, keep a tight grip of his ideas, and push them forward as firmly as possibly, but secretly, so that to one but himself (and he not too clearly) is aware of your pressure.

5. Remain in touch with your leader as constantly and unobtrusively as you can. Live with him, that at meal times and at audiences you may be naturally with him in his tent. Formal visits to give advice are not so good as the constant dropping of ideas in casual talk. When stranger sheikhs come in for the first time to swear allegiance and offer service, clear out of the tent. If their first impression is of foreigners in the confidence of the Sherif, it will do the Arab cause much harm.

6. Be shy of too close relations with the subordinates of the expedition. Continual intercourse with them will make it impossible for you to avoid going behind or beyond the instructions that the Arab C.O. has given them on your advice, and in so disclosing the weakness of his position you altogether destroy your own.

7. Treat the sub-chiefs of your force quite easily and lightly. In this way you hold yourself above their level. Treat the leader, if a Sherif, with respect. He will return your manner and you and he will then be alike, and above the rest. Precedence is a serious matter among the Arabs, and you must attain it.

8. Your ideal position is when you are present and not noticed. Do not be too intimate, too prominent, or too earnest. Avoid being identified too long or too often with any tribal sheikh, even if C.O. of the expedition. To do your work you must be above jealousies, and you lose prestige if you are associated with a tribe or clan, and its inevitable feuds. Sherifs are above all blood-feuds and local rivalries, and form the only principle of unity among the Arabs. Let your name therefore be coupled always with

a Sherif's, and share his attitude towards the tribes. When the moment comes for action put yourself publicly under his orders. The Bedu will then follow suit.

9. Magnify and develop the growing conception of the Sherifs as the natural aristocracy of the Arabs. Intertribal jealousies make it impossible for any sheikh to attain a commanding position, and the only hope of union in nomad Arabs is that the Ashraf be universally acknowledged as the ruling class. Sherifs are half-townsmen, half-nomad, in manner and life, and have the instinct of command. Mere merit and money would be insufficient to obtain such recognition; but the Arab reverence for pedigree and the Prophet gives hope for the ultimate success of the Ashraf.

10. Call your Sherif 'Sidi' in public and in private. Call other people by their ordinary names, without title. In intimate conversation call a Sheikh 'Abu Annad', 'Akhu Alia' or some similar by-name.

11. The foreigner and Christian is not a popular person in Arabia. However friendly and informal the treatment of yourself may be, remember always that your foundations are very sandy ones. Wave a Sherif in front of you like a banner and hide your own mind and person. If you succeed, you will have hundreds of miles of country and thousands of men under your orders, and for this it is worth bartering the outward show.

12. Cling tight to your sense of humour. You will need it every day. A dry irony is the most useful type, and repartee of a personal and not too broad character will double your influence with the chiefs. Reproof, if wrapped up in some smiling form, will carry further and last longer than the most violent speech. The power of mimicry or parody is valuable, but use it sparingly, for wit is more dignified than humour. Do not cause a laugh at a Sherif except among Sherifs.

13. Never lay hands on an Arab; you degrade yourself. You may think the resultant obvious increase of outward respect a gain to you, but what you have really done is to build a wall between you and their inner selves. It

is difficult to keep quiet when everything is being done wrong, but the less you lose your temper the greater your advantage. Also then you will not go mad yourself.

14. While very difficult to drive, the Bedu are easy to lead, if: have the patience to bear with them. The less apparent your interferences the more your influence. They are willing to follow your advice and do what you wish, but they do not mean you or anyone else to be aware of that. It is only after the end of all annoyances that you find at bottom their real fund of goodwill.

15. Do not try to do too much with your own hands. Better the Arabs do it tolerably than that you do it perfectly. It is their war, and you are to help them, not to win it for them. Actually, also, under the very odd conditions of Arabia, your practical work will not be as good as, perhaps, you think it is.

16. If you can, without being too lavish, forestall presents to yourself. A well-placed gift is often most effective in winning over a suspicious sheikh. Never receive a present without giving a liberal return, but you may delay this return (while letting its ultimate certainty be known) if you require a particular service from the giver. Do not let them ask you for things, since their greed will then make them look upon you only as a cow to milk.

17. Wear an Arab headcloth when with a tribe. Bedu have a malignant prejudice against the hat, and believe that our persistence in wearing it (due probably to British obstinacy of dictation) is founded on some immoral or irreligious principle. A thick headcloth forms a good protection against the sun, and if you wear a hat your best Arab friends will be ashamed of you in public.

18. Disguise is not advisable. Except in special areas, let it be clearly known that you are a British officer and a Christian. At the same time, if you can wear Arab kit when with the tribes, you will acquire their trust and intimacy to a degree impossible in uniform. It is, however, dangerous and difficult. They make no special allowances for you when

you dress like them. Breaches of etiquette not charged against a foreigner are not condoned to you in Arab clothes. You will be like an actor in a foreign theatre, playing a part day and night for months, without rest, and for an anxious stake. Complete success, which is when the Arabs forget your strangeness and speak naturally before you, counting you as one of themselves, is perhaps only attainable in character: while half-success (all that most of us will strive for; the other costs too much) is easier to win in British things, and you yourself will last longer, physically and mentally, in the comfort that they mean. Also then the Turks will not hang you, when you are caught.

19. If you wear Arab things, wear the best. Clothes are significant among the tribes, and you must wear the appropriate, and appear at ease in them. Dress like a Sherif, if they agree to it.

20. If you wear Arab things at all, go the whole way. Leave your English friends and customs on the coast, and fall back on Arab habits entirely. It is possible, starting thus level with them, for the European to beat the Arabs at their own game, for we have stronger motives for our action, and put more heart into it than they. If you can surpass them, you have taken an immense stride toward complete success, but the strain of living and thinking in a foreign and half-understood language, the savage food, strange clothes, and stranger ways, with the complete loss of privacy and quiet, and the impossibility of ever relaxing your watchful imitation of the others for months on end, provide such an added stress to the ordinary difficulties of dealing with the Bedu, the climate, and the Turks, that this road should not be chosen without serious thought.

21. Religious discussions will be frequent. Say what you like about your own side, and avoid criticism of theirs, unless you know that the point is external, when you may score heavily by proving it so. With the Bedu, Islam is so all-pervading an element that there is little religiosity, little fervour, and no regard for externals. Do not think from their conduct that they are careless. Their conviction of the truth of their faith, and its share in every act and thought and principle of their daily life is so intimate and intense as to be unconscious, unless roused by opposition. Their religion is as much a part of nature to them as is sleep or food.

22. Do not try to trade on what you know of fighting. The Hejaz confounds ordinary tactics. Learn the Bedu principles of war as thoroughly and as quickly as you can, for till you know them your advice will be no good to the Sherif. Unnumbered generations of tribal raids have taught them more about some parts of the business than we will ever know. In familiar conditions they fight well, but strange events cause panic. Keep your unit small. Their raiding parties are usually from one hundred to two hundred men, and if you take a crowd they only get confused. Also their sheikhs, while admirable company commanders, are too 'set' to learn to handle the equivalents of battalions or regiments. Don't attempt unusual things, unless they appeal to the sporting instinct Bedu have so strongly, unless success is obvious. If the objective is a good one (booty) they will attack like fiends, they are splendid scouts, their mobility gives you the advantage that will win this local war, they make proper use of their knowledge of the country (don't take tribesmen to places they do not know), and the gazelle-hunters, who form a proportion of the better men, are great shots at visible targets. A sheikh from one tribe cannot give orders to men from another; a Sherif is necessary to command a mixed tribal force. If there is plunder in prospect, and the odds are at all equal, you will win. Do not waste Bedu attacking trenches (they will not stand casualties) or in trying to defend a position, for they cannot sit still without slacking. The more unorthodox and Arab your proceedings, the more likely you are to have the Turks cold, for they lack initiative and expect you to. Don't play for safety.

23. The open reason that Bedu give you for action or inaction may be true, but always there will be better reasons left for you to divine. You must find these inner reasons (they will be denied, but are none the less in operation) before shaping your arguments for one course or other. Allusion is more effective than logical exposition: they dislike concise expression. Their minds work just as ours do, but on different premises. There is nothing unreasonable, incomprehensible, or inscrutable in the Arab. Experience of them, and knowledge of their prejudices will enable you to foresee their attitude and possible course of action in nearly every case.

24. Do not mix Bedu and Syrians, or trained men and tribesmen. You will get work out of neither, for they hate each other. I have never seen a successful combined operation, but many failures. In particular, ex-officers of the Turkish army, however Arab in feelings and blood and language, are hopeless with Bedu. They are narrow minded in tactics, unable to adjust themselves to irregular warfare, clumsy in Arab etiquette, swollen-headed to the extent of being incapable of politeness to a tribesman for more than a few minutes, impatient, and, usually, helpless without their troops on the road and in action. Your orders (if you were unwise enough to give any) would be more readily obeyed by Beduins than those of any Mohammedan Syrian officer. Arab townsmen and Arab tribesmen regard each other mutually as poor relations, and poor relations are much more objectionable than poor strangers.

25. In spite of ordinary Arab example, avoid too free talk about women. It is as difficult a subject as religion, and their standards are so unlike our own that a remark, harmless in English, may appear as unrestrained to them, as some of their statements would look to us, if translated literally.

26. Be as careful of your servants as of yourself. If you want a sophisticated one you will probably have to take an Egyptian, or a Sudani, and unless you are very lucky he will undo on trek much of the good you so laboriously effect. Arabs will cook rice and make coffee for you, and leave you if required to do unmanly work like cleaning boots or washing. They are only really possible if you are in Arab kit. A slave brought up in the Hejaz is the best servant, but there are rules against British subjects owning them, so they have to be lent to you. In any case, take with you an Ageyli or two when you go up country. They are the most efficient couriers in Arabia, and understand camels.

27. The beginning and ending of the secret of handling Arabs is unremitting study of them. Keep always on your guard; never say an unnecessary thing: watch yourself and your companions all the time: hear all that passes, search out what is going on beneath the surface, read their characters, discover their tastes and their weaknesses and keep everything you find out to yourself. Bury yourself in Arab circles, have no

interests and no ideas except the work in hand, so that your brain is saturated with one thing only, and you realize your part deeply enough to avoid the little slips that would counteract the painful work of weeks. Your success will be proportioned to the amount of mental effort you devote to it.

About the Authors

Bob Ball, Lieutenant Colonel, U.S. Army (Retired)

Bob Ball served in 1st Battalion, 1st Special Forces Group in a variety of positions. He was a Detachment Commander, Adjutant, Assistant Operations Officer, and a Company Commander. Additionally, Bob has extensive experience in Basilan and Mindanao. Bob's service continues as a Lane Manager for Robin Sage in the Special Forces Qualification Course.

Sheffield Ford III, Major, U.S. Army (Retired)

Sheffield Ford III has a notable military career serving our country. He commanded Special Forces Soldiers in Afghanistan and Pakistan, where he was responsible for special operations, combat missions, and training. During Operation Enduring Freedom in 2006, he was awarded numerous military awards, including the Silver Star for his heroic actions and courageous leadership in Afghanistan, as well as the Legion of Merit, the Bronze Star and many other military recognitions. Sheff is president and chief executive officer for Raven Advisory, LLC.

Paul LeFavor, Master Sergeant, U.S. Army (Retired)

Paul LeFavor was born in Virginia and was raised in a pastor's family. He graduated from Liberty University and received his M.A. in Religion from Reformed Theological Seminary and his M.Div. from Liberty Theological Seminary. Paul retired after serving twenty years in the U.S. Army Special Forces in 2009, is married to Becky, his wife of thirty years, and has two daughters Liane and Collette, a granddaughter, Annabel, and a grandson, Soren. He served as the Pastor of Christ Covenant Baptist Church, Fayetteville, North Carolina from 2012-2025, and is the author of several books, including the *U.S. Army Small Unit Tactics Handbook*, *Tactical Leadership*, *The Wild Fields*, and *The Five Warrior Virtues*. At present, Paul and his wife are serving as evangelists around the world, building churches, orphanages, and schools, for the glory of God and the advancement of the Christian Faith.

Bibliography

Army Doctrine Publication 3-05, Special Operations, Washington, D.C.:
Headquarters, Department of the Army, August 2012.

Army Techniques Publication 3-07.5, Stability Techniques, Washington,
D.C.: Headquarters, U.S. Department of the Army, August 2012.

Asprey, Robert B. War in the Shadows: The Guerrilla in History. Vol. 1. 2 vols.
Garden City, NY: Doubleday, 1975.

Breuer, William B. MacArthur's Undercover War: Spies, Saboteurs, Guerillas,
and Secret Missions. Edison, NJ: Castle Books, 1995.

Childress, Michael, The Effectiveness of U.S. Training Efforts in Internal
Defense and Development: The Cases of El Salvador and Honduras,
Santa Monica, Calif.: RAND Corporation, MR-250-USDP, 1995.

Coll, Steve. The Achilles Trap: Saddam Hussein, the CIA, and the Origins of
America's Invasion of Iraq. Penguin, 2024.

Coll, Steve. Directorate S: the CIA and America's Secret Wars in Afghanistan and
Pakistan. Penguin, 2018.

Coll, Steve. Ghost Wars: The Secret History of the CIA, Afghanistan and Bin
Laden. Penguin UK, 2005.

Crandall, Russell. America's Dirty Wars: Irregular Warfare from 1776 to the War
on Terror. Cambridge University Press, 2014.

Egel, Daniel, and Charles T. Cleveland. "The American Way of Irregular War: An
Analytical Memoir." (2020).

Field Manual 3-18, Special Forces Operations, Washington, D.C.:
Headquarters, U.S. Department of the Army, May 2014.

Field Manual 3-24 and Marine Corps Warfighting Publication 3-33.5,
Counterinsurgency, Washington, D.C.: U.S. Department of the Army,
December 16, 2006.

Galula, David. Counterinsurgency Warfare: Theory and Practice. Bloomsbury
Publishing USA, 2006.

Hammes, Thomas X. The Sling and The Stone: On War in the 21st Century.
Zenith Press, 2006.

Hart, B. H. Liddell. Strategy. New York: Meridian, 1991.

Hogan, David W. Jr. Special Operations in the Pacific. Washington, D.C.: Department of the Army Center for Military History publication, 1992.

Joes, Anthony. Resisting Rebellion: The History and Politics of Counterinsurgency. University Press of Kentucky, 2004.

Jones, Seth, "The Future of Irregular Warfare Is Irregular," National Interest, August 26, 2018. As of October 5, 2018: https://nationalinterest.org/feature/future-warfare-irregular-29672

Jones, Seth G. Waging Insurgent Warfare: Lessons from the Vietcong to the Islamic State. Oxford University Press, 2017.

Kilcullen, David. The Accidental Guerrilla: Fighting Small Wars in The Midst of a Big One. Oxford University Press, 2011.

Kilcullen, David. Counterinsurgency. Oxford University Press, 2010.

Kilcullen, David. Out of the Mountains: The Coming Age of the Urban Guerrilla. Oxford University Press, 2015.

Kilcullen, David. "Three pillars of counterinsurgency." U.S. Government Counterinsurgency Conference. Vol. 28. Washington, DC: U.S. Department of State, 2006.

Kiras, James, Special Operations and Strategy: From World War II to the War on Terrorism, New York: Routledge, 2006.

Koffler, Rebekah. Putin's Playbook: Russia's Secret Plan to Defeat America. Simon and Schuster, 2021.

Komer, Robert W., Bureaucracy Does Its Thing: Institutional Constraints on U.S.-GVN Performance in Vietnam, Santa Monica, Calif.: RAND Corporation, R-967-ARPA, 1972. As of May 7, 2020: https://www.rand.org/pubs/reports/R967.html

Krieg, Andreas. Subversion: The Strategic Weaponization of Narratives (2023).

Lansdale, Edward Geary. In The Midst of Wars: An American's Mission to Southeast Asia. (1972).

Lawrence, Thomas Edward. The Seven Pillars of Wisdom. Graphic Arts Books, 2020.

Loyn, David. The Long War: The Inside Story of America and Afghanistan Since 9/11. St. Martin's Press, 2021.

MacArthur, Douglas. Reminiscences. New York: McGraw-Hill Book Co., 1964.

Moyar, Mark. A Question of Command: Counterinsurgency from the Civil War to Iraq. Yale University Press, 2009.

Nagl, John A. Learning to Eat Soup with a Knife, Counterinsurgency Lessons from Malaya and Vietnam. Chicago: University of Chicago Press, 2002.

Ricks, Thomas E. Fiasco: The American Military Adventure in Iraq. Penguin UK, 2007.

Ricks, Thomas E. The Gamble: General Petraeus and the American Military Adventure in Iraq. Penguin, 2010.

Tyson, Ann Scott. American Spartan: The Promise, the Mission, and the Betrayal of Special Forces Major Jim Gant. Harper Collins, 2014.

Volckmann, R. W. We Remained: Three Years Behind the Enemy Lines in the Philippines. New York: W. W. Norton & Company, Inc., 1954.

Von Clausewitz, Carl. On War, Everyman's Library (Princeton, NJ: Princeton University Press, 1993.

Whitlock, Craig. The Afghanistan Papers: A Secret History of The War. Simon And Schuster, 2021.

Zedong, Mao. On Guerrilla Warfare. Trans., Samuel B. Griffith, II. Chicago, IL: University of Illinois Press, 1961.

Acknowledgments

This book is indeed a team effort. We are overwhelmed by the support, insight, feedback, and advice from the Green Berets of Camp Mackall and all seven Special Forces Groups. We have drawn liberally from this well of collective wisdom and experience. We would like to take this opportunity to thank everyone who has been involved in this project, especially: Major General Edward M. Reeder (Ret.), Colonel Kalev I. Sepp (Ret.), Chief Warrant Officer Three Todd Swan, and Command Sergeant Major Mike Sahms (Ret.). We are extremely grateful to everyone involved for all their help. Thanks to all.

Recommended Reading List

1. *On Guerrilla Warfare* by Mao Zedong.
2. *The Accidental Guerrilla* by David Kilcullen.
3. *A Question of Command* by Mark Moyar.
4. *FM 3-24, Counterinsurgency.*
5. *FM 3-24.2, Tactics in Counterinsurgency.*
6. *Victory Has a Thousand Fathers* by Christopher Paul.
7. *Learning to Eat Soup with a Knife* by John Nagl.
8. *The American Way of Irregular War* by Charles Cleveland.
9. *How Insurgencies End* by Ben Connable and Martin C. Libicki.
10. *Schoolbooks and Krags: The United States Army in the Philippines, 1898-1902* by John M. Gates.
11. *Honor in the Dust: Theodore Roosevelt, War in the Philippines, and the Rise and Fall of America's Imperial Dream* by Gregg Jones.
12. *The Philippine War* by Brian McAllister Linn.
13. *We Remained* by Russell Volckmann.
14. *Fire in the Jungle* by Larry S. Schmidt
15. *MacArthur's Undercover War* by William Breuer.
16. *The Jedburghs: The Secret History of the Allied Special Forces, France 1944* by Will Irwin.
17. *Operation Jedburgh: D-Day and America's First Shadow War* by Colin Beavan.
18. *Eisenhower's Guerrillas: The Jedburghs, the Maquis, and the Liberation of France* by Bejamin F. Jones.
19. *In the Midst of Wars* by Edward Lansdale.
20. *Counterguerrilla Operations* by Charles Bohanan and Napoleon Valeriano.
21. *The Huk Rebellion: A Study of Peasant Revolt in the Philippines* by Benedict J. Kerkvliet.
22. *The Pentagon Papers* by Neil Sheehan.
23. *Vietnam Declassified* by Thomas L. Ahern.
24. *SOG: The Secret Wars of America's Commandos in Vietnam* by John L. Plaster.
25. *The Salvadoran Crucible* by Brian D'Haeseleer.

26. *The CIA's Black Ops* by John Nutter.
27. *War in the Shadows* by Robert B. Asprey.
28. *The American War in Afghanistan* by Carter Malkasian.
29. *American Spartan* by Ann Scott Tyson.
30. *The Afghanistan Papers* by Craig Whitlock.
31. *The Only Thing Worth Dying For* by Eric Blehm.
32. *Operation Enduring Freedom-Philippines, Routledge Handbook of U.S. Counterterrorism and Irregular Warfare* by David S. Maxwell.
33. *America's Dirty Wars* by Russell Crandall.
34. *Success in the Shadows: Operation Enduring Freedom–Philippines and the Global War on Terror, 2002–2015* by Barry M. Stentiford.
35. *Fiasco* by Thomas E. Ricks.
36. *Endgame* by Michael R. Gordon and Bernard E. Trainor.
37. *By All Means Available* by Michael G. Vickers.
38. *Surprise, Kill, Vanish* by Annie Jacobsen.
39. *Relentless Strike* by Sean Naylor.
40. *In the Company of Heroes: The Personal Story Behind Black Hawk Down* by Michael J. Durant, Michael Durant, Steven Hartov.
41. *Black Hawk Down* by Mark Bowden.
42. *The Battle for Syria: International Rivalry in the New Middle East* by Christopher Phillips.
43. *Destroying a Nation: The Civil War in Syria* by Nikolaos van Dam.
44. *Burning Country: Syrians in Revolution and War* by Robin Yassin-Kassab and Leila Al-Shami.
45. *The Gates of Europe: A History of Ukraine* by Serhii Plokhy.
46. *Borderland: A Journey Through the History of Ukraine* by Anna Reid.
47. *Subversion: The Strategic Weaponization of Narratives* by Andreas Krieg.
48. *Three Dangerous Men* by Seth Jones.
49. *Unholy War: Terror in the Name of Islam* by John Esposito.
50. *Unrestricted Warfare* by Qiao Liang and Wang Xiangsui.

Index

Abu Ghraib Prison Scandal, 292

Abu Musab al-Zarqawi (AMZ), 285, 289, 291, 293, 298, 301, 305

Afghanistan, 220-267

Africa, 319

Al Assad, Bashar, 340, 374

Al Qaeda, 221, 226, 227, 251, 257, 301, 309, 380

Al-Zawahiri, Ayman, 256, 301

Anbar Awakening, 305-308

Angola, 321

Anstett, Robert, 129

Badr Brigade, 303-304

Barre, Siad, 322

Battle of Tongo Tongo, 329

Bay of Pigs invasion, 205

Bell, Franklin J., General, 67

Belt and Road Initiative (BRI), 337, 382

Central Intelligence Agency (CIA), 179, 184, 204, 209, 314, 337

China, 337, 382-383

Churchill, Winston, 34, 124

Clausewitz, Carl von, 5, 10, 18, 59, 98, 319, 392

Civilian Irregular Defense Group (CIDG), 178-179

Civil Operations and Revolutionary Development Support (CORDS), 187-189

Cleveland, Charles, General, 2, 20, 391, 392

Counterinsurgency (COIN), 7, 30, 30-33

Counterterrorism (CT), 8, 250, 280

Cyr, Paul, 128-129, 133

Castro, Fidel, 25, 201-206, 213, 321

Dostum, Abdul Rashid, 222-223, 266

Drones, 253, 350, 356, 361, 365

Farabundo Marti National Liberation Front (FMLN), 207, 211

Fertig, Wendell W., Colonel, 89-97

Index

Foreign Internal Defense (FID), 7

Galula, David, 3, 15

Great Power Competition (GPC), 320

Guevara, Che, 25, 201

Hart, B.H. Liddell, 10, 74

Ho Chi Minh, 161-163

Ho Chi Minh Trail, 171, 177, 183, 184

Hukbalahap Rebellion, 141

Improvised Explosive Devices (IEDs), 235, 241, 285, 291, 295

India, 257-258, 383

Information Operations, 36, 361, 378

Inter-Services Intelligence (ISI), 240

Iran, 210, 241, 303-304, 341, 345-346, 353, 371, 378-382

Iran-Contra Affair, 210

Iraq,
Elections, 294, 300, 305, 314
Invasion, 280

Israel, 381

Kaika, Battle of, 237-239

Kennedy, John F., 164, 180, 205

Trinquier, Roger, 15

Lansdale, Edward, 16, 148

Lawrence, Thomas E., 11-13

Lord's Resistance Army, 333

Malayan Emergency, 16, 32

Mao Zedong (Tse-Tung), 12-15

McChrystal, Stanley, General, 243, 297

McRaven, William, Admiral, 254

Mujahadeen, 224, 229

Nagl, John, 17, 194, 242

Nicaragua, 15, 199, 204, 206-211

North Korea, 157, 361, 369, 371, 382, 391

Operation Overlord, 127, 130, 137

Pakistan, xi, 226, 229, 239-241, 253, 257, 266, 341, 389

Putin, Vladimir, 343, 357, 369, 372

Reeder, Edward, General, 245

Rumsfeld, Donald, 220, 282

Russia, 372-378

Sepp, Kalev I., Colonel, 71, 156, 177, 394

Ukraine, 355

Unconventional Warfare (UW), 7

Viet Cong, 167

Volckmann, Russel, Colonel, 84, 97-101

Vietnam War, 160

Village Stability Operations (VSO), 245

Vo Nguyen Giap, 15, 171

Connect with Blacksmith Publishing

The PINELANDER PODCAST

www.blacksmithpublishing.com

www.ingramcontent.com/pod-product-compliance
Lightning Source LLC
Chambersburg PA
CBHW022042020426
42335CB00012B/505